TALLAHASSEE
LIBRARY
COMMUNITY COLLEGE

Appalachian Folkways

Creating the North American Landscape

Gregory Conniff
Edward K. Muller
David Schuyler
Consulting Editors

George Thompson
Series Founder and Director

Published in cooperation with the Center for American Places,
Santa Fe, New Mexico, and Staunton, Virginia

APPALACHIAN FOLKWAYS

John B. Rehder

The Johns Hopkins University Press
Baltimore and London

Printed in the United States of America on acid-free paper
9 8 7 6 5 4 3 2 1

The Johns Hopkins University Press
2715 North Charles Street
Baltimore, Maryland 21218-4363
www.press.jhu.edu

Library of Congress Cataloging-in-Publication Data
Rehder, John B.
Appalachian folkways / John B. Rehder.
p. cm. — (Creating the North American landscape)
Includes bibliographical references and index.
ISBN 0-8018-7879-9 (hardcover : alk. paper)
1. Folklore—Appalachian Region, Southern. 2. Human geography—Appalachian Region, Southern. 3. Food habits—Appalachian Region, Southern. 4. Vernacular architecture—Appalachian Region, Southern. 5. Traditional medicine—Appalachian Region, Southern. 6. Appalachian Region, Southern—Social life and customs. I. Title. II. Series.
GR108.R48 2004
398′.09756′8—dc22 2003018060

A catalog record for this book is available from the British Library.

THE TRAVELER who follows the trails of this far country, fords its rushing streams, and forces his way through thickets of rhododendron and laurel to rest upon some beech-shaded bank of moss, and who toward sunset checks his horse upon the ridge to trace the tread of smoke which signals welcome, may yet be at a loss for a name to describe the land; but when at dawn he wakes with mist rising from every cove and valley, and echoes still sounding of half-remembered traditions, folk-lore and folk-songs, recited or sung before the fire by "granny" or "grandpap," he knows that there is but one name that will do it justice—"the Southern Highlands."

JOHN C. CAMPBELL

Contents

Preface

When I was a boy growing up in the swamps and salt marshes of coastal North Carolina, my comic-book image of mountains was of smooth, steep purple triangles with snow on top. I was immensely disappointed when I saw my first "real" mountains, the Blue Ridge, when I was ten. They were rounded blobs covered with green trees in a dense blue haze—no purple rocks, no snow (it was July). Fast-forward fifty years and a real folk Southern Appalachia is within my vision. I want to share it with you. We all have an image, a perception of what we think a place is like. And every place has a personality, a character, that must be real. Our journey together is meant to discover both.

I especially acknowledge my students, graduate and undergraduate, who have explored this marvelous place called Appalachia with me in the field. Many have made worthy accomplishments from their Appalachian experience: Jim O'Malley, Mack Gillenwater, Tink Moore, the late Larry Smith, Jim Cobb, John Morgan, Joy Medford, Stan Guffey, Steve Hill, Lisa Roberts, Vince Ambrosia, Neil Cyganiak, Kurt Butefish, Tom Maertens, Ed Nichols, Todd Fox, Del Scruggs, Bill Reding, David Mann, and dozens more. I have spoken with so many people and should have interviewed hundreds more, but Carolyn White, the late Dan Jackson, Hoyt Vanosdale, Richard Strange, Tommy Reed, Emily Smith, John Kemmer III, the late Jim Russell, Sam Venable, and many unidentified informants have been more than downright helpful. A very special thank you goes to Steve Rogers and the Tennessee Historical Commission staff for the many years of data collecting and access to archives and for my valuable association with you all. Many thanks go to Terry G. Jordan-Bychkov for his exceptional insight and help with the manuscript and for his treasured friendship. I acknowledge colleagues Charles Faulkner, Henri Grissino-Mayer, Will Fontanez, and especially Andrew Wunderlich for his superb cartography. I have special appreciation for the Thomas Jefferson Prize granted to me through the University of Tennessee in 2001. I thank George F. Thompson, president of the Center for American Places, who patiently worked with me for so many years, and Alice Bennett, formerly of the University of

Chicago Press, for her excellent copyediting. My wife Judy and my grown children Ken and Karen always get special recognition. Since my last book, our family has grown with the addition of grandchildren Allen Burkhardt Rehder and Emma Caroline Rehder, who will be my field assistants in the years to come.

Appalachian Folkways

1

The Real Appalachia

Appalachia, the word and the world, speaks of a way of life in a misunderstood part of North America. Both the families who have lived here for generations and those outside looking in have a personal image of where they think Appalachia is and what life here is like. In this chapter I describe some of the ways such images define Appalachia physically, culturally, and cognitively. In later chapters we will journey to Appalachia and discover the ways people encountered the region, how they settled it, and especially how they created a rich and varied folk culture expressed in the built environment and the cultural landscape.

The real Appalachia is Southern Appalachia. Where is it? What role does folk culture play? What can we say about the image and personality of a real Appalachia? I believe we can find the real Appalachia in much the same way that cultural geographer E. Estyn Evans found reality in his classic study of his adopted Ireland: "In Ireland we have a unique opportunity of studying the dwellings, tools and customs of a venerable social order which is admittedly dying but has by no means entirely gone. It is surely here, in the activities of a fascinating countryside, that the real Ireland may be found."[1]

If the real Appalachia resides in its folk culture, as I believe it does, then why does Appalachia defy clear and simple definition? The region suggests cryptic identities. Have you ever heard anyone say, "I'm from Appalachia"? Although Appalachia does occupy a geographical space on the planet, few citizens say they hail from there. Ask people who really are from Appalachia about where they come from and they will answer with a state, county, or community, such as Kentucky, Perry County, or the town of Hazard. You might also hear subregions like southwestern Virginia, Upper East Tennessee, western North Carolina, or north Georgia.

Regionalism, at either a personal or a sociological level, has significantly more to do with people's identification with this place we call Appalachia than with its specific location. Although few people name it as a place they come from, to them it is still home—an extremely personal place where they are born and raised, get married, have children, labor, possibly retire, and ultimately die.

If adults raised in Appalachia move to other regions in the United States, are they still "Appalachian"? Yes, I believe so. Once people learn their culture, they retain some of its original traits even though they accept traits from a new host culture. Tradition-bound cultures such as those in Appalachia are more likely to preserve their traits in the face of external change. During and after World War II, Appalachian workers and their families who moved north for work carried with them enough of their heritage to create small "Appalachian ghettos" in cities like Detroit, Cleveland, and Chicago. The workers preserved Appalachian customs in language, religion, and especially food. Many maintained contact with their roots and received "care packages" of lard, dried beans, grits, cornmeal, snuff, and other things that they couldn't find in their northern enclaves. Later, grocery stores in these "Little Appalachias" began to carry these products. The stamp of culture is deep, and for some people it can be indelible.

Appalachia inspires two rather detached clusters of cognitive images: as a relic, rustic culture, and as a scene of poverty and development. In the first distinctively cultural cluster of folk images, we see old-time mountain fiddlers, banjo pickers, and a material culture well represented in single pen log cabins, double crib barns and smokehouses, fences, quilts, split-oak baskets, and a thousand and one other exemplary folk traits. In the other image cluster we see eroding hillsides with deep scars from strip-mined coalfields, polluted streams, smoke-filled valleys, and hollows awash with derelict trailers—marked signs of poverty. Also in this second cluster are well-meaning warriors in the War on Poverty who consult, plan, and theorize about developing the region. For an area caught in a vacuum, the results can be disappointing. This image cluster leaves behind a sad commentary on what seems to be everlasting poverty. David Whisnant, eloquent observer and scholar of Appalachia, says that early in his thinking on the region these two clusters emerged, with the former in the background and the latter in the foreground. But later the two began to connect. The developmental images represented government agencies such as the Appalachian Regional Commission and the Tennessee Valley Authority that

> could at last be understood better in cultural than in theoretical or technical terms. Like rotors in TVA's generators, they were whirled between twin poles of culture.
>
> One pole was their own deeply imbedded mainstream culture—a conflicted amalgam of authoritarianism and libertarianism, individualism and paternalism, elitism and populism, nostalgia and futurism, self-help and social responsibility. The other was the ancient, tattered but treasured hand-me-down fabric of untenable cultural assumptions and judgments about mountain people. The "Appalachia" that they sought to develop was, to some extent, an Appalachia created in their own minds—a mythic Appalachia, as Henry Shapiro's *Appalachia on Our Mind* has recently shown, held to as a cultural talisman by the rest of America.[2]

It is the relic folk cultural image cluster that I will write most about in this book. But I cannot ignore the image cluster of poverty, because that is what much of Appalachia is about in contemporary reality.

Geographers have a peculiar perspective with an eye for regions and regionalization—it is integral to our profession. Geographers create conceptual regions by plotting data on maps and drawing boundaries around places that express common characteristics. But Appalachia almost defies definition, especially at its boundaries, and the essence at its heart can be just as elusive. Sometimes I think Appalachia is a state of mind, a mental construct created by its inhabitants and its observers. If it has edges and boundaries, where are they? Who drew them? And for what purpose? Here we will examine criteria for bounding the Appalachian region in terms of physical geographic definitions, government agency definitions, and cultural mapping definitions. I have woven cognitive map interpretations throughout the chapter so that the mythic and real identity of the region may elude our attention but not entirely escape it.

Physical Definitions

The Appalachian System

The entire Appalachian physical system extends northeast to southwest as a linear highlands pattern for approximately 1,500 miles, from the Gaspé Peninsula in Quebec to northern Alabama. Southern Appalachia begins in Pennsylvania and extends to northern Alabama. At its widest part, the region is nearly 400 miles wide. Southern Appalachia embodies three major physiographic or geomorphic provinces: the Blue Ridge

Province to the east, the Ridge and Valley Province in the center, and the Allegheny and Cumberland plateaus to the west.

Definitions based on the physical geography of Appalachia date to the sixteenth century. The origin of the term "Appalachia" is believed to trace to Spanish explorers Alvar Núñez Cabeza de Vaca in 1528 and Hernando de Soto in 1539, who encountered a Native American area in northern Florida called Apalachee or Apalache. The explorers believed that the Apalachee people and their land in Florida possessed untold mineral riches.[3] Somehow the name Apalachee was applied to a mountain mass much farther north, where de Soto's route in 1539 took him through Cherokee lands in the mountains of northern South Carolina, southwestern North Carolina, and southeastern Tennessee. One of the earliest pieces of visual evidence is a map by John Senex published in 1719 depicting an elongated mountainous area called the "Apalitean Mountains" that extends from southern Pennsylvania to northern Florida.[4]

Nineteenth-century scholars recognized Appalachian landscapes as unique and made efforts to delimit the region. In 1861 Arnold Guyot, for whom Mount Guyot (6,621 feet) in the Great Smoky Mountains in Tennessee and North Carolina is named, was perhaps the first to establish the concept of an entire mountain range that extended from the Gaspé Peninsula in Quebec to northern Alabama.[5] John Wesley Powell, famous for his classic geological expeditions in the American West, delimited the Appalachian region in 1895 into three familiar physiographic regions: the Piedmont to the east, the Appalachian ranges in the center, and the Allegheny plateaus to the west.[6]

Guyot's and Powell's efforts were a good beginning but were not entirely correct. Guyot's region was geologically limited because the rocks that connect the entire system extend intermittently and certainly beneath the surface from Newfoundland all the way to Texas. Powell's regionalization was in error because he should have included the Ridge and Valley Province as a separate subregion, and he should have spoken of the plateau country as two units called Allegheny *and* Cumberland. The Piedmont region physically and geologically could remain in Powell's definition, but some scholars believe that the Piedmont no longer belongs to contemporary Appalachia because of economic differences.

In its largest context, the Appalachian system consists of two major divisions: the northern Appalachians and the southern Appalachians. We rarely hear the northern Appalachians considered as a unit, and Northern Appalachia does not exist in the same cultural context as

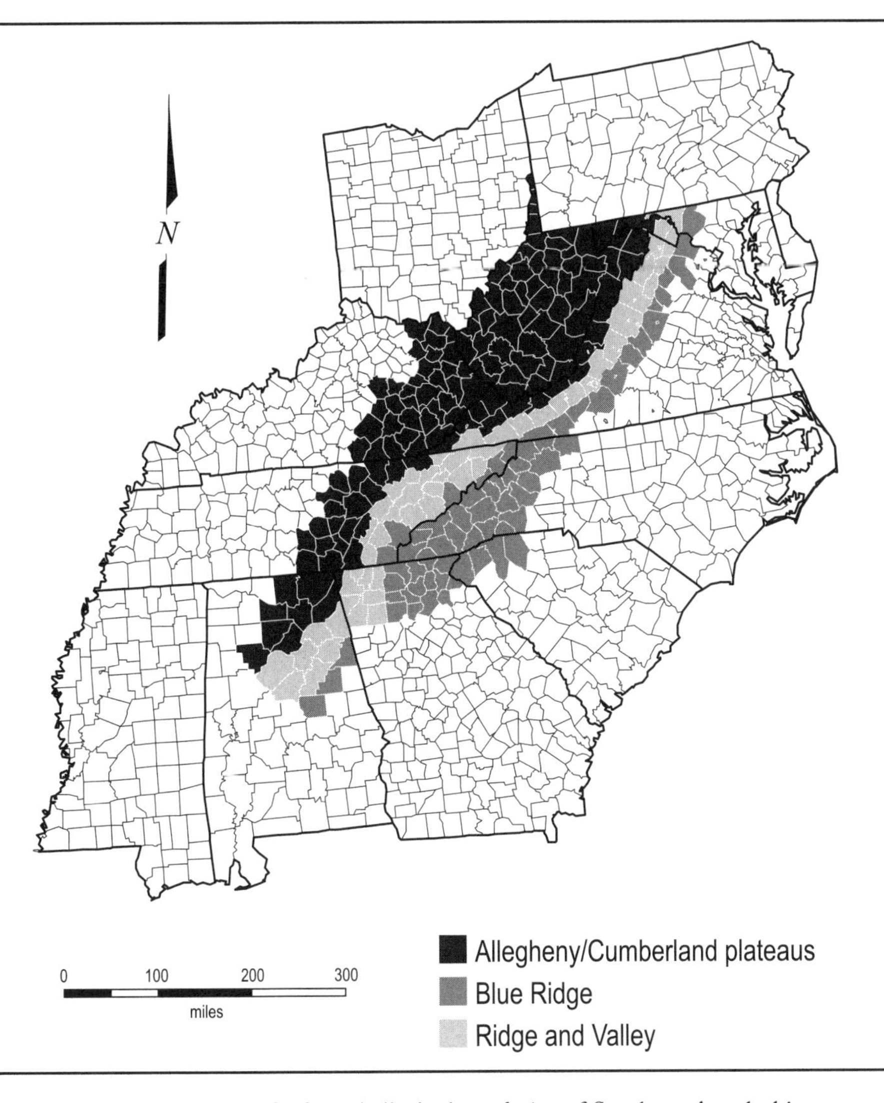

According to John C. Campbell, the boundaries of Southern Appalachia extended from southern Pennsylvania to northern Alabama. They were based on natural physiographic divisions as well as on sociological, cultural, historical, and (for southern Pennsylvania) political criteria at the Mason-Dixon Line. (Source: Campbell [1921] 1969)

Southern Appalachia. The primary ranges in the northern part of the Appalachian system contain the Notre Dame Mountains in Quebec, the Green Mountains in Vermont, the White Mountains in New Hampshire, and the Adirondack Mountains in New York. Additional smaller upland units such as the Catskill Mountains and the Pocono Mountains appear as outliers of a dissected plateau in the southeastern part of Northern Appalachia.

A valid physical boundary between the northern and southern Appalachians is the glacial border that divides glaciated northern areas from unglaciated southern ones. During the Pleistocene, the period of ice ages, continental glaciation extended to northernmost Kentucky, southeastern Ohio, and parts of Pennsylvania. Areas south of this boundary were unglaciated and thus escaped erosion by continental ice sheets. An acceptable physical boundary divides Pennsylvania into two unequal parts where the glaciated northwestern and northeastern corners belong to Northern Appalachia and the southwestern, central, and southeastern unglaciated areas are in Southern Appalachia.[7]

Southern Appalachia

Southern Appalachia begins in Pennsylvania and extends southwest to northern Alabama. Its eastern boundary includes parts of western Maryland, western Virginia, western North Carolina, northwestern South Carolina, and northern Georgia. The western boundary follows the upper Ohio River, eastern Kentucky, middle Tennessee, and northern Alabama, which is also the southern limit of the region. West Virginia is the only state entirely within Southern Appalachia.

Three large geomorphic, or landform, regions dominate the landscape. The Blue Ridge Province, a complex mountain core of ancient rocks in the eastern part, forms the cornerstone for Southern Appalachia's mountains. The Blue Ridge begins as a single ridge near the Susquehanna River in southern Pennsylvania and proceeds to Harpers Ferry, West Virginia, at the juncture of the Shenandoah and Potomac rivers. The Blue Ridge Province continues southwest through Virginia, widens in western North Carolina, and extends south to northern South Carolina and northern Georgia. Its eastern edge forms an impressively steep mountain front as it rises abruptly from the western margins of the Piedmont.

South of Roanoke, Virginia, the mountains divide into two forks that form a large oval, rejoining in northern Georgia. The eastern fork is called the Blue Ridge all the way from Virginia to Georgia, but the western fork has several names: among others, the Unaka Mountains between Roanoke, Virginia, and northeastern Tennessee, the Great Smoky Mountains in western North Carolina and eastern Tennessee, and the Unicoi Mountains. The land lying between the two forks has transverse ranges trending east-west: the Black, Pisgah, Balsam, Cheoah, Cowee,

Snowbird, and Nantahala mountains and others. Blue Ridge summit elevations range between 2,500 and 6,684 feet above sea level.

The second region, the Ridge and Valley Province, is aptly named for its corrugated surface of sedimentary rocks forming parallel ridges and narrow valleys that extend northeast to southwest from Pennsylvania to Alabama. Ridge elevations range from 500 feet to 4,000 feet above sea level. Geologically, the rocks in the Ridge and Valley subregion are younger than rocks in the Blue Ridge. Here sedimentary limestone, sandstone, and shale dominate the surface and subsurface terrain. The Great Valley, a very long and wide valley on the eastern side of the province, has served as the major route for land transportation, from Indian trails to early wagon roads to major contemporary highways such as U.S. 11 and interstate highway I-81. The Ridge and Valley Province is the major agricultural subregion and includes many of the prominent urban areas in Southern Appalachia.

The third physical subregion is the Appalachian plateau country to the west, with the Allegheny Plateau to the north and the Cumberland Plateau in the south. Both are dissected plateaus, eroded and rough in many places, with flat hilltops, steep sides, and narrow, canyonlike hollows. The plateau region originated from sediments of sandstone, limestone, shales, and coal layers that were laid down in an ancient sea, then uplifted into a tableland plateau. The Allegheny Plateau and much of the northern part of the Cumberland Plateau are almost entirely dissected, but the southern Cumberland Plateau in the central and southern parts of Tennessee and northeastern Alabama retains some of its flat, tableland appearance.

Clearly, Southern Appalachia is not just one simple mountain mass. It comprises three distinct subregions: the Blue Ridge to the east, the Ridge and Valley Province in the center, and the Allegheny and Cumberland plateaus to the west. In chapter 2 we will explore these physical subregions in detail.

Government Definitions

The government's involvement in defining Appalachia as a special region is comparatively recent. In 1864 the superintendent of the census, Joseph C. G. Kennedy, described the area that stretched from Pennsylvania, Virginia, and Tennessee to Alabama as one of seven "natural Regions of the United States."[8] In a 1935 publication for the U.S. Department of Agriculture, F. J. Marschner defined Appalachia based on geomorphology, soils, and climate. Marschner divided the region into 236 counties in nine states using the three familiar physical divisions of the Blue Ridge, the Ridge and Valley Province, and the Appalachian plateaus. He excluded counties in West Virginia bordering the Ohio River and included all of Alabama's northern tier of counties that border Tennessee. Marschner extended Appalachia south to Birmingham and then, beyond the realm of reason, southwest to Tuscaloosa. Most maps do not give Alabama this much of Appalachia.[9]

In a 1940 publication for the Works Progress Administration (WPA), Arthur R. Mangus described an Appalachian region as comprising two regions among thirty-four other rural cultural regions in the United States. The first region, Appalachia, had 154 counties in portions of Kentucky, Ohio, West Virginia, Virginia, Tennessee, North Carolina, and Georgia. The second region, Allegheny, included 125 counties in parts of West Virginia, Pennsylvania, Ohio, Maryland, and Virginia. The criteria he used were types of farming, population increase, standard of living, land value, tenancy, and race.[10] Both Marschner's and Mangus's regions were largely based on dry statistics gleaned from the gray filing cabinets of the federal government. Neither writer touched on the folk culture of the region, nor did they clearly consider poverty. When these early federal reports were produced, Appalachia, though poor, was no worse off than any other part of the United States, especially during the Great Depression. It was not until the 1960s that there was a disparity between Appalachia and other regions in terms of poverty.

With the passing of the Appalachian Regional Development Act of 1965 and the formation of the Appalachian Regional Commission (ARC) the same year, the federal government placed its greatest emphasis on Appalachia. Stimulated by earlier visits to the area and images of Appalachian poverty, President John F. Kennedy, his brother Robert Kennedy, and President Lyndon B. Johnson declared a War on Poverty, with the ARC as the ultimate army.

By 1967 the Appalachian region under the ARC contained 397 counties in thirteen states situated between southern New York State and northeastern Mississippi. The bloated region created by the ARC grew into a political region that included counties in New York and Mississippi. New York's "Appalachia" consisted of the southern tier of fourteen counties bordering Pennsylvania. Mississippi had nineteen counties in the northeastern corner of the state declared eligible for ARC funds. It was purely for political motives that Appalachia had now expanded beyond the parameters of a physical uplands environment plagued by white

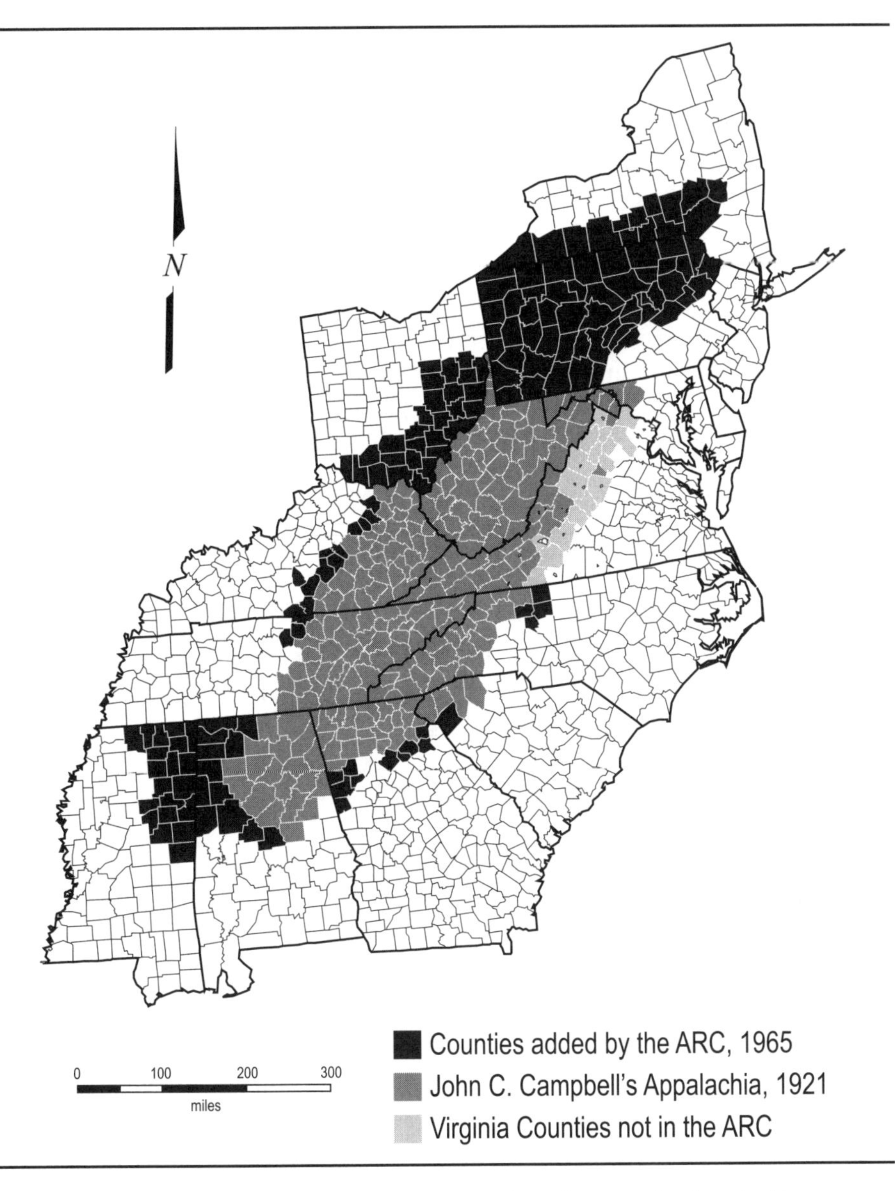

The power of politics is represented by the boundaries of the Appalachian Regional Commission, which extend from southern New York to northern Mississippi. (Sources: Campbell [1921] 1969; Ford 1967)

poverty. New York entered the picture because United States senators, especially Robert Kennedy, supported the ARC and got political support from the state as a quid pro quo. Mississippi's Senator John C. Stennis, a member of the Senate Appropriations Committee, had all along wanted his state to benefit from the ARC, and it did. Congressman Jamie Whitten, a member of the House Appropriations Committee and later chairman of the full Appropriations Committee, naturally supported the inclusion of his congressional district and other counties in northeastern Mississippi. And so these politically constituent counties became a part of Southern Appalachia, even though physically and culturally they were in the Deep South.[11] The Appalachian Regional Commission continues to define itself and Appalachia in these deceptive ways.

Cultural Definitions

The human nature of Appalachia has helped define the region for more than a century. Over the past four decades the theme of poverty has again made people aware of the region's economic struggles. In the late nineteenth and early twentieth centuries, under the watchful eyes of well-meaning philanthropists and other outsiders with missionary zeal, the region was viewed as a wild, backward place that needed to be tamed and civilized. John C. Campbell and his wife Olive Dame Campbell were excellent representatives of the missionary-philanthropist movement at the turn of the century.

Campbell's *The Southern Highlander and His Homeland,* published in 1921, remains the most comprehensive treatment of the Appalachian concept and region published before 1960. Campbell's cultural perspective was largely based on his own observations over twenty-five years as a missionary educator. In collecting information for his book, Campbell traveled through the region on horseback, visiting over seventy churches and independent schools. Unlike other writers of the time, Campbell recognized that Appalachia had many cultures rather than a single, homogeneous one.[12]

Campbell defines the boundaries of the region based on acceptable physiographic or geomorphic terms. Southern Appalachia, or the "Southern Highlands," as he calls it, begins at the Mason-Dixon Line—the state boundary between Pennsylvania and Maryland, running south to northern Alabama. In a political sense, the region consists of four western counties in Maryland; the Blue Ridge, Valley, and Allegheny counties in Virginia; all of West Virginia; eastern Tennessee; eastern Kentucky; western North Carolina; the four northwestern counties of South Carolina; northern Georgia; and northeastern Alabama.[13] The northern boundary is purely political, but the eastern, western, and southern limits follow natural physical divisions that Campbell calls the Blue Ridge Belt, the Greater Appalachian Valley, and the Allegheny-Cumberland Belt. In contemporary terms, these correspond to the Blue Ridge Province, the Ridge and Valley Province, and the Allegheny and Cumberland plateaus. After years of considering numerous definitions, I have found that Campbell's definition most closely fits my own cognitive pattern for the region.

The role that professional cultural geographers have played in defining Appalachia specifically or including Appalachia within a larger regional context is worth our attention. While most have not taken Appalachia as their primary subject, cultural geographers have created many interesting map representations of the region and surrounding areas. Fred Kniffen's pivotal 1965 paper "Folk Housing: Key to Diffusion" described the routes of diffusion for ideas and patterns in American folk house types. His map of source areas and routes of diffusion shows major streams of house types originating in the Middle Atlantic region and diffusing well into and beyond Appalachia. One of the principal routes began in southeastern Pennsylvania and, following the Great Valley, extended down the entire Ridge and Valley Province to northern Alabama, with offshoot distributary branches extending into the mountain and plateau subregions on each side.[14]

In 1968 Henry Glassie, one of America's most prolific writers on folk culture, published a map in his first book showing material folk regions of the eastern United States. The region designated "Upland South" encompasses Appalachia. Glassie's large, contiguous, loosely fitting Upland South overlies Appalachia, but it is intuitive and not based on the hard evidence of diagnostic culture traits. The "Upland South" extends well beyond Appalachia because its culture traits and people are found in extraregional locations such as in the Ozarks and other points west.[15]

For many years and in over nine editions of the *Human Mosaic,* Terry G. Jordan-Bychkov has published a map of traditional rural culture areas of the contemporary continental United States. His Midland region has origins in the southeastern Pennsylvania culture hearth, but its northern boundary extends through central Pennsylvania, northern Ohio, and westward to southeastern Illinois. The southern and eastern boundary follows the Piedmont in the Carolinas and sweeps across

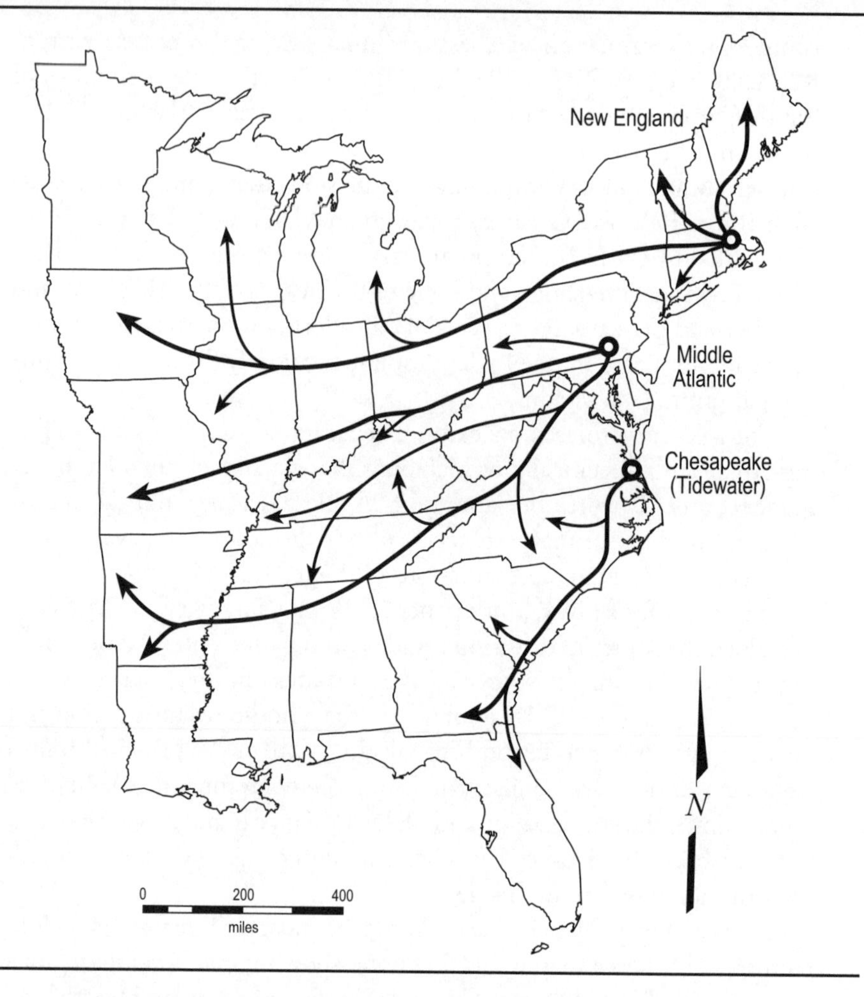

Fred Kniffen's map of source areas and diffusion routes of folk housing illustrates the development of settlement leading into and through Appalachia. (Source: Kniffen 1965, 560)

northern Alabama to include a tiny part of northeastern Mississippi. The boundary then heads north to avoid Memphis before turning west across the boot heel of Missouri. The Midland region continues across the southern Great Plains and disperses far into the western states. How does it fit Appalachia? The eastern core of the region broadly (almost too much) covers all of Appalachia and then some. Jordan's Midland region closely matches Glassie's Upland South region, except that Jordan's region is larger. Both encompass Appalachia. In 2003 Jordan-Bychkov,

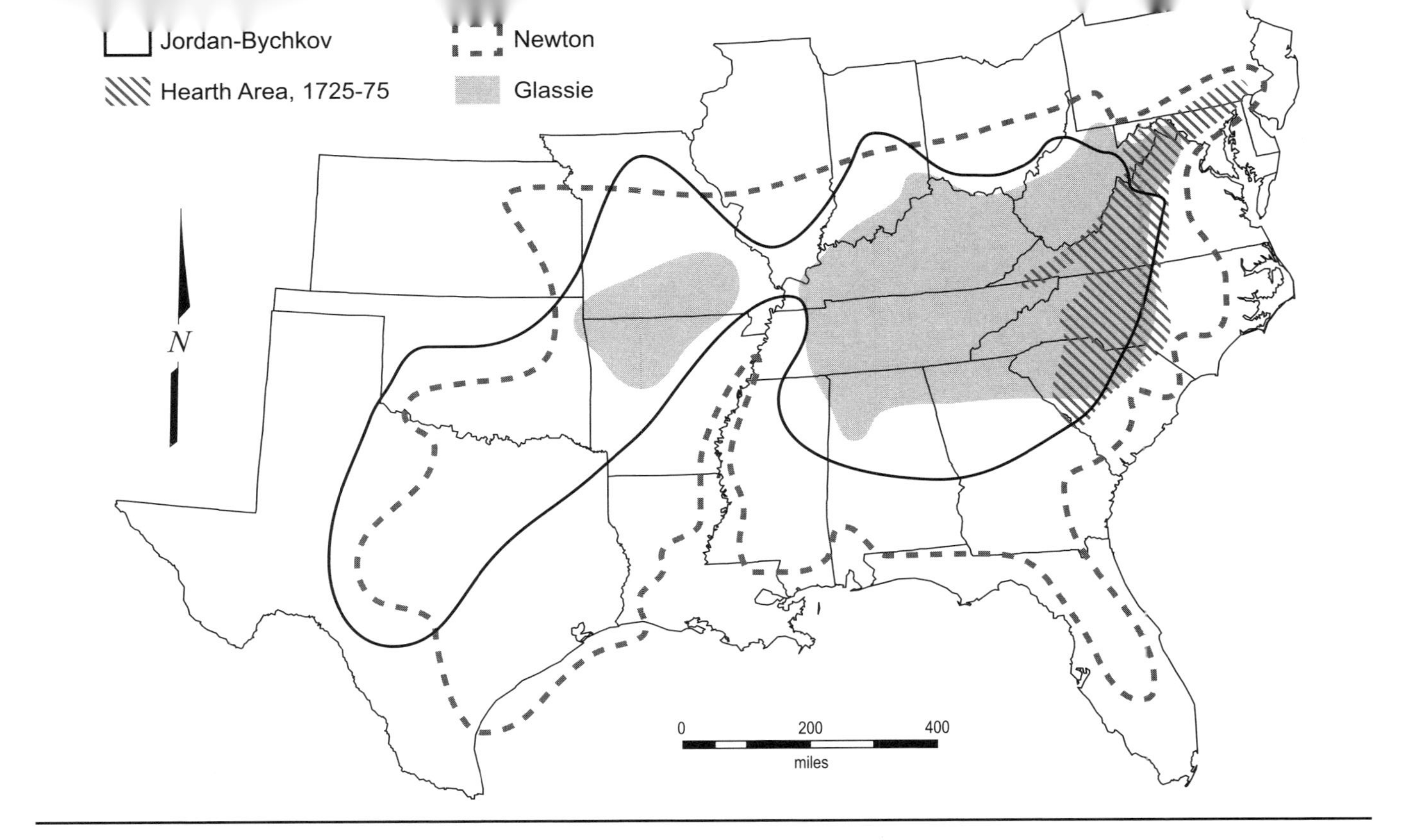

Southern Appalachia lies well within Terry G. Jordan-Bychkov's cultural boundary for the Upland South, Henry Glassie's map of the Upland South as an intuitive material folk culture region, and Milton Newton's map of the Middle Atlantic and Upland South Culture region in 1835, with its core hearth for 1725–75. (Sources: Jordan-Bychkov 2003, 7; Glassie 1968, 39; Newton 1974, 149)

in a most impressive small book, *The Upland South,* further illustrates a refined boundary pattern for the Upland South, much of which effectively encompasses Appalachia.[16]

Milton Newton in 1974 produced an interesting map of the Upland South that overlies Appalachia in two rather different ways.[17] His core regional hearth for 1725 to 1775 centers on the Philadelphia Wagon Road that runs from southeastern Pennsylvania down the Ridge and Valley's Great Valley to Roanoke, Virginia, and thence to Salem on the North Carolina Piedmont, then proceeds south into the South Carolina Piedmont at Camden before branching off to Augusta, Georgia. This may be an oddly shaped large core area, but it is an acceptable pattern for the eastern parts of Appalachia. Newton's outline of the Upland South for 1835 is by most measures exceptionally and controversially overstated. It is simply too large. He calls it the "Mid Atlantic–Upland South Culture Region," and some of it generally agrees with Glassie, Kniffen, and Jordan-Bychkov; but Newton's regional boundary covers far more territory than some scholars would agree upon. Terry Jordan-Bychkov told me recently, "According to Newton, only if your feet were standing in water [in the Gulf of Mexico] were you out of the Upland South."[18] In defense of Newton's large area, the mapped year of 1835 was a time when more of the Upland South's people and culture dispersed over broader terrain to the west and south. But not *that* broad.

Karl Raitz, a geographer at the University of Kentucky, has had the most to say about defining Appalachia, because he and Richard Ulack, with Thomas Leinbach, wrote a geography textbook on Appalachia.[19] Raitz examined different map patterns that define the region by showing the various outlines that others have drawn. Then he and Ulack chose their own definition and mapped it. The Raitz-Ulack map encompasses the Piedmont, the Blue Ridge, the Ridge and Valley, and the Cumberland and Allegheny plateaus. Some geographers do not include the Piedmont because it has become a separate subregion through economic development. Some believe that much of the culture on the Piedmont had progressed beyond a traditional folk way of life by the very nature of its industrial economy. Others would argue that the Piedmont was culturally Appalachian in origin and that despite its transformation from agriculture to industry, there are residual nodes of relic traditional culture that through time have been clearly Appalachian and certainly folk. For example, the fiddler's convention, an institution for folk music, finds many of its core elements on the Piedmont. Pioneer bluegrass musicians such as banjoist Earl Scruggs came from the Piedmont region.

Many folk traits that indeed belong to Appalachia are still found on the Piedmont.

Historian Paul Salstrom sees the three physical divisions—Blue Ridge, Ridge and Valley, and Cumberland and Allegheny plateaus—as distinct economic subregions in a temporal sense. The three physical regions were settled at different times, and each has developed and evolved into different economic subregions. Salstrom describes the subregions as Older Appalachia (the early-settled Ridge and Valley), Intermediate Appalachia (the Blue Ridge mountain core), and the New Appalachia (the late-settled plateaus of Allegheny and Cumberland).[20]

Considering these varied perspectives, we can appreciate the task of defining the region. To further complicate the issue, Appalachian culture traits can be found far outside the region. This is especially true of the Ozark culture region and parts of Texas. The same can be said of other places to which Appalachian folk have migrated, such as the Willamette Valley of Oregon, Washington State, and northern industrial cities such as Detroit, Cleveland, and Chicago. With this we have extensions of Appalachia, or Appalachian folk culture, going well beyond traditional boundaries. And that is as it should be.[21]

The Nature of Culture

If we are to understand folk culture, we should begin with fundamental concepts: big *C* culture and little *c* culture, material and nonmaterial culture traits, and processes of acculturation and assimilation. Culture is learned. All the world's people who have learned the smallest smattering or the largest dose of the stuff have culture. This encompassing worldview, or Culture, posits a universal pattern of human behavior so that everyone on the planet who has learned from his or her own culture participates in a global cultural atmosphere.[22]

Culture divides into two grand categories: Culture and culture. Big *C* culture covers broad, universal characteristics in language, belief systems, food, shelter, and other generic things that all people learn. It is the sum total of all things cultural at a global scale. Little *c* culture denotes the culture of a specific culture group or ethnic group. Examples of cultures are Navahos, Cajuns, Amish, Cubans, Finns, and thousands more. Whether one's culture is Amish or Algerian, every group has culture, whose scale is *c* culture. Specific culture groups have identifying characteristics called diagnostic culture traits that set them apart from other cultures or ethnic groups.

Culture traits are the building blocks of both Culture and culture and are classified into two types: material traits and nonmaterial traits. Material culture traits are those tangible, visible, human-made things that have form and substance. We see and touch them, as when we hold a quilt or banjo or build a log house. Nonmaterial culture traits are invisible but are present in the spoken word or concealed thought. Unwritten songs, stories, ideas, and cussing under one's breath are nonmaterial culture traits. Material and nonmaterial culture traits are possessed by groups who represent cultures, which globally and collectively add up to Culture.

The world's cultures today are more mobile than at any time in history. What happens when cultures come in contact, as they inevitably do? Two anthropological concepts describe the results. *Acculturation* states that when two cultures come together, they exchange traits but each culture retains its identity. When early Europeans in Appalachia came in contact with Native Americans, chiefly Cherokees, the groups exchanged culture traits, but their cultures remained intact. The second concept, *assimilation,* states that when two cultures come in contact, usually the stronger or more aggressive culture absorbs or destroys the other. While some residual traits may remain, the assimilated culture loses most if not all of its original identity. Assimilation can take place peacefully, too. An Appalachian example is the assimilation of the few eighteenth-century French Huguenots who made their way into Appalachia. Their numbers were small, so virtually everything French about them became assimilated except their surnames. Likewise, in Appalachia's coalfields, immigrant Italian miners quickly became assimilated into the dominant English-speaking world.[23]

The concept of a culture hearth is crucial to understanding culture regions. A culture hearth is the center of a culture region where people share common culture traits. It is a focused place, usually with a deep history and concentration of diagnostic traits. Some examples are southeastern Pennsylvania for eighteenth-century Germans; southeastern New England for the English in the Northeast; Lafayette, Louisiana, for Cajun people; and Cuzco, Peru for the ancient Incas.

Folk Culture

Earlier I suggested that the real Appalachia could be found in its folk culture. E. Estyn Evans, Ireland's most distinguished cultural geographer, focused his classic *Irish Heritage* on the relic, residual folk culture that formed the basis of his real Ireland. In Appalachia we are not in the Old World anymore, but we seek old things and old ways of life here in the New World. If we are successful, we may discover some of the culture traits that trace back to the Old World. But more important, we hope to discover diagnostic traits preserved in Appalachia's traditional ways of life.

Folk is established by tradition. Folk is traditional, it is frequently old-timey, and some people consider it old-fashioned. Folk is homemade. Folk is the self-made culture of a people. Folk is the unselfconscious constructs in the material and nonmaterial culture of a given group. Rarely does folk originate from a commercial disposition, although it may evolve into a commercial enterprise at some later time. As you read this book and examine the photographs and maps, think about the folk culture that may exist in your own setting. What are some of the traditional activities, rituals, legends, or objects and implements in your culture? They could be old tools, farm buildings, food preferences, dialect patterns, rituals, songs, or stories handed down that even in modern society serve as keystones for at least a portion of a culture that indeed may be folk.

Henry Shapiro illustrated "folk" by referring to Cecil Sharp, an early twentieth-century English folk song collector in Appalachia: "Sharp's demonstration of the 'folk' character of Appalachian culture functioned not only to establish the peculiarities of mountain life as legitimate patterns in the present, rather than merely as patterns carried over from the past, but also to define the mountaineers themselves as a legitimately distinct people in the American present. A native-born, white, Anglo-Saxon, Protestant, American folk, their very status as 'folk' defined their relationship to the rest of the American population."[24]

Tradition: "It's always been this way"

Tradition is the common denominator that makes things folk. Folk things are handed down from generation to generation. Rarely written down, these cultural commandments, rituals, sayings, songs, and ways of life form common threads woven into the fabric we call the folk culture of Appalachia. Folk people seldom know the origins of their culture or the origins of specific traits. When we ask about an object or song or ways of making a basket, the answer is, "It's always been this way."

Once you accept the idea that folk is best found in traditional settings, you are well on your way to understanding the nature of folk culture.[25] In 1916–18, Maud Karpeles, an Englishwoman working with

Cecil Sharp, collected over five hundred different folk songs in parts of southern Appalachia (see chapter 8). When Karpeles revisited Appalachia in 1951 to collect more folk songs, she made these profound observations:

> The radio, which now operates in nearly every mountain home, has let loose a flood of "hilly-billy" and other popular music, and this is gradually submerging the traditional songs. . . .
>
> Nevertheless tradition dies hard. Memory may weaken, but the love of the songs remains and with a little encouragement it springs up anew. To many a singer it was a great delight to be able to re-learn from these volumes a song that he had sung to Cecil Sharp over thirty years ago and had since forgotten. Thus, a song, originating in England and carried to America, lives there by oral tradition for some hundreds of years; it is written down and taken back to England by Cecil Sharp; then some thirty years later the song is carried back in printed form to the country of its adoption and takes on a new lease of life. Such are the devious ways of tradition.[26]

Isolation: "You cain't get there from here!"

One of the cardinal yet controversial characteristics of Appalachian culture has been isolation. Two themes separated by more than fifty years take two views of the isolation question: the belief that physical isolation in a mountain environment created cultural depravity, and the response that isolation was of little or no consequence. The first theme began in the late nineteenth and early twentieth centuries when influential people such as John C. Campbell and Olive Dame Campbell, Ellen Churchill Semple, Horace Kephart, and descriptive writers entered the region and observed that isolation was the chief culprit behind the region's backwardness. Environmental determinism, the then popular belief that all human behavior, cultural or otherwise, is rigidly controlled by environmental conditions, was widely promoted by Ellen Churchill Semple and Ellsworth Huntington. According to Semple, the mountains naturally gave people leg muscles of iron. In Huntington's view, mental capacity was attributed to favorable or unfavorable climates, and mountain climates, then considered harsh, numbed the brain.[27] Could it be true that my beloved mountains made the inhabitants slow of thought and action?

In her paper on Appalachian Anglo-Saxons in Kentucky, Semple accurately described the life of the mountaineer in eloquent prose. But her environmental deterministic views emerged when she concluded with these persuasive remarks: "The whole civilization of the Kentucky mountains is eloquent to the anthropogeographer of the influence of the physical environment, for nowhere else in modern times has the progressive Anglo-Saxon race been so long and so completely subjected to retarding conditions; and at no other time could the ensuing result present so startling a contrast to the achievement of the same race elsewhere as in this progressive twentieth century."[28]

The second theme on isolation appears in some of the Appalachian writing produced since the 1970s. In an apparent backlash, scholars such as Ronald D. Eller, Crandall A. Shifflett, and David Whisnant dismiss the notion that isolation ever had much to do with Appalachian culture or backwardness. These and other contemporary writers seem embarrassed to address the isolation issue except to minimize its significance.[29] But the more important question is, Did physical isolation or cultural isolation cause the region's inhabitants to seem culturally backward?

I believe some people became culturally isolated by choice. Even after major highways passed in front of their homes, after electricity, radio, television, satellite dishes, and other modern cultural materials were available to them, some people did not reach for the new but held fast to the old ways and appeared to be minimally affected by change agents that could have dramatically altered their lives. They remained steadfast in their comfortable old folk culture. Extending transportation to inaccessible areas was not always the solution to introducing progress to a tradition-bound society. What good would a newly paved highway do if one had no car and no ambition to take that road out to a much different way of life? Some people maintained their folk culture and folkways because they could not afford anything else. Physical isolation had been overcome through outside forces, yet cultural isolation prevailed. Mary Franks, a close friend of mine and a caseworker who knows contemporary rural East Tennessee so very well, told me of adults she knew in the early 1980s who had lived all their lives in Grainger County, Tennessee, but had never been to the "big city" of Knoxville less than twenty-five miles away.[30]

The outside world thinks we must push progressive thinking into the isolated mountain culture to bring about change. Culture change requires both an introduction to foreign ideas and acceptance of them. Here we have an archaic embedded folk culture coming in contact with the culture of modern society. Assuming that the recipient culture is open to acceptance, the cultural process of acculturation goes to work:

the two cultures make contact and exchange traits, but each retains its own identity. Acculturation, if anything, has been responsible for change in much of the region. Appalachia retains its folk identity well and at the same time has generally accepted traits from the outside world when cost, convenience, necessity, and perhaps logic were at issue. For example, when my research team and I surveyed twenty-six log smokehouses in Grainger County, Tennessee, in 1978, we found only one building still functioning as a smokehouse.[31] All the others had lost their intended purpose of smoking and storing pork and were used now for storing garden tools and assorted household junk. Rural informants told us that it was too much trouble to kill hogs and smoke meat anymore and that it was just so much easier to buy bacon and sausage at nearby stores. Change was acceptable to those who desired, accepted, and could afford the merchandise once they had access to it.

Despite good intentions to bring about change in Appalachia, we must understand the role isolation plays in preservation. In the past, isolation alone preserved much of Appalachia's folk culture. Can the strength of isolation do the same now? No. While geographical isolation still can be internalized into cultural isolation, change is arriving much too fast and too strongly for folk to get out of the way. The region needs a proactive means for preserving traditional folk culture and folk traits, and preservation must be done carefully so as not to contaminate the elements. For decades in Cherokee, North Carolina, the Native American culture as seen along the roadways was represented almost entirely by local Cherokees dressed in war paint, animal skins, and feathered headdresses—culture traits belonging to Plains Indians. Fortunately there are more authentic sides to the Eastern Band of Cherokees, the best of which is the Oconaluftee Village—a museum, collection point, and market for Cherokee crafts. Another is the outdoor pageant "Unto These Hills," which tells the history of the Cherokees along with their trials and tribulations in their contact with the White Man's culture.

Appalachia's Image

Since we see the world through the lens of culture, our personal view of Appalachia is forever colored by our own cultural composition. We do this now as people have always done it, whether or not they are conscious of it. Is there a single Appalachian culture, or are there many culture groups who live in Appalachia? There are many, but that has not always

Appalachian loggers included these men at the William M. Ritter Lumber Company at Big Branch Camp in Dickenson County, Virginia, in 1938. (C. C. Tiller Collection, courtesy of Archives of Appalachia, East Tennessee State University)

been understood. Most of America's current understanding is that Appalachia has a singular, almost uniform culture identified with white poverty. Nineteenth-century observers were convinced that Appalachian peoples were descended from a single culture of Scotch-Irish frontier settlers. While the Scotch-Irish probably constituted the majority of ethnic immigrants into Appalachia, many other cultures, both domestic and foreign, settled the region. Moreover, contemporary Appalachia has several diverse culture groups, many the descendants of older settler groups, but some of them recent arrivals. Some descendants of pioneering groups, no longer thinking of their ethnic ancestry, simply call themselves "Americans." The variety of economic endeavors extends beyond subsistence agriculture, logging, and coal mining to industry, service, and institutional patterns of economy. It is inaccurate to say that Appa-

A fading Appalachian scene has a single pen house with half-dovetail corner notches and a board-and-batten lean-to. It was home for this couple in 1976. She sees him off to work as a Ford truck with a Confederate flag in the back window arrives to pick him up. (Photograph by J. Rehder, 1976)

lachian culture is exclusively white poverty. Its cultures and economies, folk or otherwise, say much more than that.

Appalachia is a social entity, but it has many faces. Some frown under coal camp poverty; some eagerly seek progress with hopes for a better tomorrow; and some merely stare blankly in hollow psychological isolation. While it would be easy to accept the idea of a single, uniform culture, Appalachia appears to sort itself out as a crazy quilt of many cultures. The temporal issue should also be considered. Appalachian cultures have not been the same throughout the history of the region. Appalachia's peoples changed from pioneer frontiersmen, to permanent farming groups, to townspeople, loggers, coal miners, welfare recipients, industrialists, service providers, and an entire mosaic of economic identities. Setting economics aside for a moment, consider the slow evolution of language, the still slower changes in religion, and the sparse changes in values and attitudes that some Appalachian peoples still experience today. In the search for an Appalachian cultural identity, are we seeking stereotypes? Not entirely. I say this with some reservations because I believe that stereotypes are not entirely wrong. If properly understood, some stereotypical behavior can be a valid surrogate for many or most people in a culture group. If we focus on identity, stereotypical traits—if accurate and authentic—can be useful examples of a culture's identifying characteristics that lead to diagnostic culture traits. Some traits, like them or not, simply will not go away. Can we deny the existence of illegal moonshine whiskey, family feuds, snake-handling religious groups, ginseng gathering, shaped note singing in church, fiddle and banjo playing, clogging, quilting, and thousands of traits that belong to Appalachia?

How have the cartoons of Snuffy Smith and Li'l Abner or the television series *The Beverly Hillbillies* affected our understanding of Appalachian culture? Is it fair to call Appalachian people hillbillies? I abhor the term, but being politically correct does not erase one's identity, either to oneself or to outsiders who observe that a hillbilly culture is still cultivated in Appalachia. Snuffy Smith, Li'l Abner, the Beverly Hillbillies, and the Dukes of Hazzard represent Appalachia about as much as the cartoon Jetsons represent NASA's space exploration. Attempts at ethnic humor still exist in Appalachia, but nobody here calls it "ethnic," and few people seem to care. Elsewhere, the once popular Polish jokes are gone by attrition. Black jokes are strictly taboo. Southern Louisiana's Cajuns have their special in-house "coonass" identity within Cajun culture and their hilarious Boudreaux jokes that are widely told and accepted inside and outside the culture. "Redneck" southern jokes continue to be popular, largely fostered by Jeff Foxworthy's presentations of "You might be a redneck if . . ." One might argue that cultures should be able to laugh at themselves. Laughter is good medicine, and humor can be a wonderful uplift to the psyche. Can Appalachia be understood through characterizations that are the outside world's ridicule? I think not. Yet the cartoons are still published in newspapers, reruns air on television, and the hillbilly shows continue to be popular. Worse yet, a CBS reality television program is being developed to entice an unsuspecting Appalachian family to move to Hollywood—à la the "real" Beverly Hillbillies, so that the whole world can laugh at their naïveté. Such popular culture maintains the contrived sense of "innocent fun" at the expense of a poor people's rich heritage.

My role is not to debate the cultural incorrectness of stereotypes of Southern Appalachia. But I question whether ethnic or regional humor created by someone outside that culture or region can ever produce an accurate picture of the people and their place. Appalachian folk culture must be viewed apart from the images created by nineteenth-century writers and contemporary cartoons and television. The truth lies somewhere in between, but leaning ever so slightly to the right, with the left representing far-fetched feature writing and caricatures while the right is the dull conservative belief that Appalachian people are just regular folks like the rest of us except that they happen to live in the mountains. This far right position blatantly denies the uniqueness of Appalachian cultures. What makes Appalachian people a part of folk culture and what are the diagnostic traits that best identify them? These bottom-line questions are most important, and I intend to answer them in this book.

How can we put a face on contemporary Appalachia? The sociology of Appalachian life for several decades has focused on poverty—the kind of poverty reflected in photographs taken years earlier of desperately poor whites in the dust bowl days of the Great Depression. Is the face of Appalachia a woman in her middle forties whose hardscrabble life makes her look sixty? Is it the face of a deep shaft coal miner, covered with coal dust and coughing up soot from deep within his diseased lungs? Or could it be the small, round visages of dirty urchins staring blankly at the distant road that passes by their trailer up a dark hollow?

In a film documentary titled "American Hollow," Rory Kennedy has captured at least one family of Appalachian faces that may give us a glimpse of what life is like in the hollows. The film is about Iree Bowling, who came from a family of eleven who shared a single bedroom and had no electricity, no running water, and no shoes for many of the children. Mrs. Bowling went on to raise thirteen children of her own. One good measure of attainment here has been to have all of your family living nearby, close enough to visit at Christmastime. A belief in getting out of the hollow and seeking one's fortune in the big city brings disappointment when reality sets in. Neither skilled nor sufficiently educated, young people find they cannot make a living from low-paying jobs on the outside. Many return to the hollow to a family support system that seems to live on in perpetuity.[32]

The Making of an Image, 1870–1920

Until the American Civil War, the United States east of the Mississippi River was largely an agrarian world. The South had well-developed plantation economies and landscapes, while the North had family farms along with budding industries and large East Coast port cities collecting immigrants. Appalachia was surrounded by eighteenth- and nineteenth-century cultures and economies that were somewhat different but still largely rural and agrarian, and hardly more worldly. Antebellum Appalachia was not much different from any other poor rural part of America; it was just isolated and insulated from northern and southern cultures and economies. After the Civil War, both the North and the South changed economically, politically, and in some places progressively to the point that Appalachia looked almost like a mountaintop island surrounded by a war-torn but recovering America. It was as if the America of one age had shifted gears during and after the Civil War while the mountain South was left far behind. After 1870, Appalachian people and place appeared to share one definition. The mountain region had a unique physical identity, and the people appeared to be quite different from mainstream America. Appalachia was a place in but not of America. But how did this identity begin?

Color Writers Describe a Strange Land and a Peculiar People

From 1870 to 1920, Appalachia was viewed as a strange land with peculiar people. It may still be seen this way. What was Appalachia like during this fifty-year period of image making, and how were images of it formed? This is answered in two wonderful books: *Appalachia on Our Mind* by Henry Shapiro, and W. K. McNeil's *Appalachian Images in Folk and Popular Culture.* In the period 1870 to 1886, Appalachia began to be "discovered" and defined by explicit cultural characteristics. Writers such as Mary Noailles Murfree, Frances Hodgson Burnett, Julia Schayer, and John Esten Cooke were writing fiction set in Appalachian locales. They wrote for America's postwar readers hungry for stories set in interesting places, travel sketches, and illustrated pieces published in magazines such as *Harper's, Atlantic Monthly, Scribner's, Lippincott's, Southern Bivouac,* and others. During the same period, others acted as descriptive interpreters. Writers such as Will Wallace Harney, William Goodell Frost, James Lane Allen, John Fox Jr., and Josiah Stoddard Johnston described Appalachian culture and ways of life and tried to explain them. The two groups often wrote in exaggeratedly colorful ways, hence the name "color writers." Between 1870 and 1890, about 90 travelers' accounts and sketches plus 125 short stories were written to describe this strange land and to analyze the peculiar people of Appalachia.[33]

In the same vein as later cartoons like "Li'l Abner" and "Snuffy Smith," some of the early local color writers concocted unfair and prejudicial descriptions that amounted to tall tales. William Goodell Frost's influential 1899 paper "Our Contemporary Ancestors in the Southern Mountains" uses language that is both fascinating and offensive. By the title alone, Frost consigns Appalachian people who are very much alive to a dead and distant past. By terming living beings "ancestors," Frost calls them old-fashioned and out of date, implying that they are backward. As just some evidentiary traits, Frost speaks of "pioneer reminders" such as the rude Saxon dialect of the mountains, large families, a barter system, tall, gaunt, underfed folk, rough-hewn log cabins, the absence of conveniences, music in a weird minor key, fatalism, and the blood feud.[34] He describes the isolation and distance: "It is a longer journey from northern Ohio to eastern Kentucky than from America to Europe; for one day's ride brings us into the eighteenth century." He defines the region: "The mountainous back yards of nine states abut upon the lofty ridges which separate the Virginias, bound Kentucky on the east, divide Tennessee from North Carolina, and end in Georgia and Alabama. . . . The surface varies greatly in elevation and geologic structure, but as a place for human habitation the entire region has one characteristic—the lack of natural means for communication. Its highways are the beds of streams; commerce and intercourse are conditioned by horseflesh and saddlebags."[35]

In name and in content, Frost's "Our Contemporary Ancestors" is not unlike Jack Weller's 1965 book *Yesterday's People.* Both play on the image of perceived backwardness and ignorance. Both "colorize" Appalachian cultures in unnecessarily negative tones. Certainly there are people in Appalachia today who are underfed and undereducated, just as there were in Frost's late nineteenth century. There are people who have suffered the consequences of deprived indifference, some of whom come from shallow gene pools in isolated hollows and still live in backwoods isolation. But to characterize Appalachian peoples and their culture as entirely ancestral is incorrect, and to refer to them as "yesterday's people" is unfair to the total culture of the region.[36]

In a 1901 color article, "The Southern Mountaineer," John Fox Jr. described his image of mountain life: "So, in the log cabin of the Southern mountaineer, in his household furnishings, in his homespun, his linsey and, occasionally, in his hunting shirt, his coon-skin cap and moccasins one may summon up the garb and life of the pioneer; in his religion, his politics, his moral code, his folk songs and his superstitions one may bridge the waters back to the old country, and through his speech one may even touch the remote past of Chaucer. For to-day he is a distinct remnant of Colonial times—a distinct relic of an Anglo-Saxon past."[37] Fox's imagery painted from memory told of log cabins, funerals, folk speech, moonshiners, and the unusually generous nature of a people who would walk for miles over a mountain in the night to borrow cornmeal to prepare a meal for visitors. On a more somber and profound note, Fox eloquently explained what he saw as the backwardness of the culture: "The mountain dweller lives apart from the world. The present is past when it reaches him; and though past, is yet too far in the future to have any bearing on his established order of things. There is, in consequence, no incentive whatever for him to change. An arrest of development follows; so that once imprisoned, a civilization, with its dress, speech, religion, customs, ideas, may be caught like the shapes of lower life in stone, and may tell the human story of a century as the rocks tell the story of an age."[38] Fox makes an electrifying assertion about the temporal journey of a present that is already past when it finally reaches the mountain people. Is this picture true now or was it once, and are we simply refusing to believe it? Fox's writing offers us one man's interpretation of Appalachian culture.

The color writers left an indelible mark on the image of Appalachia. They were published in well-read publications with wide distributions and national exposure. Their writing created an unmistakably negative picture of backwardness brought on by mountain isolation and perceived stultifying elements in the culture. The region must still overcome these legacies that paint a region of isolation and ignorance and describe the inhabitants as yesterday's people.

Anthropomorphic Descriptions, 1869–1913

The physical appearance of a people rarely becomes an issue in the cultural geography of a folk region. But for Appalachia, we find anatomical descriptions that make us ponder why they were made. Human behavior is far more significant than appearance in understanding a folk culture. However, historical descriptions of both physical appearances and folk behavior tell us a bit more about the image of Appalachia. In 1869 Will Wallace Harney, a physician traveling through the Cumberland Mountains, observed: "The natives of this region are characterized by marked peculiarities of the anatomical frame. The elongation of the bones, the contour of the facial angle, the relative proportion or disproportion of the extremities, the loose muscular attachment of the ligatures, and the harsh features were exemplified in the notable instance of the late President Lincoln."[39]

In 1901 Ellen Churchill Semple, Kentucky geographer and environmental determinist, described Appalachian people this way:

> Though the mountaineer comes of such vigorous stock as the Anglo-Saxons, he has retained little of the ruddy, vigorous appearance of his

forebears. The men are tall and lank, though sinewy, with thin bony faces, sallow skins, and dull hair. They hold themselves in a loose-jointed way; their shoulders droop in walking and sitting. Their faces are immobile, often inscrutable, but never stupid; for one is sure that under this calm exterior the mountaineer is doing a great deal of thinking, which he does not see fit to share with the "furriner," as he calls every one coming from the outside world. The faces of the women are always delicately moulded and refined, with an expression of dumb patience telling of the heavy burden which life has laid upon them. They are absolutely simple, natural, and their child-like unconsciousness of self points to their long residence away from the gaze of the world. Their manners are gentle, gracious, and unembarrassed, so that in talking to them one forgets their bare feet, ragged clothes, and crass ignorance, and in his heart bows anew to the inextinguishable excellence of the Anglo-Saxon race.

The lot of a mountain woman is a hard one. Only the lowest peasantry of Europe can show anything to parallel it. She marries between twelve and fifteen years a husband who is between seventeen and twenty. . . . A family of from ten to fifteen offspring is no rarity, and this characterizes not only the mountains of Kentucky, but the whole area of the Appalachian system. In addition to much child-bearing, all the work of the pioneer home, the spinning and weaving, knitting of stockings, sometimes the even making of shoes and moccasins, falls on the woman. More than this, she feeds and milks the cow, searches for it when it has wandered away "in the range," or forest, hoes weeds in the corn, helps in the ploughing, carries water from the spring, saws wood and lays "stake and ridered" fences. . . .

The mountain woman, therefore, at twenty-five looks forty, and at forty looks twenty years older than her husband. But none of the race are stalwart and healthy.[40]

A final description of Appalachian people comes from Horace Kephart, an outdoorsman and writer who lived in self-imposed exile on the North Carolina side of the Great Smoky Mountains from 1904 to 1913. Kephart's disjointed yet colorful comments come from his nine years of living deep in the mountains, not merely from casually visiting the area:

Our average mountaineer is lean, inquisitive, shrewd. . . . A fat mountaineer is a curiosity. The hill folk even seem to affect a slender type of comeliness. . . .

> Spartan diet does not put on flesh. Still, it should be noted that long legs, baggy clothing, and scantiness or lack of underwear make people seem thinner than they really are. Our highlanders are conspicuously a tall race. Out of seventy-six men that I have listed just as they occurred to me, but four are below average American height and only two are fat. About two-thirds of them are brawny or sinewy fellows of great endurance. The others generally are slab-sided, stoop shouldered, but withey. The townsfolk and valley farmers, being better nourished and more observant of the prime laws of wholesome living, are noticeably superior in appearance but not in stamina.
>
> Nearly all males in the back county have a grave and deliberate bearing. They travel with the long, sure-footed stride of a born woodsman, not graceful and lithe like a moccasined indian (their coarse brogans forbid it), but shambling as if every joint had too much play. There is nothing about them to suggest the Swiss or Tyrolean mountaineers; rather they resemble the gillies of the Scotch Highlands. Generally they are lean-faced, sallow, level-browed, with rather high cheekbones. Gray eyes predominate, sometimes vacuous, but oftener hard, searching, crafty—the feral eye of primitive man.[41]

When I first read Horace Kephart's *Our Southern Highlanders,* I mistakenly thought the book was a semifictitious "color" piece generated from the fertile imagination of an outsider. Kephart was such an astute observer of Appalachian ways of life that he was clearly acculturated into becoming something of a mountaineer himself. His keen observations examine a people not of his kind. His comparisons between the southern highlanders and Native Americans and Alpine peoples as well as townspeople and valley farmers show insight from an outsider whose knowledge of the world provides a backdrop with which he compares mountain life. Kephart's astute observations were made when Southern Appalachia and its folk culture were still in splendid isolation and preservation.

Appalachia since 1960
The Social Identity of Appalachia

The social identity of Appalachia has fascinated sociologists since the 1890s. The early color writers did much to raise public awareness of a peculiar way of life in the southern mountains. But it was not until the 1960s that a major sociological thrust was made during the War on Poverty. In the 1960s, Thomas R. Ford published a significant report

Appalachian poverty has many faces. One that comes to mind all too often is a ramshackle house with a washing machine on the front porch, a yard full of junk, and an old car up on blocks. But this folk house is a double pen saddlebag house of the English pen tradition, constructed about 1920 of sawn lumber and balloon framing, not logs. (Photograph by J. Rehder, 1976)

that statistically surveyed the region's poverty and described such traits of the Appalachian people as individualism, self-reliance, traditionalism, fatalism, and religious fundamentalism.[42] In another analysis, Bruce Ergood searched twenty books and articles describing Appalachian people. Here are eleven characteristics that appeared in them, in order of most frequent citation: independence, religious fundamentalism, strong family ties, life in harmony with nature, fatalism, traditionalism, honor, fearlessness, allegiance, suspicion of government, and born trader.[43]

Loyal Jones's list of Appalachian values appears to have struck the heart of behavioral traits in the region. Jones, a prolific writer and candid scholar of Appalachia, says that the following values shared by many Appalachian people are important to their lives: religion, independence, self-reliance and pride, neighborliness, familism, personalism, love of place, humility and modesty, sense of beauty, sense of humor, and patriotism.[44]

Henry Shapiro's book on Southern Appalachia interpreted the national consciousness of the region for the period 1870 to 1920. Shapiro determined that the rediscovery of Appalachian culture by color writers, home missionaries, settlement schools, handicraft revivalists, philanthropists, and folk song collectors contributed to a mythic regional consciousness, to the point that Appalachia became to most of America "a coherent region inhabited by an homogenous population possessing a uniform culture."[45] Shapiro explained that not only was such thinking a myth, but as other intellectuals pursued reality, they began their efforts cloaked in a mythic atmosphere. Shapiro argued:

> Even the debunkers, within and without the universities, who have risen to the occasion and challenged the accuracy of particular generalizations about the mountaineers in order to paint them in a more "realistic" light, have refused to ask the central question of their craft, about the reality of the phenomenon they seek to explicate. Instead, they have begun with the assumption that the mountaineers do in fact compose a distinct people with distinct and describable characteristics. They have argued from within a mythic system about the accuracy of mythology, and attacked the generalizations of folklore which are at once so vague and so potent as to defy examination or correction.[46]

An Appalachian Ethnography

Ethnography is the time-honored anthropological subdiscipline that meticulously describes culture groups. Depicted by contemporary scholars as archaic, ethnography is purely descriptive and differs from ethnology, a somewhat newer scientific anthropological subdiscipline that analyzes cultures, probes historical developmental processes, and frequently conducts cross-cultural comparisons. In the late nineteenth and early twentieth centuries, European and American ethnographers combed the world seeking exotic culture groups. They explored, discovered, observed, and described cultures in Africa, Indonesia, South America, and other distant places with much the same intensity that biologists today bring to the study of rare and exotic plants and animals. Ethnographers intensely described the material traits and many of the nonmaterial ones in the cultures they discovered. They prepared long trait lists and cataloged hundreds or thousands of artifacts from living cultures. Their work was systematic, methodical, usually objective, and quite clinical.

In the core chapters of this book, we encounter myriad material folk traits that can be observed in the spirit of ethnography or ethnology, but especially cultural geography. As a cultural geographer, I selectively examine the folk architecture of the region, folk ways of making a living, food, folk remedies and belief systems, and music and crafts and festivals, among other things. We will observe many but certainly not all forms of Appalachian folk culture. I hope that this exploration in a cultural geographic spirit will lend the subject a soul-like grace.

2

The Shape of Appalachia

> Always enwrapped in the illusory mists, always touching the evasive clouds, the peaks of the Great Smoky Mountains are like some barren ideal, that has bartered for the vague isolations of a higher atmosphere the material values of the warm world below.
>
> MARY NOAILLES MURFREE, 1885

When viewed from space, Southern Appalachia has roughly the shape of an elongated S—moderately narrow at the northeastern end, broader in the middle, and tapering to a point in the southwestern extreme. Although the entire Appalachian region extends 1,500 miles, from Quebec to northern Alabama, the part that is Southern Appalachia covers about 1,200 miles beginning in Pennsylvania and extending southwest to northern Alabama. From east to west, three large geomorphic physical regions dominate the landscape, with the east composed of true, high mountains, the middle a wide valley filled with parallel ridges and narrow valleys, and the western region composed of dissected plateaus. The region is almost four hundred miles wide and embraces three major geomorphic provinces: the Blue Ridge, an Appalachian mountain core on the eastern flank that includes other high southern mountain sections; the aptly named Ridge and Valley Province of parallel sedimentary ridges and valleys at the center; and the Allegheny and Cumberland plateaus to the west.[1]

The Blue Ridge: Appalachia's Mountain Core

Appalachia's mountain core, composed of ancient Precambrian crystalline and metamorphic rocks, lies in the eastern part of Southern Appalachia between southern Pennsylvania and Georgia, extending for 550 miles. The Blue Ridge Province has the oldest rocks in Appalachia, dating from 600 million years or more ago. The province is a remnant of a

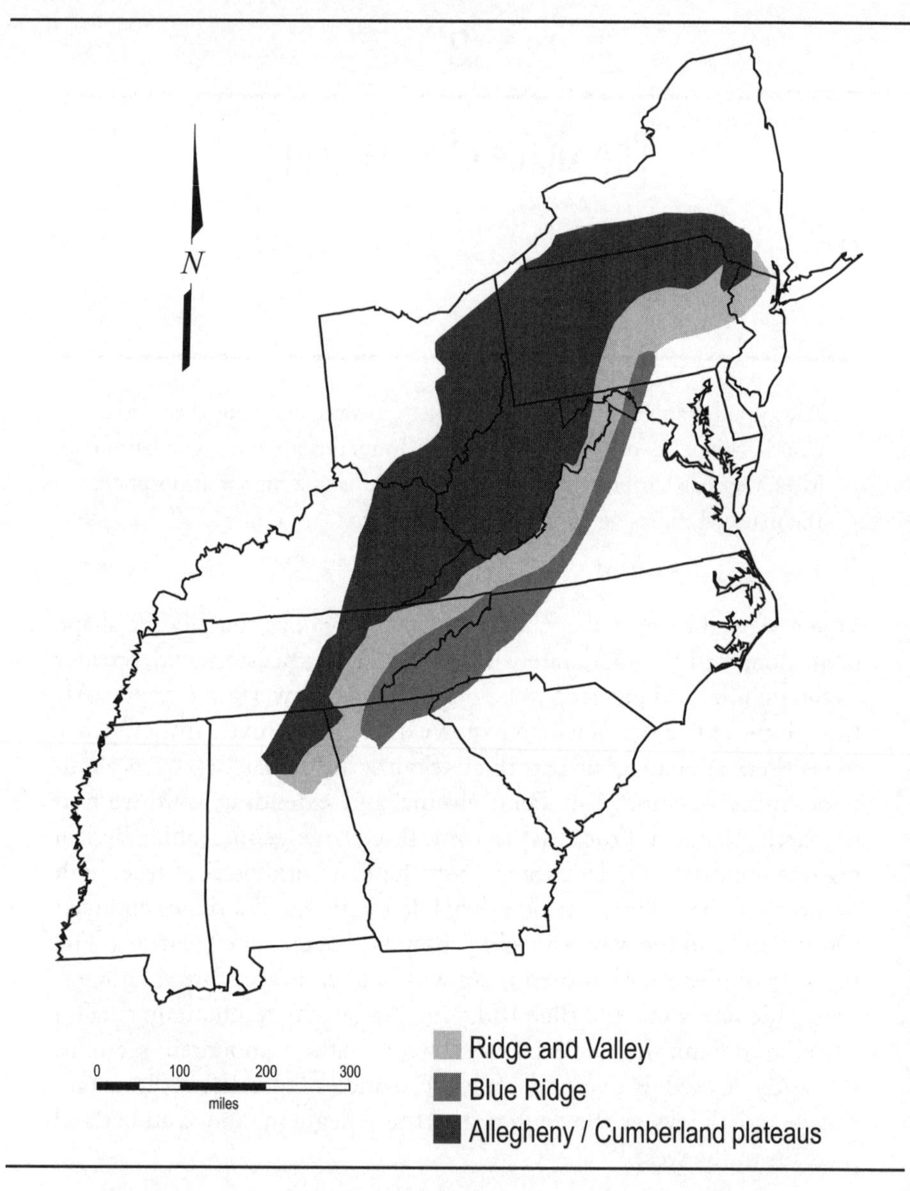

The geomorphic regions of Southern Appalachia are the Blue Ridge Province, the Ridge and Valley Province, and the Appalachian Plateau Province, better known as the Allegheny and Cumberland plateaus. (Source: Fenneman 1938)

The Great Smoky Mountains in the Blue Ridge Province are part of an ancient series of complex ranges. View south from Mount Kephart on the Appalachian Trail. (Photograph by J. Rehder, 1998)

higher mountain mass, but millions of years of erosion have worn it into the complex pattern of ranges we see today. Countless place-names for peaks, streams, and valleys can be confusing even within the same mountain range. For example, the mountains are known as the Blue Ridge in Virginia, the Carolinas, and Georgia, as the Unakas in northeastern Tennessee and southern Virginia, and as the Great Smoky Mountains in eastern Tennessee and western North Carolina, but all belong to the same Appalachian mountain core called the Blue Ridge Province. In the province south of the Potomac River in Virginia, forty-six peaks exceed 6,000 feet in elevation, and 288 peaks are above 5,000 feet. The highest mountains in the Blue Ridge are Mount Mitchell (6,684 feet) and Mount Craig (6,663 feet). In the Unakas and Iron Mountains in Virginia the highest is Mount Rogers at 5,729 feet. The highest in the Great Smoky Mountains are Clingmans Dome (6,643 feet), Mount Guyot (6,621 feet), and Mount Le Conte (6,593 feet).

The mountain core begins as a single narrow ridge near the Susquehanna River in southern Pennsylvania and proceeds as the Blue Ridge at 1,200 feet elevation near Harpers Ferry, West Virginia, at the junction of the Shenandoah and Potomac rivers. The next section of Blue Ridge in Virginia is also narrow, ranging from a mere five miles wide to fourteen, with elevations between 2,500 and 4,000 feet. The Blue Ridge continues southwestward through Virginia, widening to eighty miles to become Appalachia's highest mountains in western North Carolina. The province extends south to northern South Carolina and ends in northern Georgia. The eastern edge of the Blue Ridge forms an impressively steep mountain front as it rises abruptly from the western margins of the Piedmont. In most places along this Blue Ridge escarpment, or cliff, vertical relief patterns above the Piedmont are from 1,500 feet in Virginia to 4,500 feet in North Carolina. Actual Blue Ridge elevations are 1,200 to 6,684 feet above sea level. Mount Mitchell's elevation at 6,684 feet in western North Carolina near Asheville makes it the highest mountain east of the Rockies.

West of the Blue Ridge, other names are applied to a wider pattern of mountains within the core. Just south of the Roanoke River in Virginia, the mountains divide into two forks that separate by up to eighty miles to form a large oval, rejoining near Springer Mountain, Georgia. The eastern fork at least retains the name Blue Ridge all the way from Virginia to Georgia. However, the western fork has several names, such as the Iron Mountains and Unaka Mountains between Roanoke, Virginia, and northeastern Tennessee and the always impressive Great Smoky Mountains in western North Carolina and eastern Tennessee. Near the Great Smokies and forming a western border on this western fork, we encounter English Mountain, Chilhowee Mountain, Starr Mountain, the Unicoi Mountains, and other small mountain groups before they rejoin the Blue Ridge. The land lying within the eighty-mile-wide oval between the two forks is a series of complex east-west-trending transverse ranges that include the Black, Balsam, Pisgah, Cowee, Cheoah, Nantahala, and Snowbird Mountains, among others. The Black Mountains just north and east of Asheville, North Carolina, contain Mount Mitchell (6,684 feet) and Mount Craig (6,663 feet), the two highest peaks in the eastern United States. Just southwest of Asheville, the Balsam transverse range has Richland Balsam at 6,540 feet and Mount Pisgah at 5,721 feet in the Pisgah range.[2]

These southern highlands are among the most isolated and rugged, the highest, steepest, and most majestic mountains in all of Appalachia. In 1885 Mary Noailles Murfree captured the essence of the Great Smoky Mountains:

Gregory Bald (upper left), at 4,949 feet elevation, is only a snow-covered grassy patch in the Great Smoky Mountains. View east. (Source: U.S. Geological Survey, HAP [high-altitude photography], 1984)

Always enwrapped in the illusory mists, always touching the evasive clouds, the peaks of the Great Smoky Mountains are like some barren ideal that has bartered for the vague isolations of a higher atmosphere the material values of the warm world below. Upon those mighty and majestic domes no tree strikes root, no hearth is alight, humanity is an alien thing, and utility set at naught. Below, dense forests cover the massive, precipitous slopes of the range, and in the midst of the wilderness a clearing shows, here and there, and the roof of a humble log cabin.[3]

The Ridge and Valley Province

West of the mountain core is the second of the three regions. For nearly 1,200 miles, this serpentine region forms a series of parallel valleys and

The aptly named Ridge and Valley Province has topography to match its name. (Source: U.S. Geological Survey, Radar Mosaic, Oak Ridge, Tennessee, 1984)

ridges that trend northeast to southwest from southeastern Pennsylvania to Birmingham, Alabama. Along the eastern margin of the province, the Great Valley is the widest and longest valley of this wonderfully named landform region of North America, the Ridge and Valley Province. In northern Virginia the Great Valley is less than twenty miles wide, but at Buchanan, near Roanoke, the Great Valley constricts to a mere two miles. Farther south in eastern Tennessee, the entire province is between forty-five and seventy-five miles wide. As many as ten to eighteen parallel ridges and a like number of small valleys form a washboard pattern.

Variations in elevation are generally marked by stream patterns. Ridge elevations of 500 feet above sea level can be found in the north near Harpers Ferry and in the south near Birmingham, Alabama. Elevations of about 3,000 feet are on ridges such as Clinch Mountain and Bays Mountain. The highest parts are over 4,000 feet near Burkes Garden, Virginia, between Roanoke and Wytheville, Virginia, where a pivotal drainage divide diverts streams like the Clinch River southwest from

The Ridge and Valley Province of North America has a topography of steep forested ridges and gentle valleys, like this in Tazewell County, Virginia. (Photograph by J. Rehder, 1977)

Wytheville and sends the Shenandoah River northeast from Roanoke and Staunton before it joins the Potomac at Harpers Ferry. Valley floor elevations range from between 300 and 500 feet at Harrisburg, Pennsylvania, to 1,800 at Wytheville, Virginia, to 1,000 feet at Knoxville, Tennessee, to 400 feet in Birmingham, Alabama.

Geologically, the Ridge and Valley subregion is much younger than the Blue Ridge mountain core. Here sedimentary limestone, sandstone, and shale rocks of early Paleozoic age (Cambrian, 570 million years ago, and Ordovician, 500 mya) dominate the surface and subsurface terrain. These rocks form the basis for a variety of soils in the region, with limestone being the most favored soil type among early farmers. Sandstone-based soils were considered to be fair but still second-rate for agriculture, and shale-based soils were avoided by all but the most unfortunate, struggling farmers. In Grainger County, Tennessee, two adjacent valleys are named Richland Valley and Poor Valley. Guess which one is limestone and which is shale? The Great Valley, the longest and widest in the Ridge and Valley Province, runs along the eastern half of the province and has served an important role as the major route for land transportation dating from Indian trails to early wagon roads to contemporary highways. The Ridge and Valley Province, the oldest settled physiographic region, continues to be Appalachia's leading agricultural area and contains most of the prominent cities.

The Allegheny and Cumberland Plateaus

The third geomorphic province is the Appalachian Plateaus country, which is a maximum of two hundred miles wide along a west-to-east line at Ohio–West Virginia–Pennsylvania and a minimum of thirty miles wide in Tennessee. Commonly called the Cumberland Plateau in the south and the Allegheny Plateau in the north, both are dissected plateaus, eroded and rough, with flat hilltops and steep-sided surfaces that meet in narrow, canyonlike hollows. Layers of limestone, sandstone, shales, and coal sediments were laid down in an ancient sea over 350 million years ago during the geologic periods of the Mississippian (360–320 mya) and Pennsylvanian (320–296 mya). The original sediments remained relatively horizontal, since they were lifted over 2,000 feet above sea level into a tableland plateau with caprocks of sandstone. Once nearly all flat and level, much of the terrain is now eroded and dissected, carved by one of nature's most powerful forces—water. The Allegheny Plateau and much of the northern half of the Cumberland Plateau are almost entirely dissected, with extremely rough, irregular terrain. The term "plateau" in geomorphology tells us that the landscape is supposed to be flat on top, but on these dissected plateaus it takes imagination to visualize the earlier terrain. We see hills evenly touching the sky. Only the hilltops have equal elevations, but if we could lay a gigantic piece of plywood on the surface, we might envision a plateau tableland.

The southern Cumberland Plateau in the central parts of Tennessee and northeastern Alabama retains more of the original tableland appearance. A method for visualizing this less dissected southern Cumberland Plateau is to take a cross section or profile traveling from east to west across the region on interstate highway I-40. About forty miles west of Knoxville, Tennessee, we encounter the Cumberland Plateau Escarpment, a steep cliff one thousand feet tall at milepost 347. The escarpment marks the boundary between the Ridge and Valley Province to the east and the Appalachian Plateau Province, commonly known as the Cumberland Plateau, to the west. The terrain on top at first is rumpled into a sandstone anticline (an arched ridge of folded rocks) called the Crab Orchard Mountains, but then the sandstone caprock levels off to the

The flat tableland nature of the undissected southern Cumberland Plateau is shown at right. The Cumberland Plateau's western margins, marked by serrated shaded patterns, descend 800 to 1,000 feet to the Eastern Highland Rim, in dark tones at left. (Source: U.S. Geological Survey, Radar Mosaic, Chattanooga, Tennessee, 1984)

west as one nears Crossville, Tennessee. For about forty miles, the undissected plateau is smooth and flat until we reach the town of Monterey at milepost 300, where it becomes sharply dissected into canyons locally called gulfs. Here, orographic rainfall (rain caused by the mountains) has been much higher and erosion is greater, producing the jagged western edge of the Cumberland Plateau.[4] Beyond Monterey, the plateau descends 800 to 1,000 feet to the Eastern Highland Rim, an old limestone surface that is highly arable, between mileposts 297 and 272 and about 1,000 feet above sea level. Past this point, the profile descends another

An aerial photograph of the Cumberland Plateau illustrates patterns of fairly uniform topography with sparse farm settlement and an abundance of forest cover. Bledsoe County, Tennessee. (Source: U.S. Department of Agriculture, Agricultural Stabilization and Conservation Service, 1964)

400 feet and extends beyond the western margins of Southern Appalachia into the Nashville Basin between mileposts 272 to 186, at an average elevation of 600 feet.

A way to experience the highly dissected northern Cumberland Plateau in eastern Kentucky is to travel west to east on the Daniel Boone Parkway, a sixty-mile toll road. The parkway transects the plateau like a long, thin needle taking a biopsy of Appalachia. But this road does things that most roads do not. The road is almost cultureless. Like an elevated monorail, barely touching the ground and rarely directly touching the people, it inserts you into an Appalachian landscape of poverty in a clinical, antiseptic way. Exits and entrances come at ten- to fifteen-mile intervals; people who live somewhere in between them can only watch the road and see the privileged few who travel on it go by. The road gives us outsiders a quick glimpse—yes, a biopsy—of this part of

Kentucky's Cumberland Plateau but only distant views of its people. The primary towns and exits on the Daniel Boone Parkway are London at the western terminus, Manchester and Hyden Spur in the center, and Hazard on the east. From west to east between London and the border of the Daniel Boone National Forest sixteen miles down the road, the landscape is a bit more open, revealing Kentucky's black tobacco barns and open pastures that quickly give way to a couple of fading blue-painted barns and some small tobacco patches and gardens. Within the Daniel Boone National Forest, agriculture completely disappears, and forest, coal, and deep shadowed hollows take over until one reaches the eastern terminus.

A highly dissected plateau such as we find in eastern Kentucky's Cumberland Plateau or West Virginia's Allegheny Plateau region has two very different but similar surfaces—hilltops and hollows. The hilltops have trees and sometimes scarred steep sides, the "hollers" are ravines that hold house trailers, small shacks, and dirt tracks that pass for roads. Both are rough. Thirty years ago, the hollows in eastern Kentucky were not so congested. They had small wooden houses and the occasional house trailer, but there was daylight—some open space between residences. Hollows then had perhaps two or more occupied sites, often near the base or lower mouth. But now, from the base to the upper reaches, hollers are awash with trailers, many derelict, swathed in the trash and filth that constitutes the flotsam in this sea of poverty. Some hollows remind me of Bombay, India, where riverlike throngs of rusting, tin-roofed slums reach deep into the interior of the city with their tentacles of squalor. Appalachian hollows differ from Bombay's slums only in the color of the dwellings and the people (white), in the population density, and in the degree of poverty.

It is dark in these hollows. Sunlight, even when it is available, is limited to a few hours rather than the all-day sun on the hilltops, since the steep terrain obscures all but direct rays. "The sun wouldn't come up until about ten o'clock in the morning, and it went down at four in the afternoon," says Carolyn White, a former student and boarder at the Hindman Settlement School in Knott County, Kentucky.[5] Darkness becomes a way of life where most work traditionally has been below ground in man-made caverns called coal mines. Do people here wish they had more sunshine? Do they feel claustrophobic in these narrow, dark hollows? Some people are aware of the darkness; others don't seem to care. It's simply a matter of perspective. I prefer openness, sunshine, mountains—especially mountaintops. The hilltops here may be open and in the sun, but they are colder, windier, and certainly more costly for road access and domestic water supply. Historically, initial settlements followed creek beds, focused on spring sites, and centered on lumber and coal exploitation in the hollows. Inertia kept people in their place—in deep, dark hollows and communities named Rowdy, Dwarf, Pigeonroost, Urban, Hooker, Bluehole, Boat, Plank, Doorway, Thousandsticks, Viper, Cutshin, Talcum, Krypton, Brightshade, Busy, Butterfly, Major, Typo, Trixie, Brutus, and Hector—all real Appalachian places in Clay, Leslie, Perry, and Knott counties, Kentucky.

Rivers

All streams in Appalachia, from the largest to the smallest, follow the law of gravity—streamflow moves downward. By observing the major trends of slopes in the southern highlands, we can discern drainage patterns and usually predict stream direction, at least some of the time. While streams in Appalachia appear to go off in all directions, there are only two major trends of drainage: one flows to the Atlantic Ocean, the other to the Gulf of Mexico via the Tennessee-Ohio-Mississippi river system. The drainage divide, an elevation that separates and sets these two opposing directions, is in the Ridge and Valley Province a little south of Roanoke in the vicinity of Wytheville, Virginia. Streams north of this point flow northward and eventually east. The primary north-flowing stream is the beautiful, sinuous Shenandoah River, which begins near Staunton, Virginia, and flows north as two forks of the same stream in the Great Valley to join the Potomac River at Harpers Ferry. South of the divide near Wytheville, the Clinch and the Holston rivers flow southwesterly and join other streams that come out of the western sides of the Blue Ridge and Great Smoky Mountains, such as the Watauga, Nolichucky, French Broad, Pigeon, Little, Little Tennessee, Hiwassee, and Ocoee rivers. Waters from these tributary streams flow in concert as the Tennessee River, heading south, then west, cutting through the Cumberland Plateau near Chattanooga and ultimately flowing across the top of Alabama and then northward back through Tennessee to join the Ohio at Paducah, Kentucky.

Many streams on the easternmost slopes of the Blue Ridge begin in the mountains and flow east to the Atlantic. Predictable rivers that do this are the Catawba, Yadkin, and Broad in North Carolina and the Peedee, Saluda, and Savannah in South Carolina. However, some old east-flowing rivers such as the Delaware and Susquehanna in Pennsyl-

Mountain rivers seem to have minds of their own, like the Nolichucky River in the Blue Ridge Province's Unaka Mountains. View east. (Source: U.S. Geological Survey, HAP [high-altitude photography], 1983)

vania and the Potomac, James, and Roanoke in Virginia have headwaters much farther west in the Allegheny Plateau and the Ridge and Valley provinces. Such streams flow east but have breached the Blue Ridge Mountains. As the region's orogeny (mountain formation) went through thrust faulting and uplift, some rivers were able to keep pace by eroding downward as the mountains pushed upward, thereby maintaining their original flow direction from west to east. For some rivers the answer was stream piracy—the more aggressive eroding stream worked its way back to the west, joined a lesser stream, then captured and redirected its flow. This is what happened to the Roanoke River.

But there is a river in Appalachia that is even more bizarre. Apparently defying the laws of physics, geomorphology, hydrology, and maybe even gravity is the misnamed renegade the New River. Popular mythology calls it America's oldest river, but it has not been scientifically

Appalachia's rivers radiate from a major drainage divide near Wytheville, Virginia.

proved to be the oldest. It is believed to date from Tertiary times, about 63 mya, but it could be even older. Against all logic, the New River flows southeast to northwest, precisely across the grain of the Blue Ridge and Ridge and Valley trends. With headwaters in northwestern North Carolina at Blowing Rock near Boone in Watauga County, the New River flows northerly into Virginia before turning northwesterly and cutting across the mountainous Blue Ridge. The divide in the Blue Ridge at 3,775 feet separates the east-flowing Yadkin River from the west-flowing New River.[6] The New River then heads west, cutting across all forty

The Big South Fork of the Cumberland River flows northward across the Cumberland Plateau in Tennessee and Kentucky. (Photograph by J. Rehder, 2003)

miles of the Ridge and Valley Province trend at Radford, Virginia. Ironically, at this point the New River is near the main drainage divide between Wytheville (Gulf of Mexico drainage) and Roanoke (Atlantic Ocean drainage). But the river chooses neither northeasterly nor southwesterly flow; instead, the New heads northwest. Against rationality, the New River cuts through the Cumberland Plateau, stubbornly heading northwest through one of the deepest gorges in the eastern United States, where it joins the Gauley River at Gauley Bridge, West Virginia. From here on and beyond Charleston, West Virginia, the New River and Gauley are called the Kanawah River flowing past Nitro; ultimately the waters join the Ohio River at Point Pleasant, West Virginia. Elsewhere throughout West Virginia, in such complex dissected plateau topography, the drainage pattern for all streams is a simple northwesterly flow toward and into the Ohio River. The maverick northwesterly flow of the New, coming from North Carolina and crossing Virginia and West Virginia, makes it a river after my own heart, going against the grain, against all trends, marching to the beat of a different drummer. I like this one.

Forest Ecology

Appalachia has been blessed with a remarkable array of forest habitats, from high country boreal firs and spruce trees to mixed deciduous hardwoods, to oak-hickory patterns, to several pine species at lower elevations. No region of North America has a greater variety. The wide range of vertical elevations creates greater climatic variability than could occur latitudinally. The forests on the summits of the highest peaks at 5,000 to 6,684 feet are 98 percent different from those in the valley lowlands at 1,000 feet elevation, less than a dozen miles away.[7]

Mountain forests in the Blue Ridge Province reflect microclimatic conditions governed by orographic rainfall and mountain temperature regimes. That is to say, the higher elevations are fifteen to twenty-five degrees cooler than nearby valley lowlands, and many slopes are wetter than lower sites in the other geomorphic provinces. Precipitation ranges from eighty-five inches per year in the mountains to forty inches or less in the valleys. Mountain forests include two major forest types: needleleaf evergreen trees and deciduous hardwoods. The highest elevations host the needleleaf evergreen species Fraser fir (*Abies fraseri*) and red spruce (*Picea rubens*) at elevations from 6,684 to about 5,000 feet. The spruce-fir zone above 5,000 feet is an endangered relic forest that dates from the late Quaternary period, eight thousand years ago.[8] Spruce-fir forests reflect a microclimate similar to that of southeastern Canada. Ontario and Quebec have forests of balsam that are analogous to Fraser fir, and white and black spruce are similar to the red spruce in Southern Appalachia. I tell my students that they can take a quick, cheap trip to "Canada" this afternoon. If they hop in a car and drive about sixty miles from Knoxville, Tennessee, to Clingmans Dome, at an elevation of 6,643 feet in the Great Smoky Mountains National Park, they will experience some of the temperatures, rainfall, and vegetation—though not the culture—of southeastern Canada.

The spruce-fir zone has experienced serious environmental degradation. Fraser fir trees in particular are dead or dying from infestations of the balsam wooly adelgid (*Adelges piceae ratz*). In the 1960s the insects, which had arrived in the United States on nursery stock from Europe, reached the vicinity of Mount Mitchell in North Carolina. In the next two decades the insect infestation reached the crests of the Great Smoky

Mountains so that today 90 percent of the Fraser fir trees there are dead. While smaller Fraser firs still can be found alive, almost no adult Fraser fir trees remain in the Great Smoky Mountains National Park. The red spruce trees have been spared from this insect danger, but they are not entirely safe, since those at high elevations seem to have been damaged by acid rain.

Other needleleaf species in Southern Appalachia are hemlock and several pines. Eastern hemlocks (*Tsuga canadensis*) were once more widespread over the southern Appalachians, but they were among the primary targets during the logging boom between 1880 and 1930. Hemlocks can be found in a few grove stands and as solitary sentinels. Their habitat ranges from low elevations (1,000 feet), especially near streams, to the crests of the mountains at more than 6,600 feet. Today, hemlock trees cover approximately 5,000 acres of the 513,968-acre Great Smoky Mountains National Park. As of August 2003 there are reports that another infestation of the hemlock wooly adelgid has killed 50 to 80 percent of hemlocks in the Shenandoah National Park and the Delaware Water Gap National Park in the northern part of Southern Appalachia. Insect infestations appear in several areas in the Great Smoky Mountains National Park. A biological experiment is under way in which ten thousand tiny predator beetles have been released to kill hemlock wooly adelgids in four infested areas.[9] The pine species in the area include table mountain pine (*Pinus pungens*), an endemic Appalachian species, Virginia pine, white pine, shortleaf pine, and pitch pine.[10] Pines appear at lower elevations, from 4,000 feet to 1,000 feet, with most growing on warm, dry south-facing slopes.

Southern Appalachia's forests are better known for their deciduous stands of oak, hickory, walnut, chestnut, cherry, maple, birch, beech, ash, sourwood, and yellow poplar. These trees are mainly found at intermediate elevations of 2,000 to 4,000 feet, such as on the Cumberland and Allegheny plateaus, and below 4,000 feet in the Blue Ridge. Such valuable hardwoods attracted a major logging boom in the period 1880 to 1930, so that even here remnant climax forests can be hard to trace. It did not help that coal deposits were discovered in the plateau country and have been exploited for over 125 years (see chapter 5). Also, a blight killed virtually all the American chestnut stands between 1925 and 1938. The American chestnut was "a monarch in the Southern Appalachian forests," a favorite tree of pioneer settlers because it provided straight, lightweight tree trunks that resisted rot and insects. Chestnut delightfully bright-colored wood was invaluable for furniture making, but it was also used for building log houses. After the blight, many dead chestnut relics were split to make rail fences. This mighty tree of the southern Appalachians, which once dominated the cove (valley level) hardwoods and accounted for 30 percent of the forests in Cades Cove, Tennessee, and 40 percent of those in Cataloochee, North Carolina, was eradicated by a lowly Oriental fungus.[11] Deciduous mixed forests still bring hordes of tourists in the fall to enjoy the incredible range of leaf colors. The fall color season in October brings more visitors to Appalachia than come in any other month.

In the Cumberland Plateau country, for example, forest cover has traditionally been a mixture of northern hardwoods and southern softwood needleleaf evergreen species. Highest elevations here are not great, at less than 4,000 feet above sea level, nor is local relief, a few hundred feet from the top of the plateau to deep gorges cut by streams below. Still, in an area of ever so slight variations in elevation, the forests seem to recognize the difference. James Lane Allen, writing in 1886, gave us a cross section of vegetation in the area of southeastern Kentucky from Somerset (1,200 feet) to Cumberland Gap (2,000 feet), over a distance of about eighty miles. Between Somerset in Pulaski County and Cumberland Falls in Whitley County, Allen wrote,

> On the summits one sees the tan-bark oak; lower down, the white oak, and lower yet, fine specimens of yellow poplar; while from the valleys to the crests is a dense and varied undergrowth, save where the ground has been burnt over, year after year, to kill it out and improve the grazing.[12]

In the purer plateau land between Cumberland Falls and Williamsburg in Whitley County, some eighteen miles away, Allen continued,

> It is truly called "flat woods country," and viewed from Sellico Mountains, whence the local elevations are no account, it looks like one vast sweep of sloping, densely wooded land. Here one may see noble specimens of yellow poplar in the deeper soil at the head of ravines; pin oak and gum and willow.[13]

In the Pineville area of Bell County just north of the Cumberland Gap, Allen observed,

> Our road at times cut through forests of magnificent timbers—oak (black and white), walnut (black and white), poplar, maple, and chestnut, beech, lynn, gum, dogwood, and elm. Here are some of the finest coal fields in the known world, the one on Clear Creek being fourteen feet thick.[14]

The lowlands represented by the Ridge and Valley Province in this otherwise southern highlands region include a wide variety of hardwoods and various pines and cedars. Because the area has experienced the longest period of settlement, agriculture, logging, and other human incursions, the province is a patchwork of mixed forest types with "shared dominance." Notably there are yellow poplar and several oaks—white oak, red oak, black oak, and post oak, the trees that became the most important for early settlement and log construction. Other lowland trees are hickory, sycamore, elm, black locust, sugar maple, and numerous second-growth pines, especially pitch pine, Virginia pine, and eastern white pine.

The understory plants create environmental interest as well. Rhododendrons are found throughout the Appalachian region, from the lowlands to the highest peaks. Broadleaf and evergreen, with magnificent flowers, rhododendron plants can reach the size of small trees along streams in the mountains. Rhododendrons and mountain laurel sometimes appear in the open on mountain slopes known locally as heath balds and laurel slicks, which can become impenetrable jungles. One such is a place called "Huggins Hell" in the Great Smoky Mountains National Park, just east of Mount Le Conte and within sight of Newfound Gap, where a fellow named Huggins entered the thicket and was never seen again.[15] Elsewhere, understory trees on the lower slopes and in the Ridge and Valley Province lowlands can be dogwoods, redbuds, and small red oaks.

The natural history for any place is a never-ending story of environmental change, some of it occurring naturally but much of it man-made. Southern Appalachia has not escaped environmental degradation from both causes, yet the region remarkably still reflects many of the patterns that shaped it. One still finds virgin forests, old-growth forests, and a general range of native animal wildlife somehow enduring in a place that brushes up against one of the most populous and progressive parts of North America, if not the world. I find it astounding that at this very minute I can gaze toward the east from my garret home office in the Ridge and Valley in East Tennessee and view the splendor of the Great Smoky Mountains.[16] In one broad panorama, I can observe over seventy miles of mountain crest terrain in the Smokies and beyond. On the northern horizon, I see Mount Cammerer (4,928 feet) and Mount Guyot (6,621 feet) at the northern extremity of the Great Smoky Mountains National Park. In the center I can view Mount Le Conte (6,593 feet), which is exactly thirty-nine miles away according to my GPS—a global positioning system receiver that is accurate to within one hundred meters. Unfortunately, from my vantage point here in the Ridge and Valley, the highest peak in the Smokies—Clingmans Dome at 6,643 feet—is obscured by Keller Bend Ridge. However, toward the southeast and directly in front of me, I can watch Thunderhead Mountain (5,527 feet) and Gregory Bald (4,949 feet), both on the crest of the Great Smoky Mountains, with Gregory Bald's slope forming the southwestern terminus of the park. On the distant southern horizon, I see Straton Bald (5,341 feet) and Haw Knob (5,472 feet) in the Unicoi Mountains. Between me and those distant ranges is Chilhowee Mountain (2,696 feet), a foothills ridge of the Smokies dominating the intermediate terrain. Since I am on a limestone ridge in the center of the Great Valley, I can see numerous oak- and pine-forested ridges in the Ridge and Valley Province and the special valley that the Tennessee River flows through on its way south. My house is perched high on the south-facing brow of a ridge at about 950 feet elevation. Forty feet below me, a vacant lot with a patch of eroded red clay surrounded by wildflowers greets my eye on this day in Appalachia. In this place, I am truly blessed.

3

Ethnicity and Settlement

Southern Appalachia represents the core of a much larger culture region called the Upland South. For more than fifty years, geographers have attempted to circumscribe the boundaries of culture regions in the United States. To draw conclusive boundary lines, ideally a scholar would have to analyze all diagnostic material and nonmaterial culture traits, then map the geographic distribution of the culture's most distinctive characteristics. Using old-fashioned intuitive thinking, field experience, and historical research, you then make an educated guess. Because of the late start for study and the many years over which cultural boundaries have become blurred, the Upland South presents uncertain boundary evidence.

The South can be subdivided into two Souths based on historical development and terrain. The Deep South or Lower South, in part the vernacular Dixie, was characterized by an economy based on cotton, tobacco, and rice plantations, slavery, large African contributions to the culture, and flat to gently rolling terrain. The Upland South was characterized by subsistence farming, few slaves (resulting in a sparse black population), and land whose steeper slopes and higher elevations formed mountains, ridges, plateaus, and generally hilly terrain. Both Souths remain "Anglo" by virtue of their Scotch-Irish and English heritages, and stereotypically they still draw appellations such as Dixie, Bible Belt, redneck, and white trash. Within the cultural and geographic context of Anglo America, Southern Appalachia is the heart and hearth of the Upland South.[1]

Ethnicity

In the late eighteenth century and until the mid-nineteenth, Appalachia's European ethnicity comprised Scotch-Irish, English, and German populations. Other groups included Irish, Scots, Welsh, and French

Huguenot settlers, together with other small groups. Well-established Native Americans of the Cherokee, Shawnee, and Yuchi tribes hunted game throughout Southern Appalachia. The Cherokees claimed large territories in the southern Blue Ridge and inhabited permanent villages along river valleys. This chapter explores the early settlement of folk groups, their origins, and their dispersals, giving particular attention to routes of diffusion into and throughout Southern Appalachia. While the focus is on the Scotch-Irish, English, and German ethnic groups, who together were culturally and numerically dominant, I have included material on French Huguenots, Native Americans, African Americans, and a fascinating group called the Melungeons.

Scotch-Irish and English

Scotch-Irish people originated as seventeenth-century Presbyterian folk from Scotland who emigrated to "plantations" established by King James I in northern Ireland's Ulster Province after 1609. Most Scottish emigrants to Ulster came from the Border country of present-day southern Scotland and northern England. Others arrived at various times from the western highlands of Scotland and the Hebrides Islands. For nearly a century, these "planted" Scottish folk experienced an unknown level of acculturation by exchanging traits with northern Ireland's Ulstermen. But between 1717 and 1775, the folk that we now collectively call Scotch-Irish began major emigrations to the Americas, entering chiefly through the ports of Philadelphia, Pennsylvania, and Charleston, South Carolina.[2]

These simple facts cannot fully explain the complex events and the identity of the people who endured them. Less Scottish than most highland Scots and far less Irish than the Irish, the Scotch-lrish were still Celtic, as were other northern and western peoples of the British Isles.[3] Should we call them Scotch or Scottish, Ulster Scots or Northern Irishmen? The adjective "Scotch" identifies whiskey or plaid, but "Scot," meaning Scotsman, produces the combined term Scots-lrish. Throughout the literature and especially in the American vernacular, Scotch-lrish is the accepted term and will be used here.

It is unclear how many Scotch-Irish emigrants made their way to the Appalachian region. R. J. Dickson estimated 114,000 emigrants between 1718 and 1778.[4] Ian Graham placed the number between 102,000 and 125,000 for 1707–83.[5] It is impossible to arrive at an accurate enumeration of people with Scotch-Irish ethnicity because the first United States Census in 1790 recorded Scots and Irish as separate entries but identified no group called "Scotch-Irish." Hanna, using the 1790 census to apportion the population of 1775, estimated a total of 335,000 Scotch-Irish and Scots for the following states (north to south).[6]

Pennsylvania	100,000
Delaware	10,000
Maryland	30,000
Virginia	75,000
North Carolina	65,000
South Carolina	45,000
Georgia	10,000

Attempts to differentiate the groups from the British Isles have been controversial if not confrontational. Contemporary historians tried to separate emigrants by surname.[7] Other scholars argued that the surname method was unsound and that estimates of ethnic populations from the British Isles must come from analyses of the 1790 census,[8] which disclosed a total population of 3,277,000 in the new United States. Thomas L. Purvis calculated percentages of the 1790 census for each of seventeen states for the following groups: English, Welsh, Scotch-Irish, Scottish, Irish, German, Dutch, French, and Swedish.[9] Converting Purvis's percentages into ratios for eight states in or bordering the Appalachian realm yields a disproportionate number of English over Scotch-Irish. Such a pattern can be explained. The English-settled coastal states of Virginia, the Carolinas, and Georgia and the border states of Delaware and Maryland had large numbers of English in the Tidewater region that skewed the statewide figures. In the core of Appalachia proper, the Scotch-Irish population should have dominated the English. Perhaps coming a little closer to this pattern is the McDonalds' estimate based on Scottish, Irish, Welsh, and English surnames, which shows closer ratios of Scots plus Irish versus English.

Nothing clearly answers questions about the Scotch-Irish in Appalachia: Which Scots came from northern Ireland's Ulster Province? Which Irish came from northern Ireland? Which Scots came from highland Scotland and which from lowland Scotland? Who came from the borderlands near England? And which of the immigrants were clearly English? Or should it even matter? It would be meaningless to attempt to differentiate between groups from the British Isles in Appalachia. Obtaining a clearer picture of Scottish immigrant populations in the South could be done, particularly for the highland Scottish enclaves in eastern North Carolina, but any assessment of a pure Scottish ethnicity in Appalachia is very uncertain.[10] Finally, according to Grady McWhiney, the general pattern in each census after 1790 until 1860 was that about 50 percent of the southern population was Scottish, Irish, and Welsh while about 25 percent was from western and northern England.[11]

Table 3.1, based on Purvis's estimates, compares percentages for Scotch-Irish, English, German, Swedish, and French populations in the states of Delaware, Georgia, Kentucky, Maryland, North Carolina, Pennsylvania, South Carolina, Tennessee, and Virginia. The largest concentrations of Scotch-Irish were in Kentucky, Pennsylvania, the Carolinas, and Tennessee.[12] While the English clearly dominated every state except Pennsylvania, the Germans led there, with 38 percent of the state's population in 1790. The small number of French, who are assumed here to be French Huguenots, are concentrated in Virginia and the Carolinas, especially South Carolina, where a small residual French Huguenot population remains. I have included the Swedish category to show the disproportionate figures for these states and to show a concentration, albeit small, for Delaware, where Scandinavian log construction techniques were first effectively introduced in the seventeenth century.[13]

Eighteenth-Century German Immigrants

The early Germanic folk who arrived in Appalachia entered America chiefly through the port of Philadelphia between 1681 and 1750. The Germans arrived at different times and for various reasons. In 1681 William Penn brought the first group of thirteen German and Dutch Mennonite families, who settled in Germantown, Pennsylvania. By 1709, large-scale immigration began when thousands of Palatinate Germans from central Germany arrived in southeastern Pennsylvania by way of England and New York's Hudson River Valley. An estimated half to two-thirds of the early German immigrants were indentured servants called "redemptionists," who worked off their passage costs with four years of labor in the New World.[14]

Like so many other groups seeking asylum in colonial America, many Germans were seeking changes in their lives for economic and political reasons such as heavy taxation and punitive inheritance laws. Others came for religious freedom. Early eighteenth-century German religious groups such as Lutherans, Calvinists, Dunkers, Amish, and Mennonites—largely from Rhineland and Swiss points of origin—entered, mingled, and formed a Germanic hearth (a focused region with a concentration of cultural characteristics) in southeastern Pennsylvania. By the

TABLE 3.1
Ethnic Populations as a Percentage of State and U.S. Populations in 1790

State	Scotch-Irish	English	German	Swedish	French
Delaware	9.2	63.3	2.6	3.8	1.7
Georgia	12.2	58.6	3.5	0.4	2.6
Kentucky	16.5	54.8	4.9	0.2	1.5
Maryland	10.4	52.5	12.7	0.3	3.0
North Carolina	15.8	53.2	5.1	0.3	2.5
Pennsylvania	15.1	25.8	38.0	0.6	0.9
South Carolina	18.9	47.6	5.5	0.3	3.7
Tennessee	17.8	50.6	6.6	0.4	0.9
Virginia	11.7	61.3	4.5	0.2	2.4
All	10.5	59.7	8.9	0.3	

Source: Purvis 1984, 98.

1740s Moravians initially established settlements in Pennsylvania, and between 1753 and 1770 they migrated to the North Carolina Piedmont in the vicinity of Winston-Salem. At the samc time, Schwenkfelders settled in Pennsylvania just northwest of Philadelphia. Both groups came from a Germanic-Slavic borderland. Schwenkfelders originated in the western parts of present-day Poland, while Moravians came from the Moravian and Bohemian provinces of the Czech Republic.[15]

Thousands of Germans were attracted to the New World by land agents and promotional schemes. The "emigrant letter" or "American letter" played a profound role. The letters often contained exaggerated and even fabricated descriptions of a land free of religious and political persecution, filled with fertile thousand-acre tracts, herds of fat livestock, plentiful orchards, and vineyards heavy with grapes.[16]

Eighteenth-century German immigration reached a peak about 1750. A total of about 100,000 Germans came to the colonies during the eighteenth century. Benjamin Franklin estimated in the 1770s that Germans constituted one-third of Pennsylvania's population. Franklin's guess was accurate; scholarly analyses of the 1790 census reveal that between 33.3 and 38 percent of Pennsylvania's population was German.[17] Germans and Scandinavian groups from northern Europe were easier to identify by surname in the 1790 census, which showed that of the 3,277,000 total national population, 8.6 percent or 281,822 people were German.[18]

German immigrants in Appalachia lent much to the ethnic composition of the region. German surnames and place-names abound. The family name Körper (now Carper) has been in Craig County, Virginia, near Roanoke, since the mid-1700s. Likewise, place-names such as Sweitzer, Strom, and Bessemer appear in adjacent Botetourt County. On the North Carolina Piedmont, Moravians with names like Betz, Feldhausen, Grube, Ingebretsen, Kalberlahn, Kirschner, Kühnast, Lischer, Loesch, Pfaff, Pfeil, and Pfeiffer began settling the area around Salem between 1753 and 1770. Year by year, the Moravians colonized in increasing numbers. In 1753 the Moravians' initial occupance brought twelve men to the budding North Carolina settlement. Eight more men came in 1754, and a year later thirty-four men and women arrived. The population gained twenty-five more immigrants in 1756–58, sixteen in 1759, forty-five in 1760–65, and eighty-one between 1766 and 1770.[19] Such an organized pattern of focused colonizing was rare beyond Moravians, Amish, Mennonite, and other German religious groups. Customarily, families migrated to places where land was available. With the family as the focus and not the religious order, groups tended to form enclaves of kin. Over time clans of families with the same last name came to occupy "compounds" in sections of valleys and hollows in the region.

In 2001 I was doing fieldwork in Grainger County, Tennessee, retracing some of the historical building surveys I had done in 1978. In Dutch Valley ("Dutch" here comes from *Deutsche,* or "German"), I stopped at a familiar little country store. Mrs. Frye, an older lady who once ran the store, was sitting at a table—not behind the cash register as she customarily did. Her daughter and son now run the struggling business. After reacquainting ourselves talking about the 1978 surveys, I asked her maiden name. She said it was Wolfenbarger. I replied, "Well, now, that's a real fine German name." She replied "Is it?" I suppose my look of astonishment startled her, because here was a Wolfenbarger in Dutch Valley who was not aware of her German ancestry. I could only thank her for remembering me and my surveys from so long ago. Dutch Valley still has a significant number of German family names like Wolfenbarger, Idol, and Frye.

French Huguenots

French Huguenots were Protestants in sixteenth- and seventeenth-century France (1560–1629) who emigrated to North America in the seventeenth century because of religious persecution. The French Huguenots left first from Dutch ports and later from the French port of

La Rochelle. By the late 1600s, several thousand Huguenots had arrived in America to settle in the English colonial areas of New England, New York (New Rochelle), Virginia, North Carolina, and South Carolina. In 1687 some 600 French Huguenots settled in eastern Virginia, and in 1700 another 700 came. Between 1704 and 1710, French and Swiss Huguenots began to settle in coastal North Carolina at Bath and New Bern. Table 3.1 shows small but meaningful concentrations of French Huguenots in 1790 in the Atlantic coastal states of Maryland, Delaware, Virginia, North Carolina, South Carolina, and Georgia; the largest group at the time was in South Carolina. The French became so assimilated into English-speaking America that, other than surnames, hardly a trace of their culture remains. The only southern area that retains Huguenot identity is around Charleston, South Carolina, where French names like Chastain are found.[20] Much like the English who migrated inland following rivers and trails from the English-settled Tidewater, a few of French Huguenots likely followed similar paths to parts of Appalachia. Family names like Daignan, Neubert, Delozier, and Sevier and some place-names remain, but few if any other French traits survive in Appalachia.

Melungeons

The most mysterious, almost mythical, ethnic group in Appalachia is the legendary Melungeons. In the 1650s, English explorers in parts of Southern Appalachia reported seeing people with olive complexions, blue or sometimes brown eyes, straight hair, and European facial features who spoke broken English and claimed to be "Portyghee" (Portuguese). Unlike the Native American Cherokees and later African Americans, the Melungeons had thin lips, round eyes, and a skin color that was brown but rarely red or black. They were not like anyone in America at that time and have remained a mystery for centuries. Once a secretive, isolated ethnic enclave, present-day Melungeons live in Hancock and Hawkins counties in northeastern Tennessee and in adjacent southwestern Virginia's counties of Wise and Lee. They are also known to have lived in Scott and Dickenson counties in southwestern Virginia and in parts of western North Carolina's Ashe, Yancey, Surry, and Allegheny counties.

Who are the Melungeons? Possible origins of the term *melungeon* are interesting yet confusing. The French word *mélange* means mixed or blended. The Greek *melas* means dark or black. The African Portuguese *melungo* or *melango* means shipmate. The Turkish terms *melan* and *can,* which together sound like *melun-jun,* translate as "damned fool."[21]

Several colorful theories surround Melungeon origins, but the most plausible traces Melungeons to the dark-skinned peoples of North Africa and Iberia (Spain and Portugal), and possibly Turkey.[22] Brent Kennedy, who is a Melungeon and a contemporary expert on the subject, discovered his own genetic connection to the Mediterranean through a rare illness, erythema nodosum sarcoidosis, that he contracted in 1988.[23] Kennedy's ensuing field and documentary research pointed to Mediterranean Moors, Turks, Berbers, and Iberians. If we accept that Melungeons came from Mediterranean stock, how did they arrive in Appalachia so early—before 1700? Kennedy distills his thoughts:

> I contend that the remnants of Joao ("Juan") Pardos forts, joined by Portuguese refugees from Santa Elena, and possibly a few stray Dominicans and Jesuits, exiled Moorish French Huguenots, and escaped Acadians, along with Drake's and perhaps other freed Turkish, Moorish, and Iberian captives, survived on these shores, combined forces over the ensuing years, moved to the hinterlands, intermarried with various Carolina and Virginia Native Americans, and eventually became the reclusive Melungeons. I as strongly contend that the Turkish/Moorish element was at least in the beginning the predominant one, explaining why the probable Turkish self-descriptive term "Melungeon" came to be associated with various populations regardless of their location.[24]

Kennedy's statements require considerable explanation. In 1566, Spanish officer Captain João Pardo, probably of Portuguese descent, brought two hundred Iberian soldiers to Santa Elena, a Spanish colony in the vicinity of Beaufort, South Carolina. Pardo sent groups of soldier-settlers into the interior to build four or five forts in northern Georgia, western North Carolina, and eastern Tennessee. Kennedy believes that the Pardo forts and the established settlement at Santa Elena formed some of the more plausible entry points for the first Melungeons.[25] A second source for entry is assigned to Sir Francis Drake, who in 1586 captured about five hundred Moors, Turks, some South American Indians, and a few Iberian soldiers. It is said that Drake *may* have deposited this human cargo on the North Carolina coast at Roanoke Island, the site of the English settlement known as "the Lost Colony." Uncertain at best—but if true, this would have been a good source and entryway for a dark-skinned Moorish/Turkish/Iberian population to have gained another foothold on the East Coast and to have migrated into the interior.[26] Author Eloy J. Gallegos does not accept Kennedy's assertion that Drake may have deposited a Moorish or Turkish cargo in North Carolina.[27]

The Drake connection is still unproven, and so are others in Kennedy's list of possibilities—Dominicans, Jesuits, and Acadians who fit neither the physical descriptions nor the temporal patterns for the region.

In the context of physical anthropology, however, the Melungeons are genetically different from other dark-skinned peoples in Appalachia. Early unscientific interpretations erroneously declared them to be a "triracial isolate," meaning they might be a combination of Caucasoid, Mongoloid, and Negroid, or a white-Indian-black mix.[28] However, in a 1969 study, anthropologists Pollitzer and Brown studied physical traits in gene frequencies, skin color, hair color and texture, and cephalic index of 177 Melungeons in Hancock County, Tennessee, and Lee County, Virginia. They concluded that the Melungeons' physical traits were more associated with Caucasoid populations and had far less connection with Mongoloid (Native American) and Negroid (African American) traits.[29] In a 1972 analysis, Pollitzer reported that his samples of Melungeons were 90 percent Caucasian and only 10 percent Mongoloid and that there were no Negroid traits according to gene frequency data.[30] From these and more recent studies, we can assume that the Melungeons are a dark-skinned Caucasoid racial group in Appalachia that has genetic connections to peoples of the Mediterranean region and North Africa.[31]

Melungeon surnames are similar to Portuguese and Spanish names. Common ones are Bell (Bela), Carrico (Carico), Caudill (Caudillo), Collins (Colina, Colinza), Goins (Goinza), Martin, Mullins (Molina), Perry (Perez), Reeves (Rivas), Gibson, and Goodman. First names, differing from neighboring Scotch-Irish and German ones, include Selena, Helena, Eulalia, Mahala, Louisa, Salvadore, Sylvester, Alonso, Canara, and Vardy/Vardeman/Navarrh.[32]

Tennessee's Melungeon populations appear to have come from North Carolina and possibly from Virginia. Just before 1790 Vardy Collins and Shep Gibson became the first Melungeon settlers on Newman's Ridge, the hearth for contemporary Melungeons in Hancock County. Vardy Collins was born in 1764 in North Carolina and was still living on Newman's Ridge in 1850; Gibson died on the Ridge in 1842. By 1820 Hawkins County (which then included Hancock County) had 310 Melungeons identified in the census by the category "other free persons of color." Ten years later the 1830 census showed 331 Melungeons in Hawkins County, represented by fifteen Collins families, thirteen Gibson families, three Bowlins families, two Goins families, two families each of Minor, Moore, and Mullins, and one family each of Goodman, Mosley, Nichols, and Williams.[33]

Eloy Gallegos argues that John Sevier, Tennessee's first governor, was actually Juan Javier or Xavier and was probably a Melungeon.[34] To most scholars the name Sevier looks like a French Huguenot name, so that there is uncertainty about the ethnic origin of the governor's name and places named for him, such as Sevier County and the county seat at Sevierville.[35]

Throughout most of their history, the unique Melungeon populations in Appalachia sought shelter from racially charged discrimination. Neighboring outside groups within Appalachia attempted ethnic cleansing. Knowing that the sun would make their dark skin darker, Melungeons stayed covered up in an effort to minimize the discrimination based solely on skin color. Their physical isolation in the Appalachian terrain and their own cultural and social denial of who they really were led to an ethnic amnesia. Through the efforts of Brent Kennedy, Eloy Gallegos, and other students of the culture, the Melungeons are becoming better understood as a unique ethnic group with an unusual heritage in Southern Appalachia.

Native Americans

Native American tribes surrounded the core of Appalachia without thoroughly occupying the center of the region. At the northern edges Iroquois, Susquehanna, Conestoga, and Delaware Indians held territories in present-day Pennsylvania, Maryland, and Delaware. To the south, Cherokees, Shawnees, and Yuchis occupied portions of Southern Appalachia but not the core areas of eastern Kentucky and West Virginia. Indian hunting and war parties did, however, travel widely through the region, and it was on these established hunting and war trails that Europeans later entered and traveled through the region.[36]

If a single Native American tribe can be associated directly with Southern Appalachia it would be the Cherokees. At the time of European contact, which was probably the height of Cherokee occupation, there were an estimated 25,000 Cherokees living on and claiming 40,000 square miles of land in what would be parts of today's Tennessee, North Carolina, South Carolina, Georgia, and Alabama. Three grand divisions marked early Cherokee settlements: the Lower Towns was an area along the upper part of the Savannah River in South Carolina; the Middle Towns were in western North Carolina; and the Upper Towns, better known as the Overhill Cherokees, occupied the Little Tennessee River Valley in eastern Tennessee and part of western North Carolina.[37] In the ensuing years, European and Cherokee contacts and conflicts over more than thirty treaties reduced the tribe's population and territorial holdings in Appalachia. Cherokees' acculturation with Europeans, however, led to their acceptance of traits like European clothing, log housing, political organization, and literacy. In the 1820s the Cherokee chief Sequoyah created a Cherokee alphabet, and a written language was born; by 1827 the Cherokee newspaper the *Phoenix* was being published and circulated in the region. A federal census of the Cherokees in 1835 listed the following populations: 8,946 in Georgia, 3,644 in North Carolina, 2,528 in Tennessee, and 1,424 in Alabama, making a total of 16,542 Cherokees in the region.[38]

The darkest times for the Cherokees came in 1837–38, when the United States government began a systematic ethnic cleansing. Gold had been discovered in Dahlonega in northern Georgia, and greed fed a gold rush to confiscate remaining Cherokee-held lands. The treaty of 1835 called for an official Cherokee Removal. Between the spring of 1837 and the fall of 1838, thousands of Cherokees were rounded up and driven in wagons and on foot over 1,200 miles to Oklahoma. Approximately 4,000 men, women, and children perished on this Trail of Tears. Cherokee survivors in Oklahoma came to be known as the Cherokee Nation. The 1,400 Cherokees who were able to escape and hide in Southern Appalachia's mountains came to be known as the Eastern Band. In the Qualla Boundary, a Cherokee reservation in western North Carolina, this Eastern Band today has about 12,000 members who live in communities such as Cherokee, Snowbird, Birdtown, Wolftown, Yellowhill, Painttown, and Big Cove.[39]

African Americans

African American populations were extremely limited in Appalachia before the Civil War. However, in Virginia's Shenandoah Valley and other parts of the Great Valley of Virginia, it had been reported that there were plantations that raised few crops but many slaves. They were commercial breeding centers whose sole purpose was to produce slaves for cotton and sugar plantations much farther south beyond the Appalachian realm.[40] Appalachia was not built on a plantation economy, so many areas of the region lacked an African American slave population. Some contemporary Appalachian counties originally settled by white subsistence farmers have few or no black residents today.[41]

In the post–Civil War years black populations entered the region as wage laborers for logging and coal mining enterprises. The largest movement of African Americans to Appalachia occurred during the early coal mining period between 1900 and 1920. For example, in 1907 35 percent of the coal miners in West Virginia's Pocahontas Coal Field were black. By 1920, 43 percent of all black coal miners in the United States were living and working in West Virginia.[42]

Migrations into Southern Appalachia

Before the nineteenth century, Scotch-Irish and German migrations dominated the European settlement history of Appalachia. Routes for all settlers were governed by such location factors as ports of entry, land already occupied by other European groups, established trails and wagon roads, territories held by Native American tribes, and lands that remained open to settlement. New England did not welcome additional immigrants because of economic problems there, and few Atlantic ports other than New York, Philadelphia, and Charleston were attractive to large-scale immigration. Between 1720 and 1770, lands adjacent to port cities were already filling fast, and the abundance of cheap open lands to the west pulled settlers to venture into the backwoods frontier.[43]

Early transportation routes into Appalachia followed a succession of wild game trails, Native American hunting and war trails, Europeans' horse and foot traffic traces, and wagon roads. Most covered the same stretches of ground, evolving and widening as more traffic was compelled to reach distant destinations. Wagon roads became the principal way for settlers to reach new lands.

The pattern of migration was not a uniform westward movement from the Atlantic seaboard toward the southern Appalachian highlands. Quite the contrary, the migrations took the form of two widely separated streams that entered through the ports of Philadelphia, Pennsylvania, and Charleston, South Carolina. From these cities, migrations west and south from Philadelphia and west and north from Charleston formed a pincers movement that created an arc of Scotch-Irish and German settlement in colonial America.

The larger and more important first stream began in southeastern Pennsylvania at an established culture hearth that formed around a gathering place and staging area in the Susquehanna River Valley west of Philadelphia in 1710–30. Between this somewhat younger developing German hearth area in Pennsylvania and a much older Swedish settlement (1638–75) in the Delaware River Valley to the east, a Midland culture hearth emerged that would supply the American landscape with architectural traits based on corner-notched log buildings.[44] This first

An Appalachian view from 560 miles up shows the all-important Cumberland Gap and Wilderness Road trend. The Ridge and Valley Province at center extends from upper right to lower left. The Blue Ridge Province's Great Smoky Mountains are at the lower right. (Source: U.S. Geological Survey, NASA Landsat imagery, 1978)

stream coming out of Pennsylvania proceeded approximately one hundred miles west of Philadelphia until it encountered the Allegheny Front, a steep, high plateau escarpment that naturally hindered further westward movement for a time and effectively deflected settlers southward. In the vicinity of Gettysburg, the migration indeed turned southward following the Great Valley to reach southwestern Virginia by 1735.

The Philadelphia Wagon Road or Great Wagon Road

The route called the Philadelphia Wagon Road or the Great Wagon Road became the most important route in the settling of Appalachia be-

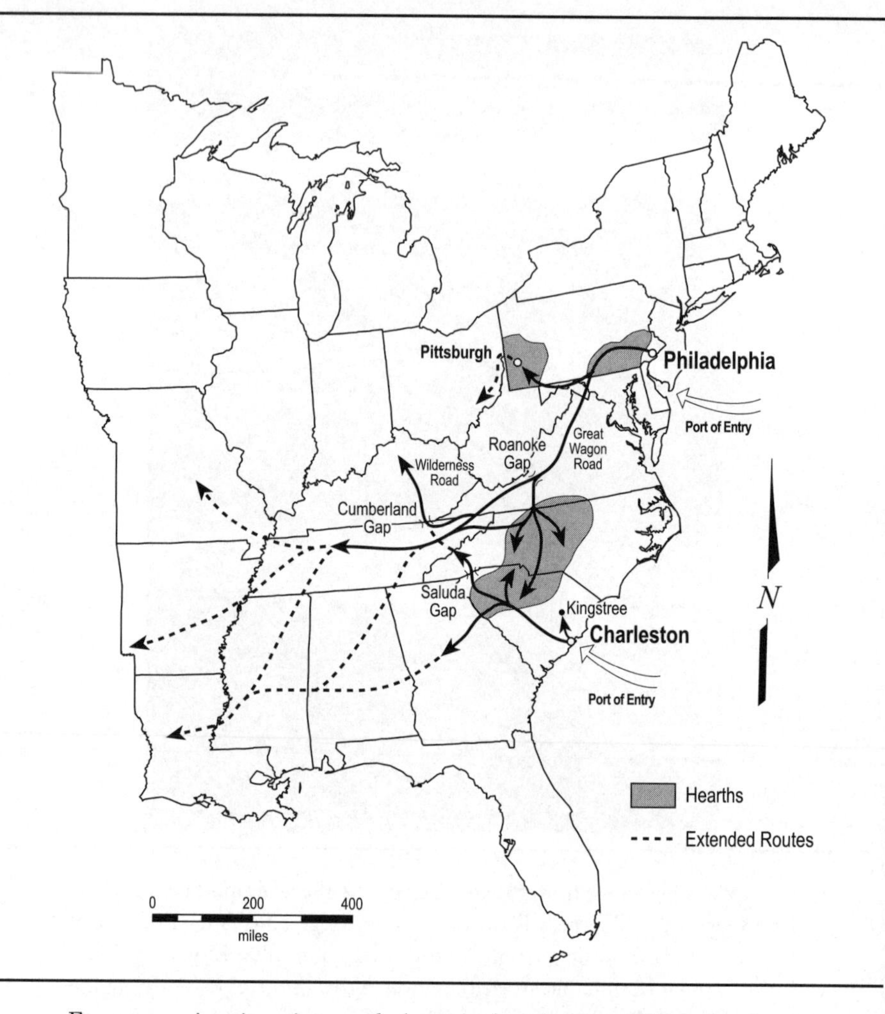

European migrations into and through Appalachia established culture hearths between 1720 and 1790. (Sources: Rehder 1992, 98; Newton 1974, 149)

tween 1740 and 1774. The Great Wagon Road covered an impressive distance of eight hundred miles from Philadelphia via Roanoke, Virginia, and Salem on the North Carolina Piedmont to its terminus in Augusta, Georgia. Beginning at Philadelphia, the road went west to Lancaster, York, and Gettysburg, Pennsylvania. Turning southward and following the Great Valley geomorphic trend, it went on to Hagerstown, Maryland, to Harpers Ferry at the Potomac River, and thence to Winchester, Harrisonburg, Staunton, and Lexington, Virginia. Continuing south in the Great Valley, the road crossed the James River at Buchanan and proceeded to Roanoke, a settlement then called Big Lick.[45] Today's interstate highway I-81 parallels highway U.S. 11 and closely follows the direction, if not the actual roadbed, of the Great Wagon Road. U.S. 11, however, does rest on the bed of the Great Wagon Road north of Roanoke.

At Roanoke, the main trunk of the Philadelphia Wagon Road went directly south and easterly, passed through the Roanoke Gap, followed the Roanoke River, crossed the Dan River, and deposited immigrants onto the Piedmont region of North Carolina. In the 1740s and 1770s, the lands between the Catawba and Yadkin rivers became a secondary yet important culture hearth for German Moravians and Scotch-Irish settlers.[46] Eventually the Great Wagon Road extended southward quite far beyond the North Carolina Piedmont settlements of Salem, Salisbury, and Charlotte. The road passed through Camden, South Carolina, and reached its southern terminus at the English fur-trading outpost of Augusta, Georgia, founded in 1735.[47]

This relatively continuous settlement took time, even along well-worn routes like the Great Wagon Road. The migration from Pennsylvania to the North Carolina Piedmont lasted two or three generations, with each new generation venturing farther and building a European landscape in the Appalachian wilderness. Isaac Weld Jr., writing in 1796, offers a glimpse of settlement in the Great Valley in Botetourt County, Virginia, twenty miles north of Roanoke and just west of the Blue Ridge:

> This town [Fincastle] . . . was only begun about the year 1790, yet it already contains sixty houses, and is most rapidly increasing. The improvement of the adjacent country has likewise been very rapid, and land now bears nearly the same price that it does in the neighbourhood of York and Lancaster, in Pennsylvania. The inhabitants consist principally of Germans, who have extended their settlements from Pennsylvania along the whole of that rich track of land which runs through the upper part of Maryland, and from thence behind the Blue [Ridge] Mountains to the most southern parts of Virginia. These people . . . keep very much together, and are never to be found but where the land is remarkably good. It is singular, that although they form three fourths of the inhabitants on the western side of the Blue Ridge, yet not one of them is to be met with on the eastern side. . . . They have many times, I am told, crossed the Blue Ridge to examine the land, but the red soil which they found there was different from what they had been accus-

> tomed to. . . . In Bottetort [Botetourt] County, on the contrary, the soils consist chiefly of a rich brown mould, and throws up white clover spontaneously. . . . A bed of [limestone] runs entirely through the country, so that by some it is emphatically called the limestone county.[48]

Weld speaks volumes in this short passage. He tells us that Germans here are the principal inhabitants, that they choose lands based on their knowledge of soil quality, that land prices are as about the same here as in Pennsylvania, and that the Germans have established an almost continuous line of settlement that extends back to the Pennsylvania German culture hearth.

The Wilderness Road

The traveler wishing to go south and west from Roanoke branched off from the Great Philadelphia Wagon Road and took the Wilderness Road (1774–89). It was a leading road that traveled southwest through the Great Valley and linked the settlements of Christiansburg, Fort Chiswell (1768), Wytheville, and Abingdon, Virginia, to connect two very important sites: Cumberland Gap on the west side of the Valley and the Watauga settlements, a tertiary Scotch-Irish culture hearth, on the east side in northeastern Tennessee.[49]

The juxtaposition of the Wilderness Road with another trail coming from western North Carolina meant that the Watauga settlements could ultimately flourish and develop into a political unit of considerable cohesiveness for the short-lived State of Franklin, a seceded temporary politico-hearth of frontiersmen in 1784. From the Watauga settlements, the Wilderness Road made a beeline to the most renowned gateway to an expanded settling of Appalachia's west and north—Cumberland Gap.

Situated at the junction of the present-day states of Kentucky, Tennessee, and Virginia, Cumberland Gap was the preeminent focal point granting access to points north and west of the plateau country; it was also a milepost landmark for foot and wagon traffic. First used by large game animals in their seasonal migrations and their search for salt licks, the Gap became a major hunting trail followed by Native Americans. In 1750 Dr. Thomas Walker "discovered" the passage that offered an opening through the thousand-foot-high Cumberland Plateau escarpment. It was not until 1775 that Daniel Boone blazed the Wilderness Trail for frontier settlement to push far into the interior of Kentucky. Boone's Wilderness Trail through Cumberland Gap linked settlements in western North Carolina and Watauga settlements in northeastern Tennessee

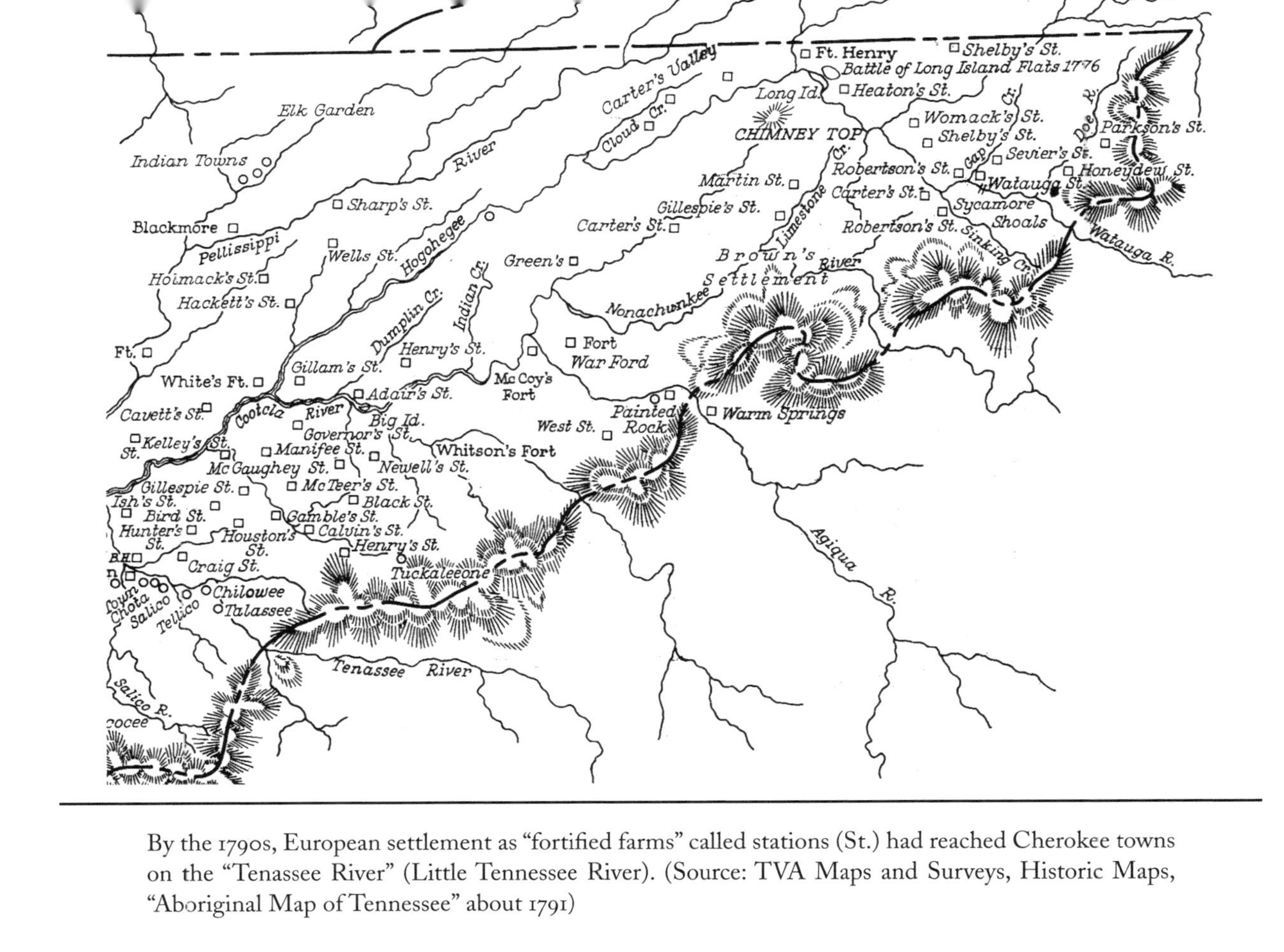

By the 1790s, European settlement as "fortified farms" called stations (St.) had reached Cherokee towns on the "Tenassee River" (Little Tennessee River). (Source: TVA Maps and Surveys, Historic Maps, "Aboriginal Map of Tennessee" about 1791)

to Kentucky settlements at Boonesborough and Harrodsburg in Kentucky's Blue Grass Basin. Passage through Cumberland Gap stood as a symbol of human endurance and accomplishment in the settling of Appalachia in general and Kentucky in particular as larger settlement components followed the flow of tens of thousands of settlers who took this "new" route to the West. Between 1775 and 1810, an estimated 300,000 people passed through the Gap en route to Kentucky and points north and west.[50]

Entry from Charleston, South Carolina

The second arm of the pincers was a stream of immigrants that entered the South Carolina port of Charleston beginning in 1732 and pushed seventy-five miles into the interior at Kingstree. As more and more immigrants arrived, the migration stream moved deeper into the South Carolina up-country. Here on the Piedmont, a South Carolina culture hearth emerged and pushed smaller streams of settlement southwestward to Georgia and northeastward to link up with the North Carolina Piedmont hearth.[51] From the South Carolina hearth, settlers moved northward through the Blue Ridge Mountains through Saluda Gap, to forge into the Asheville Basin and eventually spread settlement westward into Tennessee.

Secondary loops of migration settlement continued throughout the southern region. Heading west from the North Carolina Piedmont hearth in 1771–72, Daniel Boone, James Robertson, and John Sevier led settlers across the Blue Ridge Mountains, followed the Watauga River, and settled the Watauga settlements in northeastern Tennessee.[52] From the Watauga settlements that became Kingsport, Bristol, and Johnson City, additional migration went westward overland and circuitously via the Tennessee River, as the Donalson party did in 1779–80 to the Nashville Basin. The availability of land grants and the drive to migrate quickly through dangerous Indian-held territories led to an established outpost of European settlement in the then distant Nashville Basin.[53] Still following trails and wagon roads down the main course in the Great Valley of Virginia, thousands of settlers continued to turn northwest through Cumberland Gap, where they followed the Wilderness Road into the Blue Grass Basin in east-central Kentucky. South of Cumberland Gap, settlers continued down the valley toward areas that were to become Knoxville, Chattanooga, and Birmingham, as the threat of Indian attacks lessened and those lands were opened for permanent settlement after 1838.

Parts of Southern Appalachia were the homelands of Cherokees, Shawnees, and other aboriginal groups, so much of the land was closed to European settlement until after the 1790s. From the vicinity of the site of Knoxville, Tennessee, Indian-held territory prevented major European incursions until after 1830. In 1837–38 most Cherokees were physically removed from Appalachia in the Trail of Tears cultural atrocity, but that single awful event opened the southernmost regions of Southern Appalachia to even more permanent European settlement.[54]

The English, meanwhile, made slow progress inland, following river valleys from their Tidewater hearths along the Atlantic seaboard. The English Tidewater had became a region whose early focus was not toward the western frontier and the interior. The English in the Tidewater were content to develop a narrow colonial English string of coastal ports and plantations in tobacco, rice, and cotton that looked to the English homeland for social and economic alliances. Before 1760, from a well-established English-dominated hearth in the Chesapeake Bay area, additional settlers of various cultural and economic statuses began to move up-country, following the Potomac, the James, and other eastern rivers and eventually linking with the Shenandoah Valley and Great Valley settlements of the Scotch-Irish and Germans. From the English-dominated southern coastal towns of Wilmington, Charleston, and Savannah, additional English settlers moved westward toward Appalachia. By 1800, European settlement had expanded to embrace the core of the Appalachian region.[55]

By the first quarter of the nineteenth century, Cumberland Gap had fewer settlers passing through. Another gateway much closer to East Coast population centers was being constructed in 1811–18. It was a new road called the National Road, but also given the confusing name of Cumberland Road, which put immigrants on the fast track to the West. Today we know it as U.S. 40. The route began in Maryland and went up the Potomac River to Cumberland, Maryland, and across a narrow divide in the plateau to cross the Youghiogheny River. The road continued westward across West Virginia to Wheeling, where it linked with the Ohio River. This was a momentous development, making it possible for immigrants to disembark from ships in East Coast ports, travel overland to the Ohio River, and settle points farther west. In time the National Road connected Wheeling with Indianapolis and thence the Mississippi River at St. Louis, Missouri.[56]

4

Folk Architecture and the Cultural Landscape

The Settlement History of Appalachia

Folk architecture is important because the structures people build tell us much about the personality of places and the human character behind them. We will examine the Appalachian cultural landscape in terms of first effective settlement, the imprint of the English metes and bounds land survey system, log construction, notch types, house types, barns, small outbuildings, fences, and other landscape features. This excursion will be neither spatially complete nor temporally comprehensive, and for that I apologize. But it will illustrate the types of folk architecture that represent the settlement history of Appalachia.

When you visit Southern Appalachia, don't be disappointed if you fail to see a log house or barn or a split-rail fence anyplace but in outdoor museums and parks. Southern Appalachia has its share of endangered species, and among them are components of the material culture expressed in folk architecture.

Appalachian cultural landscapes have changed over time and under different economic conditions. The Appalachian subsistence farmstead of the eighteenth and nineteenth centuries produced a different landscape than does the hobby farm or the rural nonfarm settlement of the late twentieth century. Dairy farms, with their big barns and silos, have always looked different from small tobacco farms. Coal camps and logging settlements in the late nineteenth and early twentieth centuries treated the countryside indifferently and created a unique landscape of temporary settlements and permanent scars. There can be no single Appalachian settlement type; there are many.[1]

Initial Occupance and First Effective Settlement

In any study of a people and their landscape, the initial occupance or first effective settlement, seen as a conceptual stratigraphic layer, is paramount to understanding successive settlements.[2] Much like the tells of the Middle East, each settlement might have several layers. The first one is important to understanding site selection and how this initial layer relates to subsequent layers. By comparison, the concept of first effective settlement puts more emphasis on the success of a first layer or perhaps adjacent layers. If the very first contact is slight and impermanent, the concept assesses the significance of subsequent layers until an effective settlement emerges. For example, initial European forays into Appalachia by explorers, trappers, and military expeditions created no permanent settlements. They may have been the first, but they left no lasting evidence of permanent settlement. It was not until European settlers arrived to build blockhouses, clear land, and establish farmsteads and stations that the first effective settlement could be realized. That first permanent layer created by eighteenth-century Scotch-Irish, German, and English settlers in Appalachia established patterns that have endured through more than two centuries of land occupation.

Metes and Bounds

The cadastral or survey system by which property was measured engraved a network of settlement evidence into the landscape. Land surveys relied on the English survey system of metes and bounds. Parcels were allocated by the claimant's choice of boundary lines to encompass selected soils, water, and woodland.[3] The cadastre, or tax register, followed a complex method of sightings to and from objects such as rocks, trees, water features, and wooden or iron posts. Colorful boundary descriptions included "the missing chestnut stump" or "the place where Will Smith's cow drowned in the river." Land in Appalachia today is still sold with an imprecise disclaimer: "sixty acres, more or less."

As generations of families occupied a land that had such uncertain boundaries, "walking the farm" became a tradition. At a selected time of the year, usually in early winter when the leaves were off the trees, the father or grandfather—the patriarch of the clan—would take his family along the bounds of their land and point out landmarks and boundary points along the way. To the youngsters it was a joyous walk in the woods, but to those in line to inherit, the walk was meant to imprint the critical landmarks that marked the farm's property lines, no matter how inconspicuous they might be. My first experience with walking the farm came in my early adult years as my father-in-law walked his North Carolina farm with us. He began with clearly defined markers: iron posts, large trees known to surveyors as witness trees, a railroad tie planted vertically in the middle of a field that separated his farm from his mother-in-law's farm next door. As we walked deeper into the woods near the back of the property he stopped and said, "We own the land that from here goes two hollers into the swamp." I looked puzzled, so he repeated the mantra but added that "on courthouse plat maps the land goes to the run of the swamp," meaning that the legal bounds just might be a bit different from his two hollers, but he was satisfied with the "run of the swamp," where the main streamcourse was the official boundary.

As land was claimed, surveyed, and settled, the map became a complex crazy quilt blanketing the terrain. The nineteenth-century landscape reflected circuitous landownership patterns in property boundaries marked by fences and tree lines, in the irregular shapes of cultivated fields, and in the network of trails and secondary roads that linked holdings. Settlement became widely dispersed. Individual farms, though often within sight of each other, were clearly separate. Unlike the European homeland, the rural agricultural landscape contained no villages, few hamlets, and few other rural agglomerations.

The Farmstead

The arrangement of buildings on the Appalachian farm followed some patterns from the Old World.[4] The house was near a water supply and a means of access such as a path, creek, or road. But even in the Upland South the house was perched on slightly higher ground to avoid local flooding. A hilltop site let the occupants observe their small world if not the larger one around them. Deciduous oaks, hickories, or maples shaded the house site. In the absence of lawnmowers, most yards were cleared of grass, exposing clay, sand, or rock. My late friend Milton Newton, a scholar of the cultural geography of the South, once asked an elderly lady why she was sweeping her clay yard. He was expecting some ancient African folk explanation, but she answered, "Dangit, to keep it clean!"[5]

Beyond the house, a variety of outbuildings were spread out as if they too had been swept out back or to the sides of the main house. A barn, sometimes as large as the house, was the focal point for myriad much smaller outbuildings such as cribs and coops. Each building had a function, sometimes just one job. The barn provided storage for farm implements, hay, and corn and protection for oxen, horses, and mules. A complete farmstead maintained corncribs for storing corn; a smokehouse for curing and storing meat; a springhouse to protect the main water supply at the spring and a "dairy" to keep perishable foods in cooling troughs under the roof; an apple house or root house for storing fruits and root

On Appalachian farms like the John Sevier plantation in Knox County, Tennessee, the farmhouse occupied the center of the site, with outbuildings and animal enclosures distributed around it. (Source: Green and Peden, Photographers. Photo no. 180 from the Jim Thompson Collection, Thompson Photo Products, Knoxville, Tennessee. Used by permission.)

crops; a chicken coop, pigsty, and other animal enclosures; and an outhouse. Beyond the house site and outbuildings stretched fields and pastures arranged in no particular size, shape, or alignment, bounded by fences of split wooden rails or posts and wire. Bottomlands—flat, rich soils along streams—were devoted to the principal crops of corn and tobacco. Treelines followed fences and property lines, but woodlands on steeper terrain and in larger units joined to fill in the landscape.

The Southern Mountain Log Cabin as a Symbol

The log cabin represents Appalachian shelter in powerful ways. To some it is the ultimate symbol of frontier life as a pioneer family struggles with harsh conditions on the frontier. It also serves as a "dirt floor" symbol of poverty existing long before commercial logging and coal mining entered the picture. The popular image of Appalachian folk housing takes us first to the one-room log cabin, the house we call a single pen. It is a small and simple place, probably with few if any windows and a dirt floor.

While the ramshackle appearance of the southern mountain log cabin conveys the image of folk housing in Southern Appalachia, there are at least eight notable house types and variations that are meaningful to landscape interpretation. Appalachian folk house types follow patterns established largely by the English pen tradition: the single pen; double pens called Cumberland, saddlebag, and dogtrot; I-houses; and four pen dwellings. The region's folk house typology also includes three-room German houses and some twentieth-century floor plans in bungalows and box houses. I will discuss these later.

Early observers sometimes described Appalachian dwellings in unflattering ways. In southeastern Kentucky in 1886, James Lane Allen wrote:

> The dwellings—often mere cabins with a single room—are built of rough hewn logs, chinked or daubed, though not always so. Often there is a puncheon floor and no chamber roof. . . . The low chimneys, made usually of laths daubed, are so low that the saying, inelegant though true, is current, that you may sit by the fire inside and spit out over the top. The cracks in the walls give ingress and egress to a child or a dog. Even cellars are little known, their potatoes sometimes being kept during winter in a hole dug under the hearth-stone.[6]

Ellen Churchill Semple described cabins in Kentucky's mountains in 1901 thus:

> The home of the mountaineer is primitive in the extreme, a survival of pioneer architecture, and the only type distinctly American. It is the blind or windowless one-room cabin, with the rough stone chimney on the outside. The logs are sometimes squared with the hatchet, sometimes left in their original form with the bark on; the interstices are chinked in with clay. The roofs are covered with boards nearly an inch thick and 3-feet long, split from the wood by a wedge, and laid on, one lapping over the other like shingles. The chimneys, which are built on the outside of the houses, and project a few feet above the roof, lend a picturesque effect to the whole. They are made of native rock, roughly hewn and cemented with clay; but the very poorest cabins have the low "stick chimney," made of laths daubed with clay. In the broader valleys, where the conditions of life are somewhat better, the double cabin prevails—two cabins side by side, with a roofed space between, which serves as a dining room during the warmer months of the year. Sometimes, though rarely, there is a porch in front, covered by an extension of the sloping roof. In some of the marginal counties of the mountain region and in the sawmill districts, one sees a few two-story frame dwellings.[7]

Allen and Semple describe the single pen cabin, the fundamental house type in the region. The single pen log cabin is mistakenly thought to be distinctly American; it actually has European origins. Semple also describes a double pen type—the dogtrot—based on the open passage and roof. In 1899 color writer William Goodell Frost interpreted the log cabin as a stage in life, as if it were akin to a stone age dwelling, implying that the people living in single pen cabins might evolve to live in double pen cabins or even larger and more substantial two-story dwellings.[8] Horace Kephart described the mountain settlement of forty-two houses where he lived in 1904: "Two or three, only, were weatherboarded frame houses and attained the dignity of a story and a half."[9] The other forty houses in the Hazel Creek area of the Great Smoky Mountains in western North Carolina were log cabins.

Contrary to these observations, in 1913 John H. Ashworth argued for house values and images in Virginia's mountains:

> The "typical home" of the "mountain whites," the rough and ill-kept cabin, which appears so often in current literature, is preposterous. Such may be a typical cabin, although the unrepresentative, the worst cabin of all, is usually shown; but cabins constitute only a small percentage of the homes. Beautiful dwellings, worth from two thousand to five thousand dollars, are in most every community, while residences costing from five thousand to twenty thousand and even thirty thousand dollars are not at all uncommon.[10]

In these few excerpts we find descriptions of the single pen cabin and one type of double pen. But we also see an evolutionary pattern emerging in some of the material, and this is as it should be. Contemporary historian Ann E. McCleary, writing in 2000, interprets three stages of development in vernacular architecture in the Shenandoah Valley of Virginia:

> The first phase occurred from the 1780s until about 1810, when the first group of permanent houses began to appear. Although few in number, these dwellings introduced architectural ideas which fed into subsequent rebuilding phases. In the second phase, from around 1810 through the 1830s, local families experimented with a variety of traditional ethnic ideas and popular new pattern book designs, creating tremendous diversity in architectural forms. During the 1830s and 1840s, in the third phase of this rebuilding, the ideas introduced in the previous thirty or so years congealed, by the 1850s, to create what would become the classic Shenandoah Valley home—a brick, central-passage plan, often called an "I-house."[11]

Log Construction

Folk house types in Appalachia are much more than dirt-floored one-room cabins, and they reflect respectable diagnostic traits through their form, function, and construction. Houses do not have to be built of logs to be folk. Folk structures may be built of stone, brick, or sawn, framed lumber. I emphasize log construction here because it is intrinsic to understanding Appalachian folk architecture. Log construction from the eighteenth and nineteenth centuries and part of the twentieth becomes the gold standard for folk architecture in Appalachia. Virtually anything built of logs before 1960 is a potential folk structure. We will explore the fundamentals of building a simple log house and examine the old folkways of log construction that were common from about 1700 to 1880, a period of home-built, folk architecture mostly using logs. When milled lumber became widely available after 1890, the use of logs for home construction declined. There was a short resurgence of log building in the 1930s when homeless rural folk built "Depression houses" using small-diameter pine logs with saddle notches. While these temporary dwellings were folk houses, the workmanship and the quality of available materials were frequently poorer than in the log structures built in the nineteenth century. Since the 1970s, commercial companies have been erecting sparkling new Appalachian log houses in the region. These buildings have never been folk either by design or by intent; they are commercially constructed with spruce and lodgepole pine from western states. The only adherence to tradition is that companies create half-dovetail notches on hewn logs.

Publications on Folk Housing

Scholars investigate genuine American log buildings to help us identify a culture's history and diffusion patterns. We seek to trace authenticity, origins, and dispersals of the Appalachian log structure because its morphology reflects a way of life. In 1939 Harold Shurtleff wrote *The Log Cabin Myth,* dispelling historians' belief that seventeenth-century English settlers had built log cabins as their first homes in America.[12] Another myth is that Native Americans built log cabins and taught European immigrants how to do it. Both notions are nonsense! Scholars of American folk architecture universally agree that the log house in America came from European sources, particularly from Scandinavian regions and German-speaking Europe.

Scholarly interest in American log buildings remains strong, especially among cultural geographers and folk life specialists. In 1963 Henry Glassie published an article appropriately called "The Appalachian Log Cabin," showing us a young scholar's exploration into folk housing and its identity with Appalachia. The journal also printed Glassie's early illustrations in pen and ink.[13] In 1965 Fred B. Kniffen, the father of folk geography in the United States, published his most definitive paper, "Folk Housing: Key to Diffusion," which set the stage for a widely accepted typology for American folk architecture.[14] In 1966, Fred B. Kniffen and Henry H. Glassie published their pivotal work on notch types, "Building in Wood in the Eastern United States."[15] In 1978 James R. O'Malley and I published a paper on the two-story log house in the Upland South. Our aim was to illustrate the rarity of such structures, but our research revealed the complexity of construction and variations of multilevel log folk dwellings.[16] In 1985 Terry G. Jordan, one of America's best-known scholars on the geography of folk architecture, published a wonderful book, *American Log Buildings: An Old World Heritage,* which sought European origins. Jordan's book is based on extensive fieldwork in Europe and contains one of the best comparisons of log buildings in the Old World and the New.[17] In 1990 John Morgan, a former student of mine and the best field worker in historic building surveys, wrote *The Log House in East Tennessee.* Morgan used data from his doctoral dissertation for the book, which focuses on log houses in Blount, Grainger, Morgan, and Union counties, with the greater emphasis on Blount County, Tennessee.[18] In 1989 Terry G. Jordan and Matti Kaups published another pivotal piece called *The American Backwoods Frontier,* which tells us that the origins of log construction in North America trace to Karelian sources in Scandinavia. In 2003, Terry Jordan-Bychkov published the excellent *Upland South,* which condenses four decades of research on the region.[19] Other publications on folk architecture may be found in the journals *Pioneer America* and *Journal of Material Culture,* among others.

Origins of Log Construction in America

In eighteenth- and nineteenth-century Appalachia, log construction was the primary building method, in which structural walls consisted of horizontal rounded or square-hewn timbers notched at the corners. In folk architecture, the typology for corner notches includes saddle notch, saddle V notch, V notch, half-dovetail notch, full dovetail notch, square notch, half notch, diamond notch, and half log or semilunate crown notch.[20]

Corner notching has been suggested as a way to diagnose cultural identity. Which culture groups were responsible for the corner notching types found in Appalachia? Aboriginal Indian groups did *not* originate the technique, because they built wooden shelters with vertical posts and woven saplings. Using horizontal logs with corner notches was a European introduction.

Despite the widespread use of log construction by the Scotch-Irish in Southern Appalachia, the technique did not originate with them either. Before coming to America, neither Scotch-Irish nor any other peoples from the British Isles had a history or an apparent knowledge of using wood in this way.[21] In northern Ireland's Ulster Province, Scottish settlers in 1611 built structures of stone and thatch, just as they had in Scotland, while English immigrants at the same time and place were cutting and squaring timbers for structures built with half-timbering techniques.[22] Log construction did not exist in Britain at the time they left. Moreover, on arriving in America, initial Scotch-Irish settlers from Ulster still did not know about log construction. In 1732 at Kingstree, South Carolina, about seventy-five miles north of Charleston, the first homes of Scotch-Irish settlers were crude pits dug in the earth and roofed with pine saplings and sand.[23] English settlers in Virginia at first built cruck structures (like an A-frame but with curved support timbers) with walls of wattle-and-daub—woven saplings with mud packed in the interstices.[24] As early as 1610–12 and for centuries thereafter, the English in the Tidewater region continued building structures of framed, sawn timber with clapboard siding.[25]

Not one but two European cultural sources have been suggested for American log construction. The first, championed by Jordan and Kaups and supported by the works of Wright, Mercer, and Weslager, posits a Scandinavian origin.[26] The second, as put forth by Kniffen and Glassie with support from Bucher, Wertenbaker, and others, argues for a German origin.[27] Supporting the Scandinavian origin is considerable field evidence collected by Jordan and Kaups. Notch types are more than coincidental, and cultural links between south-central Sweden and the Delaware Valley are firm for the 1630s albeit based on few Swedish populations.[28] The German evidence, based on larger immigrant populations, suggests a connection between German immigrant builders of log structures in Pennsylvania and their homelands in Switzerland, southwestern Germany, and areas of Moravia, Bohemia, and Silesia settled by eastern Germans.[29] The argument favoring a German origin can be based on log building cultures with immigrant diffusions from Germanic-speaking Europe to eighteenth-century southeastern Pennsylvania. But the argument favoring the Scandinavian origin is based on recent field evidence of log structures and notch types found in the Fenno-Scandian region and the historical evidence of Swedish immigration into the Delaware Valley adjacent to southeastern Pennsylvania in the 1630s.

I suggest a modest but no less evidentiary interpretation. I agree with Jordan and Kaups about the groups who introduced log construction to America—they were quite obviously the Swedes and Finns who settled the Delaware River Valley below Philadelphia in 1638–75. Jordan and Kaups's book is filled with arguments for this initial Scandinavian introduction, and I believe they are correct.[30] Notwithstanding, later-arriving German-speaking immigrants, coming from European areas with traditions of log buildings with corner-notching technologies, surely must have influenced other folk in the region. The question of who brought log construction here first is settled—Swedes and Finns. But the Germans should be credited with reintroducing, reinforcing, and maintaining log construction in the hearth. And more important, Germans and perhaps others actively educated much larger and more mobile populations such as Scotch-Irish immigrants and others who lacked the knowledge of building with horizontal logs and corner notching.

For linkages to Appalachia, the decisive presentation of log construction techniques came with European immigrants who settled in southeastern Pennsylvania in the first half of the eighteenth century. Within this Pennsylvania Midland hearth, cultural linkages were formed between German users of notching types and Scotch-Irish recipients. As both groups migrated throughout Appalachia, they exchanged culture traits through acculturation, and Scotch-Irish settlers and others learned the methods of log construction and notching types so closely identified with Appalachia. Either the Germans were good teachers or the Scotch-Irish and other British Isles immigrants were good students, for it is from all of these groups that log construction techniques diffused from the Midland hearth in southeastern Pennsylvania southward and westward into Appalachia and beyond through the Upland South and across a young and vigorous eighteenth- and early nineteenth-century America.

How to Build a Log Cabin

The following method comes from the first volume of *The Foxfire Book,* edited by Eliot Wigginton, and from my thirty-six years of observing log structures in Appalachia. The Foxfire instructions on how to build a log cabin were collected in the field by high school students and their teacher Eliot Wigginton in Rabun Gap, Georgia, in the late 1960s and early 1970s.[31] At that time, only a few old men in the Blue Ridge Province in northern Georgia still remembered how to build a log house; fewer still could summon the energy to demonstrate authentic techniques in log construction.

We are building a single pen log house with dimensions of sixteen by eighteen feet. Most single pens in Appalachia are rectangular rather than square, and their dimensions hover around sixteen feet, with variations ranging from fourteen to twenty feet.

As simple as it might appear, a log house is a puzzle whose pieces end up as a complete structure. Every building has a foundation, walls, and roof. The steps in building a log house consist of preparing the *foundation:* stone foundation, sills, sleepers (floor joists), and puncheon floor; the *walls:* notched horizontal logs and loft (ceiling) joists; and the *roof:* rafters, purlins, ridgepole, and roof covering. Additional elements are chimney, doors, and windows. Despite my immediate and impatient attraction to discussing notched logs in walls for log houses, we must begin with the foundation. Not all Appalachian log houses had wooden floors—many had dirt floors—but all required a foundation made of either stone or sill logs.

Stone Foundation

The foundation begins with large flat stones either stacked in a series of piers or placed to form an entire stone-walled rectangle at least eighteen inches high. If piers are used—and this is the more common method—we must pile three to five large flat stones to at least that height. This dimension is important, because a folk belief claims that termites will not go above eighteen inches to get into a wooden house. I have not measured compliance with this component of folk wisdom, but I have observed many old log structures with noticeably tall stone foundations.

The next step is to measure and cut the log sills—two long beams that rest on the long axis of the foundation. Sill logs, twelve to twenty inches in diameter, are the largest and strongest logs in the cabin and must bear all the weight of the structure. If the house is to be more than twenty feet wide, we will need a middle sill placed on stone piers near the center and parallel to the long axis to lend internal support. Most sill logs are hewn on all four sides, but some are hewn flat only on the top or bottom or both.

Sleepers, or floor joists in modern terminology, are floor beams that are laid at right angles across the sills to form a stable base for attaching the puncheons or floor planks. Sleepers are placed at regular intervals with three- to four-foot centers; intervals greater than six feet will let the floor sag. The sleepers are square-hewn beams six to twelve inches thick and can be quarter notched into the sills to create a stable flush joint. Sometimes a builder used round log sleepers but still had to hew the tops of the logs with a foot adze to maintain a flat surface to accept the next layer—the puncheon floor.

Puncheon floors have rough sawn or hewn boards that must be flat on top, straight, and laid with a tight fit so that the wind will not blow up from under the house. Making puncheons was time consuming and labor intensive, requiring considerable skill. Some areas had puncheon-making enterprises before sawmills began to emerge in the late nineteenth century. One such place was Puncheon Camp, Tennessee, in the shadow of Clinch Mountain in Grainger County, where puncheon planks were produced for flooring houses and other buildings throughout this end of the county.

Walls

The next step is building the external walls, the very essence of a log house. Enter the world of the true log cabin—one with log walls that have corner notched timbers. After we get this cabin built, I will discuss corner notch types in greater detail. Logs meant for walls may be left in the round or hewn on two or all four sides. The tree species, the log's width and length, and the learned traditions and skills of the builder determine the size of the building and the type of notch used. Each log is measured, cut, trimmed, usually hewn flat, and notched to fit into the log below it. Hewing a log to trim the bark and create flat sides requires a broadax. A typical broadax has an eight- to ten-inch cutting edge and a narrow butt. Its unusual handle is curved to the side so the user can stand on top of the log and hew the sides with more direct force. Hewing logs this way is much safer and easier than using a standard logger's ax. The job looks simple but takes great skill, especially for producing lockable notch types such as V notches, half-dovetail, full dovetail, and diamond notches. In corner notching, each shaped cut on the top of a log must have a corresponding groove or cut on the bottom; but the bottom cut must be at right angles to the top cut. Trust me, it's best to look at the illustrations, because describing an intricate notch is almost as difficult as shaping one with an ax. As the walls go up, we must allow for doorways and possibly window openings. Early log houses had only one or maybe two doors, no windows, and as few openings as the builder and the occupants would allow. Windowless cabins were called blind cabins. Some early houses had shutters; windows with glass panes came later. Not only was cutting doors and windows laborious, but such openings let in too much cold air. It was drafty enough anyway.

Chinks are the cracks between logs in a log-walled house, and we will want to fill them with chinking. Chinking can be mud, clay, stones, bits of wood, rags, paper—just about anything that might slow the winter wind. Del Scruggs, a former student of mine, was hired to "house sit" a log cabin rent-free. He said, "The house had plenty of chinks but very little chinking. It was so airish that I had to pitch a camping tent *inside* the house in order to sleep without freezing."[32] But builders of log barns and many small outbuildings left the chinks open so corn, grain, hay, and livestock could get ventilation.[33]

Roof

We're now up to the roof, and a lot of work remains. First, we must cut loft joists and fit them into the uppermost log, called the wall plate. A well-made house will have mortised loft joists, square grooves about four by four inches and two to three feet apart cut in the top logs or two or three logs below the top so that the joists fit snug and flat. Loft joists are analogous to the floor joists (sleepers) because they support a flat surface, only this one is the ceiling; hence loft joists are now called ceiling joists. The loft is then floored with the same kind of puncheon boards that made up the ground floor flooring. A small stairway or ladder leads to an opening into the loft, a sleeping place for children.

The framework of a ridgepole-and-purlin roof consists of ridgepole, rafters, and purlins, arranged to support the wooden shingles (shakes) that cover the roof. Rafters are inclined lateral poles whose bottom ends rest in notches in the top log/wall plate and whose top ends are attached to the ridgepole at the apex of the roof. Each set of rafters and loft joists forms a triangle; these are spaced about three feet apart and held together by horizontal purlins extending the full length of the house from gable to gable. The purlins walk up the roof slope like ladder rungs and terminate at the ridgepole. They are one to two feet apart depending on the length of the shingles. This gridded frame is now covered with wooden shakes split from white oak boards. "Riving shakes with a froe" means splitting shingles (shakes) with a froe—an odd implement with a knife-like blade mounted at right angles to a handle. You hold the blade sharp

"Riving shakes with a froe" was the way to make roof shingles from white oak. Swain County, North Carolina. (Photograph by J. Rehder, 1974)

side down against the butt end of a white oak log, hit the dull side from above with a mallet, then wiggle the handle back and forth until the wood splits into a long, thin shake. A white oak shake roof should last fifty years or more before rotting through; a corrugated tin roof lasts twenty to twenty-five years before rusting through. The log cabin is now roughed in, and the only woodwork that remains is to add doors and optional window treatments, tasks I won't describe here.

A log house is not a house unless it has a chimney and fireplace; most are built of stone or brick. The earliest log houses in Appalachia had mud and stick chimneys—suggesting an alarming risk of fires. But in the firebox the thick mud was mixed with sand and stones, and outside the house the sticks and mud were high enough on the chimney that fire was not as likely as you might imagine. Few mud and stick chimneys survive.[34] Later came chimneys made of flat fieldstones, and then chimneys of brick or quarried and dressed limestone. Fireplace hearths were generally twenty-nine to forty-eight inches wide at the floor. A proper fireplace had a throat, a constriction above the firebox to prevent rain and debris from falling directly into the fire. But most important, the throat of the fireplace determined whether the unit would "draw"—that is, pull smoke up the chimney and out of the house. The small orifice at the throat was six to thirteen inches front to back. The throat could be almost any size as long as the spaces below and above it were larger. Furthermore, the opening at the very top of the chimney where the smoke came out had to be about the same size as the throat or slightly larger. If the opening was too small, smoke built up inside the house;[35] too large, and drafts became too great, heat escaped, and the fire consumed too much fuel. It took a good chimney mason to do this job. A house without a working fireplace had no heat and no place for indoor cooking. (See chapter 6 for hearth and fireplace cooking.)

Corner Notching and Notch Types

Notched corner timbers are the essence of a log structure. Notch types cover a range of possibilities for securing horizontal logs in a wall. Forget for a moment your Lincoln Logs. Instead, follow the progression of notch types through saddle notch, V notch, saddle V notch, half-dovetail notch, full dovetail notch, diamond notch, square notch, semilunate crown on half logs, and double notches.[36]

Saddle notches are simple rounded cuts made on a round log so that adjacent logs will overlap. The round cuts can be on either the top or the bottom of the log, but the smart builder notches only the bottom side to keep rainwater from collecting in the notch and creating a weak point for wood rot. My father-in-law built his first double pen log house in 1941 with saddle notches on round pine logs. When I asked why he chose this notch type he replied. "I didn't know of any other way to make it. I just did what came naturally." Saddle notches in Appalachia are exclusively found on round logs, frequently round pine logs or small recent poplars. In historic buildings data from seventeen east and middle Tennessee counties that I analyzed, the saddle notch is represented in fourteen counties by from one to twenty-one occurrences. Out of 1,119 structures, the survey recorded 99 with saddle notches, accounting for 8.8 percent of the notch types. As a "poor man's notch," the quick-to-make, utilitarian saddle notch is commonly found on outbuildings built after 1880 and on houses built between 1930 and 1950.

The V notch is named for the inverted V at the top of the log at the butt end. It is best to look at logs "end on" from the butt to determine the type, especially for the rest of these notch types. From an end per-

Left: Saddle notch. (Photograph by J. Rehder, 1978). *Right:* V notch. (Photograph by J. Rehder, 1978)

spective, a V-notched log looks like the gable end of a small house. The V notch in Southern Appalachia is an old notch type when found on large hewn oak or yellow poplar logs. Frequently the V on square-hewn logs is on oak timbers. In the same seventeen-county data set from Tennessee, the V notch is represented in twelve counties with one to twenty occurrences. The sample survey had eighty-four V notches, accounting for 7.5 percent of the notch types. It does appear more frequently in the northern half of Appalachia, especially in Pennsylvania, along the Ohio River, and in western Virginia.

A variation subtype of the V notch is the saddle V that appears on unhewn round logs. From end on the saddle V is pear shaped. The top retains the V shape while the bottom of the round log remains rounded. Even though the V notch and saddle V notches are both Vs and are derived from the same European origin, those in Appalachia are quite different in timber type, in age, and in quality of workmanship. Is a saddle V a V notch or a saddle notch? Originally it comes from the same Scan-

Saddle V notch. (Photograph by J. Rehder, 1978)

dinavian origins as the V notch on hewn logs, so technically the saddle V is a subtype of the V notch type. But in the Deep South and Southern Appalachia the saddle V has more identity with saddle notches and as such can be found on outbuildings built after 1880, and is more frequent on pine log buildings built between 1900 and 1950. There were 53 saddle V notches in my seventeen-county data set from Tennessee; they account for a mere 4.7 percent of the 1,119 buildings examined.

The most common notch in the southern half of Southern Appalachia is the half-dovetail notch. It is found on all types of log structures but is more common on houses and on substantial barns and small outbuildings built with yellow poplar and oak. The half-dovetail is an exceptionally good but simple locking notch that sheds water well. When viewed from end on, the half-dovetail has its upper side sloped at a thirty to forty-five degree angle. The lower part of the log is straight horizontal, and the two sides are straight vertical. All of these shapes combined, but especially the sloped upper cut, form the diagnostic traits of the notch. In those same seventeen East Tennessee and Middle Tennessee counties, the half-dovetail notch is represented by at least one in every county, with 90 occurrences in Johnson County and 116 in Union County. Out of 1,119 structures, the sample survey had an extraordinary number of 517 log buildings with half-dovetail notches, accounting for 46.2 percent of the notch types. In some counties, half-dovetail notches accounted for as much as 68 percent of the notches.

The half-dovetail notch apparently became a favorite in many parts of Appalachia, but especially in areas beyond the more strongly German-influenced northern portions. I am cautious in saying this, but there appears to be an affinity for the notch in the Upland South and particularly in the southern parts of Southern Appalachia, where Scotch-Irish outnumbered German settlers. We must remember that no culture group from the British Isles knew anything about building with horizontal logs and using corner notching techniques. We must credit the early Swedes and Finns in the Delaware River Valley for introducing corner notching to the American landscape and particularly for using half-dovetail notching here long before German settlers arrived. As settlement expanded, eighteenth-century Germans in the Midland culture hearth in southeastern Pennsylvania taught Scotch-Irish and other folk the skills needed to hew logs and to lock them down with corner notches. Thus a connection exists between the dominance of half-dovetail notching and the dominant culture group in the area, people who did not bring any corner notching technique with them from Europe. This is as good a case for acculturation as could be seen in folk architecture.

One of the rarest of notches in Southern Appalachia is the full dovetail notch. There are only 4 occurrences out of 1,119 in my seventeen-county analysis; this is a mere 0.35 percent. I believe this complex notch is simply the most difficult to cut. From end on, the full dovetail notch has a thirty- to forty-five-degree angle on top and another thirty- to forty-five-degree angle cut on the lower portion of the log, but pointing down. When we try to match both slopes to the adjacent logs above and below, then the builders cuss and the fireworks start, because the cuts must be precise so the fit is exact.

Several notch types that are rare but represented in Appalachia include diamond notch, square notch, semilunate crown, and double notch. One of the rarest notches in America is the diamond notch; it is almost as rare as real diamonds. From end on, the diamond notch looks exactly like the diamond symbol in a deck of playing cards. The geographic distribution of this "ace of diamonds" notch is mostly limited to

Left: A half-dovetail notch cut into a yellow poplar log was one of the most common old timber-notch combinations in the southern half of Southern Appalachia. (Photograph by J. Rehder, 2003). *Right:* Full dovetail notch. (Source: Tennessee Historical Commission, 1990)

the both sides of the fall line in Virginia and North Carolina, and it is largely found on tobacco barns and other small outbuildings there. My seventeen-county Tennessee survey lists only one house (destroyed in 1978) in Grainger County and one single crib barn in Marion County; hardly worth calculating a percentage.

The square notch is perfectly square as seen from end on. It is a weak connector, subject to collapse, and a poor substitute for intelligence, since it requires so much labor to hew it yet it cannot lock logs in place. In my sample of 1,119 log buildings, 37 structures in thirteen counties have square notches. They range from one to six occurrences per county and as a group represent 3.3 percent. Some buildings are entirely square notched, and others have mixed notches of squares, half-dovetails, and others. It is quite possible that in later years, as the craftsmanship in notching declined, builders became less precise and often accidentally produced a square when they were aiming for a half-dovetail.

Semilunate crown notches on half logs. (Photograph by J. Rehder, 1978)

A semilunate crown appears on half logs. The builder takes a round log, splits it in half, then cuts small curved notches in the bottom of each log and assembles them much the same way as he would in saddle notching. The semilunate crown, so named because it looks like a half moon, is rare in Appalachia. There are a total of five in my seventeen-county data; four are in Grainger County, and one is in Johnson County.

The double notch is also quite rare on the Appalachian landscape, though it is familiar from children's Lincoln Logs. I have only two occurrences in the Tennessee data, and they are on "suspect" late-model log buildings in and near a state park recreation area built in the 1930s. I conjecture these were camp structures either built on private land while the parks were being established or bought later and moved off park land. In either case they are *not* folk houses here, but they resemble the double-notched log buildings so common in Scandinavia and Alpine Europe.[37]

Geographic Patterns of Notch Types

Corner notching in Appalachia is at best a mixed collection of notch types with varied temporal and geographical patterns. So where are they? In the eastern United States but looking beyond Appalachia for a moment, we find saddle notches with a broad sweep of dominance in the Deep South, south and east of the fall line along the inner coastal plain of the Carolinas and Georgia and clearly dominating the inner Gulf Coastal Plain of Alabama, Mississippi, and non-French interior Louisiana, and going well into Arkansas and Texas. There is a strong correlation between saddle notches and pine trees, because the saddle notch is always cut on round logs and most pine trees remain slender, with small diameters. The geographic distribution I just described for the saddle notch is identical to the distribution of the southern pine forests. To the north of Appalachia, the V notch finds its largest geographic pattern beginning in its hearth in Pennsylvania and clearly dominating in a broad westward sweep in the hardwood forests of the Ohio River Valley through southern Ohio, Indiana, Illinois, and out across Missouri. The less common saddle V notch is used with round pine and poplar logs, but because of its slightly greater complexity and its affinity with the northern pattern of V notches in particular, the saddle V does not follow the same geographic distribution as the saddle notch. The diamond notch, though rare in all parts of the eastern United States, has a concentration along both sides of the fall line in the Virginia and North Carolina Piedmont. Its rare appearance forms a quasi boundary to the east of Appalachia.[38]

Now that we have Appalachia surrounded at least on three sides, what notch if any dominates the region? The consensus would be the half-dovetail notch, especially in the southern half of Southern Appalachia, that is, roughly south of the New River in Virginia. As for the northern half of Southern Appalachia, up-valley from the New River especially beyond Roanoke, the dominant notch is the V notch, but closely followed by half-dovetail notches and even full dovetails in Pennsylvania. In spite of these simple general patterns, Appalachia has examples of all the types described here. In my analysis of log structures in Tennessee, notch frequency patterns, in rank order for the top five notches are: half dovetail, with 517; saddle, with 99; V notch, with 84; saddle V, with 53; and square notches with 37. The four remaining notch types each had five or fewer occurrences.

How Old Is It?

"Hits about a hundert year old," or "Built before the Civil War," or "This house were here when this part of Tennessee was still in North Carolina!" (meaning before 1796). These are the answers I get when I ask people in the field about the age of a log structure. Determining the age of a folk building is tricky. Since I do not always accept vernacular answers as accurate, I must rely on other means for dating log buildings.

My thirty-six years of experience in the field lets me make educated guesses based on knowledge of the landscape and confirmed dates on a few selected well-known (by me) buildings. As I inspect a particular log house in East Tennessee, I examine the timber, and if it is poplar or oak I know that the structure is probably old. Hewn poplar or oak logs thicker than twelve inches are usually older. If the timber is small-diameter pine and is left in the round, not hewn, it is probably newer. Next, the corner notches may indicate age. Saddle notches are typically newer than V and half-dovetail notches. Saddle notches on round pine logs tell me that the house was probably built no earlier than 1880 and is likely to have been built between 1930 and 1940 or possibly as late as the 1950s. Stone chimneys are often older than brick ones; stovepipes and cinderblock chimneys are newer still. The latter, however, may have been replacements, so we have to be careful in dating a house by the chimney alone.

I try to determine the family name attached to the place or the old owner's full name. If I am successful, I can often analyze a nearby family cemetery to match the former occupant's name with the property. Let's say the George R. Wolfenbarger place is our subject log house. In the family cemetery nearby, I find the man's gravestone, and his dates are born in 1820, died in 1877. Assuming that young George built the place at age twenty, that could have been about 1840. It is entirely possible that George may have built it in his late teens, but hardly before he was fifteen. It is also possible that he built it when he was in his thirties or perhaps his forties, but less likely while he was in his fifties. So we have a range of years between 1840 and 1877 for the Wolfenbarger place based on the working lifetime of its initial owner. We don't want to estimate too late, because during the Civil War years between 1860 and 1864 it's unlikely that George was home to build a log house. If I lean toward the 1840 mark when George was twenty and already the father of several children (based on other gravestones in the family plot), the 1840 to 1850 bracket is a bit more plausible. Since the wood is yellow poplar and V-notched, I am thinking that this is an old place. By now I am definitely leaning toward 1840 and certainly no later than 1860 for this two pen dogtrot house.

Dendrochronology

The odyssey of the mind described above is one I have often repeated in the field and with courthouse records and other archives. But another way to date log buildings is through dendrochronology, the science of using tree rings to study climatic variations as well as to determine the age of wooden objects. How many of us, perhaps as children, remember counting the tree rings on a stump in the forest to calculate the age of a tree when it was cut? I have had the pleasure of working with geography faculty colleague Henri Grissino-Meyer and two fine graduate students, Bill Reding and David Mann, at the University of Tennessee. Using techniques of dendrochronology, Reding and Mann have determined the dates for five log buildings in East Tennessee. Bill Reding examined four log buildings, one in each of four contiguous counties: Union, Grainger, Jefferson, and Hamblin. They had been surveyed in the late 1970s by teams of historic buildings surveyors for the Tennessee Historical Commission. I had been the principal investigator on the Grainger and Union County survey teams in 1978 and 1979. With our twenty-four-year-old estimates as a base, Reding took core samples from the logs and established dates for when the trees were cut. By comparing the rings with a master pattern of known tree ring chronology in the lab, Reding could match rings and confirm dates for his core samples. Here are his findings. A Union County blacksmith shop with yellow poplar logs and half-dovetail notches had a dendrochronology date of 1773; the surveyors' (Rehder and Morgan) estimate based on interviews and intuitive fieldwork was 1770. In Grainger County, a smokehouse with yellow poplar logs and half-dovetail notches was dated to 1860 from dendrochronology and given an estimate of 1850 from the survey team of John Morgan and John Rehder. In Hamblin County, a log I-house with yellow poplar logs and V notches was dated to 1812 from Reding's dendrochronology and to 1815 by the surveyors. In Jefferson County, another I-house with half-dovetailed yellow poplar logs, now a complete ruin, had a dendrochronology date of 1827; the surveyors' estimate was 1835.[39] I am pleasantly surprised that the surveyors' estimates came so close to the actual date when the tree was cut as verified by dendrochronology. But I am especially proud that we now have an accurate method for determining the age of log structures in Appalachia.

David Mann's work centers on the Swaggerty Blockhouse, an unusual log building in Cocke County, Tennessee, constructed of red oak timbers with half-dovetailed notches. A blockhouse was an early fortified log building that could be the corner unit in a military fort or, as in this case, a stand-alone fortified house. This one was believed to be very old, part of the "first effective settlement" phase in the region's settlement. The Swaggerty building, one of only two log blockhouses remaining in Tennessee, has a large metal historical marker stating that the structure was built by James Swaggerty in 1787. With the owner's permission, David

Mann cored the logs, took thirty-seven samples, and dated the structure with dendrochronology techniques at an alarming date of 1860! Then, using archaeological techniques and an analysis of materials excavated from forty-two shovel test pits on the site to verify his findings from dendrochronology, Mann arrived at average dates of 1860 for nails and 1864 for glass shards on the site. Here a young cultural geographer, using dendrochronology and archaeology, has proved that a site was marked in error concerning both the date of the structure and its use. The Swaggerty Blockhouse turned out to be a cantilevered springhouse and storehouse built on the farm by Jacob Stephens in 1860.[40]

Folk House Types

For the geography of house types and the interpretation of their cultural significance, we owe much to Fred B. Kniffen. Through Kniffen's works and those of his students and supporters, the study of folk housing has contributed much to our understanding of the American landscape.[41] Folk houses reveal a lot about the personality of a culture region. As representatives of the folk who built them, they allow us to use dwellings as diagnostic traits to identify the cultures involved. They allow us to trace the migrations and dispersals of folk cultures. Meanwhile, they stimulate us to seek answers to questions about source areas, points of origin, and exact routes and times of diffusion, as well as variations in types, styles, construction materials, and degrees of diagnostic reliability.

Appalachian folk house types follow patterns established largely by the English pen tradition, but the region also has early three-room German houses and twentieth-century bungalow and box house floor plans. The most common house types of Appalachia are in the pattern of the English pen tradition, with a pen representing an individual room unit. I like to say "pens are fer people and cribs are fer critters," meaning that house units are pens and barn units are cribs. In the pen typology, houses evolve from single pen houses to double pen houses (two rooms) to I-houses (two stories tall, two rooms wide, and one room deep) to four pen houses (four rooms over four rooms). Appalachian folk houses may be built of horizontal rounded or square-hewn logs with corner notches, framed sawn lumber, brick, or stone.

Single Pen Houses

The single pen house is the basis from which other houses in the pen tradition evolve. The typical one-room cabin measures about twenty by

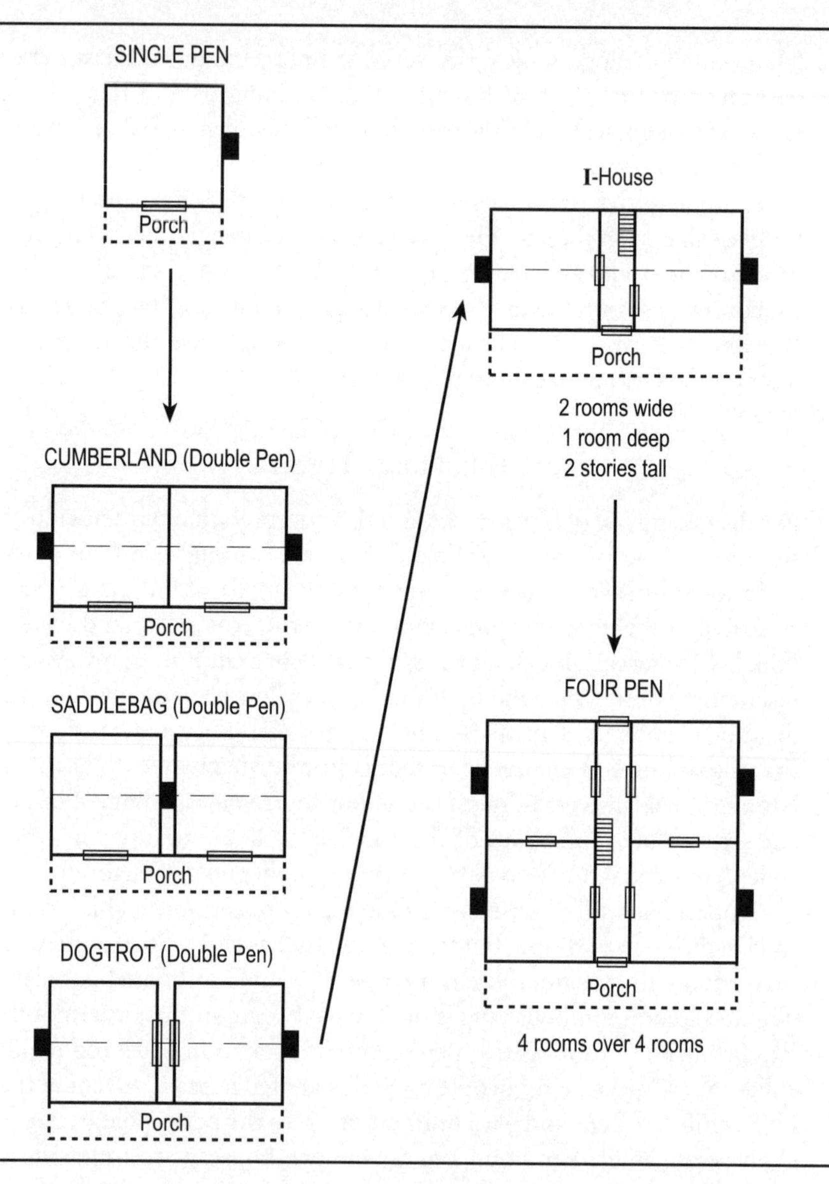

House types that followed the English pen tradition evolved from the single pen to three types of double pens to the I-house and ultimately to the four pen house. (Sources: Kniffen 1965; Jordan and Kaups 1989; Rehder 1999)

The single pen house is the initial one-room cabin. This one in Grainger County, Tennessee, once had a limestone chimney and front porch stairs leading to the loft. (Photograph by J. Rehder, 1978)

seventeen feet and has a single outside chimney at the end of one of the gables. Usually the dwelling has a single story, but some are built with a story and a half, with a sleeping loft supported by mortised loft joists in the ceiling above the main room. An interior ladder or a boxed stairway leads to the loft where the children sleep. A single front door, single back door, and optional windows provide access, air, and light. Appendages—additional structural attachments not considered in the typology of the single pen—may include front and rear porches and a kitchen in back.

Single pens in the nineteenth century tended to remain much the same size despite variations in time, place, timber, and notch types. In his investigations of log single pen houses in Alabama, Gene Wilson measured thirty-eight examples and determined that their outside dimensions averaged twenty feet, nine inches by seventeen feet, one inch. Built between 1815 and 1895, the log single pens used three types of corner notching: twenty were half-dovetailed, nine were square-notched, six were V-notched, and the rest were either mixed or unknown.[42] Elsewhere in the South, Wilson examined seventeen other log single pens dating from 1726 to 1875 in Maryland, Virginia, Kentucky, and Tennessee and found that they measured an astonishingly close average of twenty feet, nine inches by seventeen feet, two inches.[43]

Two members of my historic structures survey team, John Morgan and Joy Medford, analyzed fifty-four log single pens in Grainger County, Tennessee. The houses had average dimensions of twenty-two by eighteen feet and were built between the 1790s and 1930. Twenty-four of the houses were built before 1860. Thirty-seven of the fifty-four were notched with half-dovetail notches, and twelve were V-notched. The single pens were built from three major timber types: twenty-one houses were of pine logs, seventeen were of yellow poplar, four were of oak, with the remaining ones built with mixtures of timber types.[44]

In a related inventory in adjacent Union County, Tennessee, Vincent Ambrosia, another member of my survey team, compiled data on forty-nine log single pens. Forty-two, or 85.7 percent, were built with half-dovetail notches. The few remaining houses were two with V notches, three with saddle notches, one with saddle V notches, and one with square notches. In terms of timber types, thirty-five houses were built of yellow poplar timber, six of oak, and six of pine. The rest were mixed.[45]

John Morgan's work in Blount County, Tennessee, in 1982–84 revealed a population of 123 log single pen houses of which 30 percent were rectangular and 70 percent were square. Popular dimensions of single pens were: twenty-six by twenty feet, twenty-five by twenty, twenty-six by eighteen, twenty-four by eighteen, twenty-five by seventeen, twenty-four by sixteen, twenty-two by sixteen, and twenty by fifteen. In his analysis of *all* surveyed log houses in Blount County, Morgan discovered that 58 percent had half-dovetail notches and 24 percent had V notches. Moreover, representative timber types were pine at 68 percent, oak at 11 percent, and an astonishingly low 5 percent for yellow poplar.[46] The forest ecology here in the shadow of the Great Smoky Mountains National Park, which occupies a third or more of the county's area, partially accounts for the difference. Morgan's 123 inventoried log single pens largely come from the warmer and lower elevations in the Ridge and Valley portions of the county, which are dominated by pine and oak, with far fewer yellow poplar trees. The greater availability of pine habitat both inside and especially outside the park would account for the higher number of pine structures there.

In 2002, I analyzed and reviewed survey data on historical buildings for seventeen Tennessee counties located, from west to east, on the Eastern Highland Rim, Cumberland Plateau, Ridge and Valley, and Blue Ridge provinces. From the Tennessee Historical Commission data, I selected counties that had complete data sets that were computerized and for which building locations could be plotted. There were 265 log single pen houses in the collection, with a range of 6 to 41 at the county level. Single pens numbered 265 out of 658, or 40.2 percent of the log houses. The single pens far outnumbered the 138 Cumberland houses, the 101 saddlebags, and the 51 dogtrots in the double pen group of house types (these types are explained below).

In the Piedmont region of North Carolina, 44 log single pen houses were inventoried in the rural parts of Guilford County.[47] Corner notching included 11 houses with V notches, 10 with half-dovetail, 2 with full dovetail, and 1 square-notched. The rest were listed simply as log houses. Excellent photographs of the inventoried structures, maps, and detailed descriptions of their locations were made, but measurements and timber types were not included.

What can we conclude from these inventories and data analyses? First, single pens are not perfectly square houses; they are small rectangles measuring approximately twenty to twenty-two feet by sixteen to eighteen feet. Such dimensions represent an unusual uniformity for dwellings constructed over a period of a century and a half and built by folk builders with no blueprints. One could surmise that the dimensions are governed by tree type and size or by the weight of the logs. Measurements could be based on *x* number of ax handles, on pacing the distances, or on some other means of measurement. Or they could be based on traditional measurements carried from Europe. In early Britain a standardized measurement was the sixteen-foot rod for land measurement, for house bays or room units, and for animal stalls, which were sixteen feet square, the space needed to accommodate two pairs of oxen.[48] This standard dimension came to the American colonies as house bays in cruck structures in the earliest Virginia settlements.[49] In time, cruck structures gave way to timbered houses, but the sixteen-by-sixteen-foot bay became the basis for the single pen, bay, or square cabin of the English pen tradition.[50] Henry Glassie, in his study of the southern mountain cabin, correctly interpreted the square cabin as English and the rectangular cabin (twenty-two by sixteen feet) as a Scotch-Irish introduction. The rectangular cabin represented a direct descendent of dwelling dimensions from western Britain and northern Ireland.[51]

To summarize corner notches for single pen log houses, the dominant notch type is the half-dovetail, represented by 53.6 percent of samples in Alabama, 68.5 percent in Grainger County, Tennessee, 85.7 percent in Union County, Tennessee, and 42 percent for the single pens with identified notch types in Guilford County, North Carolina. The V notch accounted for only 15.7 percent of the houses in the Alabama survey, 22.2 percent in Grainger County, Tennessee, and somewhat more than 1 percent in Union County, Tennessee, but it dominated the identified notch types, with 45.8 percent in Guilford County, North Carolina. Temporally, for sample areas that included dated structures, the Alabama study had houses with half-dovetail notches ranging in date from 1820 to 1890.[52] In the Grainger County, Tennessee, study, half-dovetail-notched houses ranged from the 1790s to 1915.[53]

Timber types indicate changing availability and wood preferences. In Union County, Tennessee, yellow poplar timbers were used to construct 61.4 percent of log dwellings.[54] For Grainger County, Tennessee, a temporal pattern emerged as three phases of the log building period. Yellow poplar timbers dominated in 53 percent of log single pens in the early first phase, dating from 1790s to 1850. In the middle phase of 1850 to 1880, oak and mixtures of oak, poplar, chestnut, and pine timbers were used. For the last phase, until the end of the log building period 1880 to 1930, pine timbers dominated in 81 percent of the single pen houses.[55]

This means that within the ecological context of this part of Appalachia, yellow poplar trees were the trees of choice in the initial occupance and early settlement period. Of the timber types in Appalachia, yellow poplar is unquestionably superior for log construction because the trunk grows tall and straight, with only a few limbs in the uppermost part of the mature tree. The wood is soft, easy to cut and hew, lightweight and easily moved, insect and rot resistant, with considerable durability. How could a folk builder go wrong with such a marvelous tree? In the second phase, oak and mixtures of oak, chestnut, poplar, and pine indicate a change to some hardwoods but also reflect changes in the availability of timber suitable for log house construction on specific sites. In the third phase, pine is an imperfect choice because it is knotty, resinous, and decays easily. But by the late 1880s to the 1930s and thereafter, pine was the only sizable timber remaining for log construction.

On June 8, 2002, I was doing fieldwork on the Cumberland Plateau in Grassy Cove, an uvala or flat area of collapsed sinkholes in karst topography, surrounded here by the Crab Orchard Mountains. It is the most beautiful sight and site in Cumberland County, Tennessee, for limestone soil at 1,500 feet elevation. I spoke with John C. Kemmer III, the fifth-generation German American owner of a magnificent farm that covers about half of the 4,000-acre cove. The Kemmers were among the first families to settle the cove just before the Civil War, and they es-

The Cumberland house, also called the "basic or standard" double pen, is prevalent on the Cumberland Plateau. Diagnostic traits are two rooms, two front doors, and two outside chimneys. Fentress County, Tennessee. (Source: Tennessee Historical Commission, 1982)

tablished a store there in 1886.[56] Although he had no remaining log buildings on the property, we began talking about timber types for log structures. Before I could get a word in edgewise, Kemmer said, "You know, yellow poplar is the best wood for building with logs." I couldn't agree more.

Double Pen Houses

Common traits of double pen houses are two room units or pens placed side by side, one-room depth, and single-story height. Chimney placement, numbers of chimneys, and the spacing of pens are the diagnostic traits that distinguish subtypes. Double pen houses appear in three subtypes: the Cumberland house (basic or standard double pen), the saddlebag house, and the dogtrot house.

Cumberland House. The Cumberland house takes its name from the Cumberland Plateau, where in the 1970s anthropologists Riedl, Ball, and Cavender found it to have remarkable dominance in TVA's Normandy Reservoir Project area in Coffee County, Tennessee. My very close friend and colleague, the late Bert Riedl, an Austrian cultural anthropologist, coined that name for this rather ordinary-looking double pen house. It is quite fitting that Bert gave it the name, and it is in his honor that I use the term in my teaching and writing.[57]

The Cumberland house has two pens joined at the center, a chimney at the end of each of the two gables, and usually two front doors. Optional windows, interior doors, and appendages of front and back porches and an attached kitchen in back complete the structure. The house may be constructed of logs or of sawn planks, but like other small folk houses in Appalachia, it is almost never built of brick or stone.

In sampled surveys in Alabama, the Ridge and Valley Province in Tennessee, and parts of western North Carolina, the Cumberland house is poorly represented. Gene Wilson's study in Alabama examined eight such houses that represented various construction methods: three had half-dovetail notches, two were V-notched, one had square notches, and two were of frame construction.[58] In the Ridge and Valley region of Tennessee, only two log Cumberland houses were found in Union County. In adjacent Grainger County, where Morgan and Medford's analytical work focused only on log single pen houses,[59] I found only four log Cumberland houses in Grainger County in my later survey analysis in 2002. The Guilford County, North Carolina, surveys revealed ten houses of the Cumberland type, with three log houses built with V notches and seven houses using frame construction.[60] The dwelling is scarce in the surveys because Cumberland houses are more commonly found on the western margins of Appalachia in the Cumberland and Allegheny Plateau regions. From there the house type can be traced farther west into the Nashville Basin, to Arkansas and beyond; it is also well documented in Kentucky in the Bluegrass Basin, Pennyroyal, and points west. Another reason for the scarcity is that the Alabama and Tennessee survey compilations focused on log houses.

In 1992 I stated that, based on my experience at the time and on Gene Wilson's, the Cumberland house appeared more frequently in some areas of the South, but at later periods when balloon framing with sawmill lumber had become popular.[61] Let us examine the seventeen-county sample taken in 2002 to see where the Cumberland house fits into the scheme of things today. In the inclusive group of 658 log houses, the Cumberland house represents 138 log houses, or a respectable 21 percent. The Cumberland Plateau landform region appropriately holds 106 occurrences of the house out of the 138 total log Cumberlands, an enormous 76.8 percent of the log houses of this type in my sample. I rest my case.

The saddlebag house, also a double pen type, has a single central chimney and a pen on either side. Grainger County, Tennessee. (Photograph by J. Rehder, 1978)

Some of the plateau counties are outstanding. Half of Overton County is on the western edge of the Cumberland Plateau and the other half is on the Highland Rim on the next topographic level down and to the west. Overton has 51 log Cumberland houses out of a total of 150 log houses, or 34 percent. White County, south of Overton and almost entirely on the plateau, has 11 houses of 47, or 23.4 percent. Grundy County, entirely on the southern part of the Cumberland Plateau, has 9 log Cumberland types out of 22 log houses, or 40.9 percent. My analysis examined only log houses, not framed ones, of which there are perhaps hundreds more remaining on the landscape. Whether of log or frame construction, however, Cumberland houses are clearly identified with the plateau subregions in Southern Appalachia.

Saddlebag House. The saddlebag double pen house characteristically has a single central chimney with a pen built on each side. Like other multipen structures, the house in its formative stages could be built first as a single pen and have the second pen added later.[62] The saddlebag house gets its name from the imaginative interpretation of the chimney as a horse and the pens on each side as saddlebags. This house type, too, is not well represented in sampled surveys other than in the plateau country. In Alabama, Gene Wilson found thirteen saddlebag houses represented by four of log construction (full dovetail, half-dovetail, and V notches) and nine of frame construction.[63] The Union County, Tennessee, survey listed six saddlebag houses, all built with half-dovetail notches.[64] In North Carolina's Guilford County, the three saddlebag houses surveyed had V and half-dovetail notches.[65] In eastern Kentucky, Carlisle's detailed study of the Paintsville Lake Dam area indicated that three log saddlebags were built between 1836 and 1865. Indicative of the evolutionary process in pen construction and changing timber choices, one saddlebag had an older pen built in 1860 with half-dovetailed yellow poplar logs while the second pen was built after 1865 with pine logs in a mixture of half-dovetail, V, and square notches.[66]

My 2002 analysis of 658 log houses in the seventeen-county Tennessee sample area shows 101 log saddlebags, or 15.3 percent of the log houses, a rather respectable number. The houses have a focused distribution because 63 of them (62.3 percent) are found on the Cumberland Plateau and Eastern Highland Rim. Overton County with 24 and White County with 15 cornered the lion's share of log saddlebags. In my experience and John Morgan's, saddlebags of framed, sawn wood usually outnumbered log saddlebags in Appalachia's Ridge and Valley and Blue Ridge provinces, in landform regions other than the plateau country.[67] If all saddlebag houses—frame, log, or otherwise—were analyzed, there would be much better representation throughout Appalachia but with special concentrations in the southern Cumberland Plateau and Highland Rim country. They would also appear in convincing patterns in the adjacent Nashville Basin and Kentucky's Bluegrass Basin and Pennyroyal regions to the west.

DOGTROT HOUSE. The dogtrot is one of the most interesting and easily recognized folk houses on the American landscape. A dogtrot house has two pens separated by an open-air passage, with a common roof covering both. The house is described as having "two pens and a passage." Outside chimneys match the outside gable ends. Doors may be on the front but are commonly inside the open-air passage called the dogtrot. Unlike other folk houses named by scholars, the dogtrot is well known by name to folk in the regions where it is found.

In thirty-six years of examining southern mountain landscapes, I have not seen a dogtrot north of Virginia. The house is also rare east of the Blue Ridge Mountains, absent east of the Piedmont, and surprisingly rare in East Tennessee. The house has a spotty distribution in Ken-

The dogtrot house has two pens and a passage, with chimneys on the outside ends. The open passage serves as a cooling breezeway. Jackson County, Alabama. (Photograph by J. Rehder, 1978)

tucky; it is very weak in eastern Kentucky but has good distribution in central Kentucky in the Bluegrass Basin, especially southern Kentucky in the Pennyroyal.[68] The dogtrot is well represented in middle Tennessee on the Cumberland Plateau and Highland Rim and in the Nashville Basin, and it becomes increasingly frequent westward into Arkansas and the Missouri Ozarks. Dogtrot distributions sweep solidly southward from middle Tennessee throughout much of Alabama, Mississippi, northern and western Anglo Louisiana, and well into eastern and central Texas.

On the Southern Appalachian landscape, however, the dogtrot house appears to be an afterthought. If we closely examine Jordan and Kaups's detailed map of dogtrot house distributions, the log dogtrot has only four occurrences in Pennsylvania, four in Virginia, two in West Virginia, and three in South Carolina. In East Tennessee, John Morgan reported sixteen log dogtrots in Grainger, Blount, Morgan, and Union counties.[69] I know of only two observed examples to add to Morgan's total and one documented one in East Tennessee's Ridge and Valley and Blue Ridge provinces.

As we approach the Cumberland Plateau, dogtrots come into their own. In my data analysis of 658 log houses in seventeen East and Middle Tennessee counties, there were 51 log dogtrot houses. Of these, 43, or 84.3 percent, are on the Cumberland Plateau and Highland Rim, with concentrations in a straight-line north-to-south pattern in the counties of Overton (12), Putnam (4), Sequatchie (13), Grundy (5), and Marion (6). There can be no doubt that the log dogtrot house has spatial strength in numbers in the interior Cumberland Plateau and Highland Rim country. Moreover, slightly beyond the bounds of Appalachia proper, the log dogtrot gains strength in the Nashville Basin and forges ahead through the Deep South all the way to Texas. The dogtrot appears to follow the Natchez Trace from Nashville to Natchez that funneled settlers and settlement forms toward Texas. The log dogtrot also follows the routes of the 1838 Trail of Tears from the Cherokee hearths in southeastern Tennessee to Arkansas and to eastern Oklahoma. At least that is what the map pattern suggests.

The origin of the type is surprisingly uncertain, but more has been written about the dogtrot house than has been understood. Some investigators attribute it to Scandinavian origins.[70] Kniffen observed that the dogtrot house first begins to appear in southeastern Tennessee.[71] With all due respect to my major professor and mentor, I disagree. I think that southeastern Tennessee lacks the physical evidence, and its late place in the temporal development of settlement in the region after the 1830s simply does not fit with exemplary, though scattered, map evidence of old dogtrots that date from 1698 in Pennsylvania, 1739 in Maryland, 1783 in Virginia, 1784 in Middle Tennessee, and 1795 in North Carolina.[72] Gene Wilson found such a preponderance of dogtrot examples in northern Alabama that his book on Alabama folk houses could have been titled *The Dogtrot and Other Folk Houses in Alabama.*[73] Richard Hulan, looking at dogtrots in Middle Tennessee, suggested seeking the origin of the house type in the Bluegrass Basin of Kentucky, settled by "old-fashioned Virginians."[74] I like Hulan's thinking here, but he is speculating like almost everyone else. In 1968, Henry Glassie felt that the dogtrot was an Appalachian variation of the hall-and-parlor house from the English-settled Tidewater region.[75] I cannot agree with Glassie here, because if the hall-and-parlor is the progenitor house, we should be seeing many more true dogtrots in the Lowland South, especially in the English-settled lowlands of Tidewater Virginia, the Carolinas, and Georgia. We do not! Arguments for a Karelian or Scandinavian origin prevail based on Martin Wright's early work and especially on the finely crafted arguments in Terry Jordan and Matti Kaups's book.[76] Other than Jordan and Kaups, no one has studied the dogtrot house sufficiently

The I-house is North America's most ubiquitous folk house—it is two stories tall, two rooms wide, and one room deep. Grainger County, Tennessee. (Photograph by J. Rehder, 1978)

to reach an accepted resolution about its origin. So why can't we simply accept a Scandinavian origin and leave the subject alone? I think we can accept a Karelian/Scandinavian point of origin for a dogtrot house, and we can connect it to the Swedes and Finns in the Delaware River Valley. But beyond that the trail grows cold. The dots we would *like* to connect along the way through accepted avenues in and through Appalachia are few and far between. Perhaps the idea of dogtrot architecture diffused more directly, faster, and farther than the physical evidence. The jury may still be out on this house type, and as folk structures disappear from the landscape, the future work of the scholar who accepts this most important task will be made more difficult.

I-House

The evolution of houses in the pen tradition continues to the I-house, a structure two full stories high, two rooms wide, and one room deep. If any folk house can be called ubiquitous, it is the I-house because its distribution extends so widely across the United States. The house type extends from Pennsylvania to northern Florida, to southwestern Louisiana, to Texas, and all over the Midwest. The house appears sparingly in Utah, Washington, Oregon, and even in California.[77]

How did the I-house get its name? In 1936 Kniffen, observing houses of that type in the southwestern prairies of Louisiana, found that the farmer-builder-owners of these houses were descendants of midwestern grain farmers who had migrated to southwestern Louisiana in the 1880s from *I*ndiana, *I*llinois, and *I*owa. Without much fanfare, the house type was named for the first letter of these midwestern states; the name stuck and remains today.[78]

In Appalachia, I-houses have a special meaning on the landscape. Kniffen suggested that the I-house was a symbol of economic attainment in the Upland South, a region well known for its poverty.[79] He was proved correct by my former student James R. O'Malley, whose study of I-houses in northeastern Tennessee found that 86 of the 113 sampled I-houses could be directly correlated with large landholdings, the best soils, and land with gentle slopes.[80]

I-houses may be built of hewn logs, rounded logs, sawn lumber, brick, or stone. In a cross section of fifty I-houses in Grainger, Knox, and Sevier counties, Tennessee, Karen Rehder found forty-one I-houses constructed with framed sawn lumber, eight of brick construction, and one of stone; none were built of logs.[81] In Union County, Tennessee, four log I-houses, all with half-dovetail notches, appeared in the comprehensive survey made in 1979.[82] In 1978 James O'Malley and I wrote "The Two-Story Log House in the Upland South" to draw attention to a little-known and limited pattern in Appalachian folk architecture.[83] At the time, we did not expect to find many log I-houses. However, in my later seventeen-county survey of log structures in Tennessee there were an astonishing 83 log I-houses out of the 658 houses, or 12.6 percent of the group. In rural Guilford County, North Carolina, a total of 116 houses were surveyed, of which 8 were brick, 23 were built with logs but were covered with weatherboard (siding or clapboards), and 85 were frame construction.[84] Such inventoried examples indicate the frequency, general periods of construction between the 1790s and 1920, materials, and survival of I-houses in the Appalachian region.

Thousands of I-houses appear on the landscape in parts of Appalachia, and most have all wood frame and sawn lumber construction. Brick I-houses outnumber log I-houses. But there are several reasons why the log I-house is still relatively rare. Once the I-house became the symbol of rural attainment, families were unlikely to build a log I-house when sawmill lumber or brick was available. For people moving up socially and economically, a log house was not the proper symbol of opulence. Furthermore, two-story log I-houses were difficult to build. The massive weight of huge oak logs and the height they had to be lifted required more workers than one family could provide. Much like the concerted community involvement in Amish barn raisings, large log I-houses had to be constructed either in stages or all at once with gang labor.

Stone I-houses are common in Pennsylvania but are rare in Southern Appalachia. I know of four in East Tennessee. One, in northeastern Tennessee near Johnson City, is purported to have been built by early German immigrants. Another stone I-house in Tazewell in Claiborne County was built in 1814–18 by Irishman William Graham, who was Tazewell's first permanent merchant. The existence of this well-preserved limestone I-house raises two significant issues. That an Irishman was its builder and first owner suggests a direct link to an Irish heritage of stone construction. I believe that Graham was following tradition when he built the stone I-house. Second, Tazewell, Tennessee, was an important gateway, only a day's journey by wagon from Cumberland Gap. Graham thus had ample opportunity to start a trading business that let him establish permanent residence in the large, substantial limestone I-house. The Graham house is now occupied by John Kivett, whose family of lawyers has owned it since 1906.[85]

Jordan found common distributions of I-houses in the Fenno-Scandian source area. But since none predated 1700, he proposed a probable British origin for the I-house in America.[86] Outside end chimneys are an English trait, both for the English Tidewater region and for England proper. For the I-house in Appalachia, form, dimensions, end chimney placements, and other diagnostic features point conclusively to influences from the British Isles.

A Three Pen House?

Scholars of American folk housing do not seem to have identified a three pen house or used terms like "three pen" or "triple pen" in pen typology. On June 21, 2002, in the Farragut, Tennessee, city hall, I examined a faded photograph of a combination saddlebag and dogtrot house. The log structure had been built in 1810 by David Campbell (1763–1813) on Harvey Road in west Knox County. The half-dovetailed log building had three cabins joined together. The left and middle cabins were joined in the center and had a central stone chimney between them; this part was clearly a saddlebag. The middle pen and right side unit had a dogtrot passage between them and a common roof; this was the dogtrot part, even though the right side had been converted into a single-crib barn.

The four pen house completes the evolution of houses of the English pen tradition. Augusta County, Virginia. (Photograph by J. Rehder, 1997)

Another three pen house in Grainger County, Tennessee, is reported to have one or more log pens; none can be seen because the house is weatherboarded. Three pen houses are so infrequent that they do not constitute an established house type. However, the convertability and flexibility of the English pen tradition led builders of log houses and barns to create an evolving pattern of vernacular architecture in Appalachia.

Four Pen House

The final house type in the development of pen structures is the four pen house. Just as the name implies, the four pen has four rooms on each of two floors, with room pairs separated by central halls. The house is the ultimate, the pinnacle of the house type evolution for the region. As the largest house on the vernacular landscape, the four pen also is the most opulent house in this otherwise poor rural part of America. Consequently four pen houses are scarce. Four pen houses are built of framed sawn lumber or brick. They are relatively rare in stone and are almost never built of logs. While it is feasible to build I-houses and smaller structures with horizontal logs, to build a four pen, two-story log house would be a difficult undertaking and highly unlikely.[87] In the more richly settled areas in the Great Valley, particularly along the route of the Great Philadelphia Wagon Road, now U.S. 11, there are dozens of four

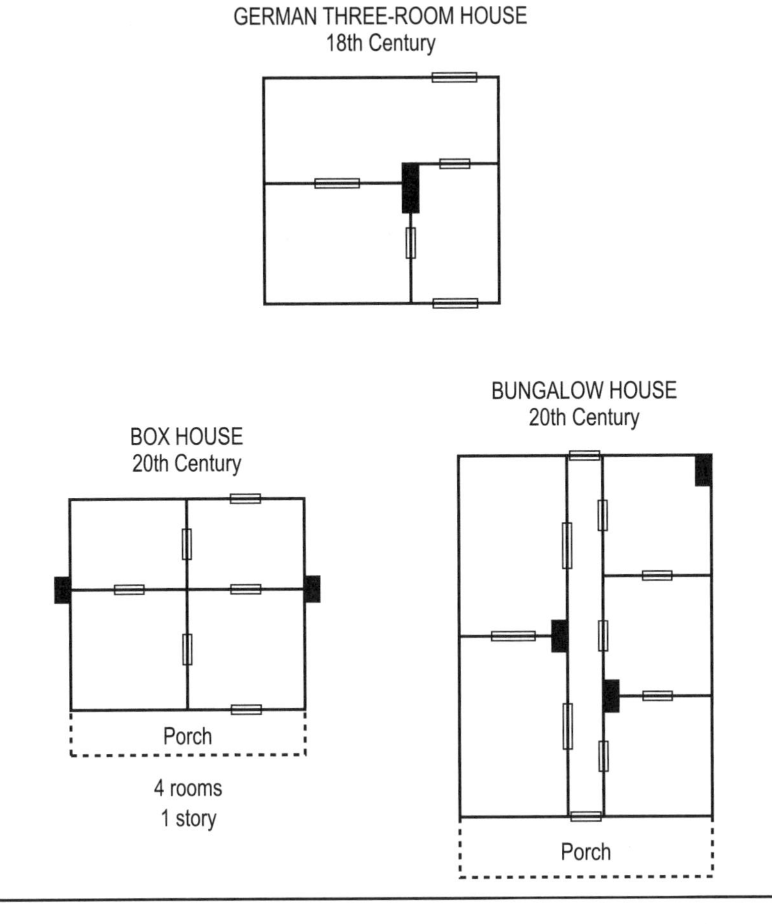

Floor plans of the German three-room house and twentieth-century bungalow and box houses reveal further folk architectural variations beyond the English pen tradition.

pen houses that once served as inns for travelers. The homes are enormous and truly represent the supreme stratum for folk housing in Appalachia.

The German Three-Room House

At this point we depart from the English pen tradition and the evolutionary pattern of pens in folk housing. The German three-room house, also called the continental house, is a rectangular to square log house that from the outside looks like a single pen dwelling. The floor plan reveals three distinct rooms—two small rooms and one large one—with a focus on a central chimney. One of the best surviving examples of a German three-room house is a log house in the restored Moravian village of Old Salem in Winston-Salem, North Carolina. The dwelling was built in 1768, and its initial occupant, blacksmith George Schmidt, lived in it from 1771 to 1789.[88] The three-room house is an oddity and is scarce except for the few remaining examples in German-settled Pennsylvania and in the German Moravian settlement of Salem on the North Carolina Piedmont.

Twentieth-Century Bungalows and Box Houses

Between 1870 and 1910 there was a decline in continuous log construction of folk house types in Appalachia, then it ceased altogether. The pen tradition, at least in logs, ceased as well. Two house types, the bungalow and the box house, emerged from a rural revolution in material culture that we might call an "age of lightweight sawn lumber."[89] Bungalow houses have a diagnostic front-facing gable. Floor plans may have three or more rooms arranged in a double line along the long axis of the house, but they are *not* shotguns. The chimney or stovepipe can be placed almost anywhere the owner or builder chooses. Bungalows are usually built with framed balloon two-by-four studs with double walls and are weatherboarded with horizontal sawn lumber. While bungalows are simple framed, easily built homes, box houses can be even simpler, cheaper, and much flimsier. For both houses, the roof can be pyramidal or even a high hip design, but most are saddle roofs, with the roof ridge running along the long axis of the house. The tin roof, a misnomer, is really made of overlapping corrugated panels of galvanized (zinc-coated) steel nailed to rafters.

By my definition, a box house is a square to slightly rectangular four-room house with all four rooms on one floor, or "four on one." If rectangular, the gables face to the sides. The chimney can appear almost anywhere. I do not agree with some other writers about the identity of box or boxed houses. John Morgan and Michael Ann Williams identify them by their thin board-and-batten construction. I maintain that it is not the construction technique alone that identifies the house but the floor plan and gable position.[90] It is true that box houses were typically of board-and-batten construction; but I have seen them balloon framed and weatherboarded, and I have seen them built with logs.

Board-and-batten houses have a weak, thin frame of sills, plates, and corner posts covered with inch-thick vertical planks. Boards eight or ten inches wide were nailed up first, then two-inch wide strips called bat-

Top: A well-preserved German three-room house from about 1768 can be seen at the museum community of "Old Salem" in Winston-Salem, North Carolina. Notice the full dovetail notches, a rare sight. (Photograph by J. Rehder, 1977). *Bottom:* Bungalow houses became the popular twentieth-century folk houses of choice in many parts of Appalachia. Hawkins County, Tennessee. (Source: Tennessee Historical Commission, 1987)

A box house of the Great Depression era was built in the 1930s with small round pine logs and notched with saddle V and saddle notches. These were among the last log folk houses built in Appalachia. Grainger County, Tennessee. (Photograph by J. Rehder, 1978)

tens were nailed over the cracks between the boards. The term "batten" has much the same meaning as in the nautical saying "Batten down the hatches!" Between 1880 and 1940, board-and-batten construction became the principal method for cheap buildings, whether they were box houses, sheds, barns, or houses and other buildings in logging and coal camps. Board-and-batten construction was cheap, easy to work with, and arrived at an ideal time, when the rural landscape was being logged, with portable sawmill operations scurrying about, and when agriculturally settled areas were running out of suitable logs for construction.

While it is remarkably unusual, some bungalow and box houses were built of logs. In my seventeen-county Tennessee survey data, there are ten log bungalows and nine log box houses. Together they represent less than 1 percent of the 658 log houses in the group. Seven out of ten log bungalows and six out of nine log box houses were on the Cumberland Plateau and Highland Rim. Dates of construction for some of the log bungalows were 1910 for two houses and 1924, 1927, and 1930 for others. The log box houses were built in the 1930s. Most were "Depression houses" that were built with round pine logs and saddle or saddle V corner notches; on some of the houses the small-diameter logs were painted. The old saying "necessity is the mother of invention" applies here in a modified way. Building log houses during the Great Depression years meant that folk had to revert to an old, likely forgotten, construction technique that used whole pine logs and notched the corners with saddle notches. Necessity bred the reinvention of old folk traits, albeit weak ones in log construction that afterward disappeared again when times got better.

The popularity of sawn lumber for building in twentieth-century Appalachia cannot be overestimated. Both twentieth-century houses—bungalows and box houses—took advantage of sawn lumber that allowed for expanded house dimensions with certainly more rooms under one roof. Houses were no longer restricted by the length and weight of logs, and floor plans were no longer under the constraint of log pens. Thin walls allowed for more interior and exterior openings. The raw materials in twentieth-century houses—window glass, cheap sawn planks, machine-cut nails, and tin roofs—meant a revolution in folk housing. It was to be the last.

Folk Architecture in Materials Other Than Log

Whether the materials are logs, stones, bricks, or sawn lumber with balloon or board-and-batten construction, genuine folk buildings are folk first, regardless of the materials they are constructed from. I know what you're thinking: Why did Rehder just spend so much time building the case for folk log structures? I wanted to demonstrate that a significant share of the Appalachian landscape over time has been dominated by buildings constructed of logs. I explained that log structures are an exceptionally strong diagnostic indicator of folk architecture. But I don't want readers searching for folk houses to be disappointed, thinking that a framed dogtrot or a stone I-house is meaningless because it is not log. The form and the floor plan make buildings culturally meaningful. The forms discussed throughout this chapter are indeed culturally important, from house types to outbuildings and other structures. Many are not built of logs, but they are still folk and still important to our reading of the landscape.

Barns

The vernacular landscapes of Appalachia are characterized by rustic barns and small outbuildings. Barns, much like houses, have an evolutionary pattern that proceeds from single crib barns to double crib barns,

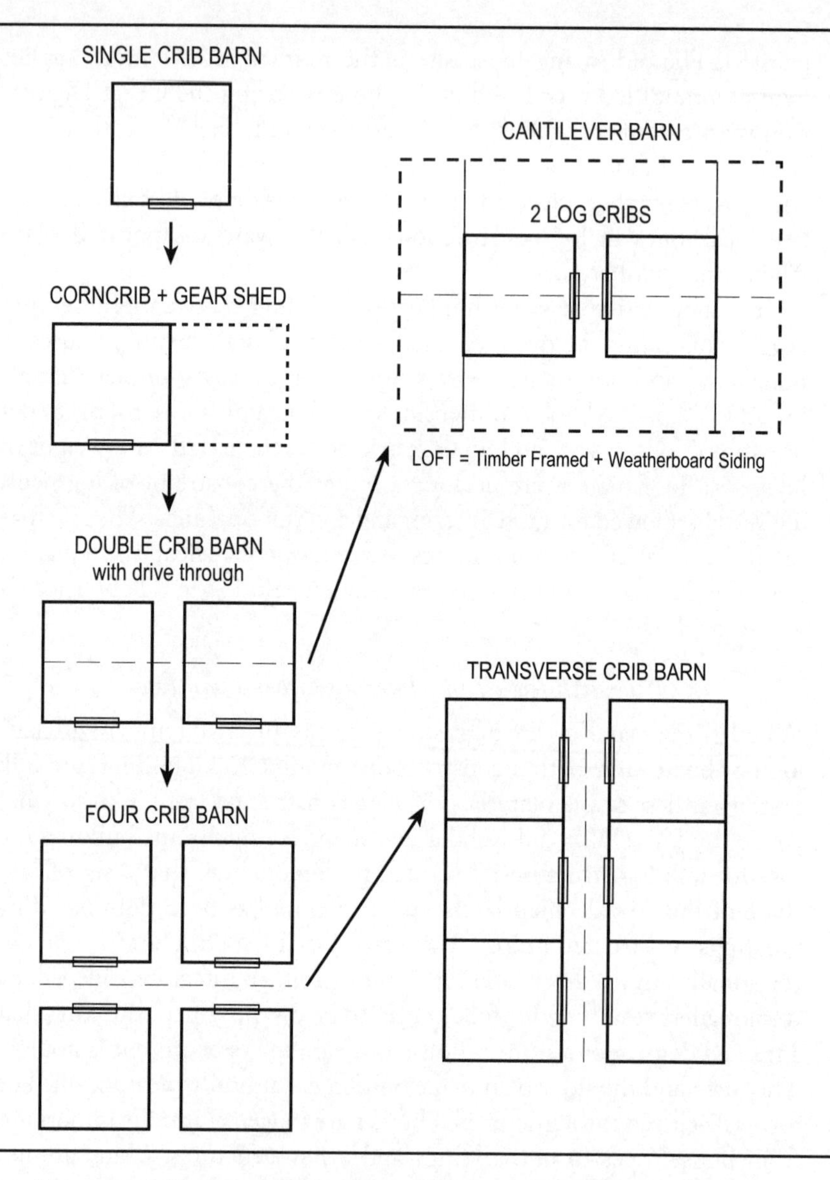

Barn types follow the evolution from single crib to double crib to the cantilever barn and the combination of four cribs to transverse crib patterns. (Sources: Kniffen 1965; Glassie 1968; Jordan and Kaups 1989; Jordan-Bychkov 1998; Rehder 1999)

to four crib barns, to transverse crib barns.[91] The modular units are called cribs instead of pens, for good reason: remember, "Pens are fer people and cribs are fer critters." In 1901, Ellen Churchill Semple described a typical mountain farmstead's outbuilding arrangement:

> The usual home of the mountaineer is a one-room cabin. Near by is the barn, a small square log structure, with a roof projecting from 8 to 10 feet, to afford shelter for the young cattle or serve as a milking-shed. These vividly recall the mountain architecture of some of the Alpine dwellings of Switzerland and Bavaria, especially when, as in a few instances, the roofs are held down by weight-rocks to economize hardware. Very few of them have hay-lofts above, for the reason that only a few favoured districts in these mountains produce hay.[92]

Single Crib Barn

The single crib barn is a one-unit structure that may be either square or rectangular. Single cribs have dimensions ranging from six feet on a side for some of the smaller buildings to twenty-five feet for larger cribs. The crib's function is to store things—usually corn, but it can also be used as an animal stall. Since most single crib barns are built for corn storage, in the folk vernacular they are corncribs. Some single cribs may also provide adjacent shelter for farm implements with an oversized overhanging roof called a gear shed. The gear shed, a cantilevered appendage, not only adds protection to the side(s) of the building, it also adds character.

The seventeen-county survey data from Tennessee is rich in single crib barns. Of 461 log barns and outbuildings, 172 are single pens, representing 37.3 percent. The largest concentrations are in the Ridge and Valley and Blue Ridge provinces, with 62 in Union County, 21 in Grainger County, and 47 in the mountain county of Johnson in the northeast corner of the state, in the Watauga River Valley and old settlement hearth area. Appalachian farmers quite early decided that a single crib barn could be doubled by adding another crib to the side of the first one.

Double Crib Barn

The double crib barn has been expanded to two cribs separated by an open passage but with a common ridgepole roof. Double crib barns look like dogtrot houses, but the two structures do not positively correlate in geographic distribution in the Upland South. However, Jordan has found close correlations between double pen houses and double crib

Top: A single crib barn with its gear shed in the shadow of Clinch Mountain in Grainger County, Tennessee, still stands but does not look like this anymore since the owners remodeled the roof. (Photograph by J. Rehder, 1978) *Bottom:* Sometimes a single crib barn has two gear sheds that have been cantilevered with large crossbeam timbers. Meigs County, Tennessee. (Source: Tennessee Historical Commission, 1979)

A double crib barn with the drive-through passage between the cribs. The "Alpine" roof projection that is so common to true European Alpine outbuildings can still be found on small outbuildings in the central and northern parts of Appalachia. Bland County, Virginia, near Wytheville. (Photograph by J. Rehder, 1977)

barns in Finland.[93] The double crib barn displays several variations in shape, size, function, and roof design. Double cribs can be square or rectangular. They can vary from four by eight feet per crib to eighteen by twenty-four feet but are rarely any larger.

Different roof orientations are interesting grist for the mill of geographic and cultural interpretation. In the northern half of Southern Appalachia, especially in the Ridge and Valley Province, many two crib barns have a roof that is oriented front to back over the short axis of the structure. They also have curious and quite Alpine-looking overhanging roof projections. In the southern half of Southern Appalachia, double crib barns have a roof ridge that is oriented along the long axis of the structure, very much like a dogtrot house. These observations suggest that the saddle roof with "Alpine" roof projections, found in the northern half of the region, is associated with places where early German settlers were more numerous than other settler groups, whereas in the southern portion of the region the double crib barns with long ridgepole roofs are found where fewer Germans but more Scotch-Irish settlers and other non-Germanic Europeans had established themselves. There is a

A double crib barn with the drive-through passage between the cribs. What differences do you see between this barn in Bradley County, Tennessee, near Chattanooga, and the one on page 119? (Source: Tennessee Historical Commission, 1984)

temporal component here as well, because settlement history tells us that the Ridge and Valley Province was settled largely from north to south. Thus older landscape features are found in the northern half while comparatively "younger" settlements and their corresponding material culture are found in the south. Such speculations coming from an old-time observer like me open the door, challenging me or someone else to prove or disprove a rather simplistic interpretation of a very complex set of landscape circumstances.

In my 2002 data analysis of seventeen Tennessee counties, 117 of 461 barns and outbuildings were double crib barns. This represents 25.3 percent of the structures. While these numbers may be representative and acceptable for the total, I should explain some of the variances between counties. Northeast Tennessee's Johnson County dominates with 72 double crib barns; Union has 12, Bradley has 6, and Morgan has 5. There are three more counties with 4 each. The overabundance of double crib barns in the Johnson County survey reflects three things. First, the cutoff dates for surveyed structures were in the 1940s. Double crib hay barns, sometimes described as "pole barns," became very popular in the 1930s, 1940s, and especially after World War II. Second, John Morgan and his crews from Emory and Henry College were extremely thorough

The basic four crib barn has log cribs on each of its corners. Hancock County, Tennessee. (Photograph by J. Rehder, 2001)

in their survey work. Their catch rate was exceptionally good because they looked inside lumber-sided barns and found double cribs serving as foundations. Most surveyors would not be so thorough. Third, I can be blamed for interpreting a number of barns as double cribs because of the crib foundations. If I were to reexamine the data, perhaps some double crib barns might indeed be reclassified as cantilever barns (see below). So there you have it. Not every surveyor is so thorough, not every data analyst can be accurate, and all data sets have flaws. In spite of this, Johnson County still would dominate with single and double crib barns; it is simply *that* rich with barns and outbuildings.

Four Crib Barn

The four crib barn represents further barn evolution; it is a barn with a crib on each of the four corners. Open passageways crossing at the center separate the four cribs, and a gabled roof covers the entire structure. Henry Glassie described the development of the four crib barn and its transformation into a transverse crib barn this way: "The Tennessee Valley farmer found himself in need of a large barn. . . . He built a pair of double-crib barns facing each other, roofed the whole, and had a four-crib barn. As it was built, the four-crib barn developed its own symmetry, one of its two passageways was blocked off to provide additional stabling, and this painfully neat evolutionary sequence resulted in the transverse-crib barn."[94] This journey in imagination on the part of a young Henry Glassie can be neither denied nor proved based on our landscape evidence. But I have observed that in Appalachian four crib barns all four cribs were usually built at the same time and covered at the same time. Temporally, some are not particularly old, dating from the late 1880s and even perhaps to the 1930s and 1940s.

The log four-crib barns in my seventeen-county Tennessee data disappointingly accounted for only 12 examples out of 461 log outbuildings. Union County had the most, with 5 log four crib barns. Bradley and Overton counties had two barns each; Johnson, Marion, and White counties had one each. Beyond the survey data, I know of a few more, with one each in Sevier and Hancock counties in Tennessee and one in Haywood County, North Carolina. But even with this experience, I suspect there are or have been more out on the landscape.

Transverse Crib Barn

Another method of expansion is to join three or four cribs on each side of a central passage and create a transverse crib barn. This enlarged barn, which may be sixty to ninety feet long, has a single central passage, but the diagnostic traits are the multiple cribs on either side. Some are corncribs to store livestock feed, others are used as stables for mules, horses, and other farm animals. As Glassie noted in the quotation above, the transverse crib barn evolved in concept and in actuality from the four crib barn. Both the four crib and transverse crib barn types are considered original American barn types but with links to western Europe through the single and double crib barns.[95]

The seventeen-county Tennessee data also appear sparse, with 15 log transverse crib barns out of 461 structures. Somehow I expected more, and there is no significant major concentration. Johnson and Marion counties have 3 barns each; Grainger and Union counties have 2 each, and Bradley, Morgan, Polk, Putnam, Rhea, and Sequatchie each have one. Perhaps other four crib and transverse crib barns were covered with weatherboarding and thus escaped detection. It is entirely possible that other transverse crib barns exist either in counties that have not yet been surveyed or in counties where the surveyor's lack of thoroughness allowed some structures to slip by. Ideally, unless we were to have complete and thorough surveys of each county in Appalachia during the 1970s to 1990s, I doubt we would ever know the true nature of all of Appalachia's folk architectural patterns.

This transverse crib barn in Grainger County, Tennessee, has three cribs on each side; it is constructed of round logs with saddle notches and has a shake roof. (Photograph by J. Rehder, 1978)

Forebay or Pennsylvania Barn

Beyond the evolution of cribs is the forebay barn or, as Robert Ensminger calls it, the Pennsylvania barn.[96] The diagnostic feature is an overhanging loft or forebay that allows feed to be delivered from the loft through a trapdoor to the barnyard below. The forebay overhang also provides livestock with a little shelter in bad weather. Ensminger, unquestionably the expert on this barn type, wrote:

> In the early eighteenth century, a two-level barn appeared in southeastern Pennsylvania. This barn was larger and more substantial than preceding pioneer farm structures in North America. It was more versatile than the one-level ground barn (*Grundscheier*) being built at the same time. This multipurpose barn could house various livestock in the basement (lower level) stable, while on the second floor it could store hay and straw and accommodate the threshing of feed grains. These were then stored in the bins of a granary, which was usually located in the forebay. The forebay, or "overshoot," is the second-floor extension, which projects over the front stable wall for a distance that varies from four feet to even twenty or more feet in some barns. The forebay, this

> extra space, or bay, on the fore side of the barn, is the distinctive and diagnostic feature of the Pennsylvania barn.
>
> Access to the upper level of the barn is created by banking the barn (building it into a hill) or by constructing a gentle ramp or bank leading to the second level, allowing farm machines and wagons to be driven up into the barn.[97]

Forebay bank barns, or Pennsylvania barns, dominated southeastern Pennsylvania by the end of the eighteenth century and diffused from this important culture hearth to the west and south in the nineteenth century. The barn's diffusion pattern took it in a very broad westerly direction across Pennsylvania, through the Ohio River Valley, and particularly throughout the farm country of Ohio, Indiana, and Illinois. A smaller but no less important route took it down the Great Valley of Virginia to eastern Tennessee. Forebay barns in the southern Appalachians are smaller than those in the northern part of the region, but they trace to the German-settled areas of rural Pennsylvania, where huge forebay barns still dominate the landscape. Furthermore, they can be traced to European source areas in central and eastern Switzerland, especially to Canton Graubunden.[98]

Cantilever Barn

The cantilever barn diagnostically has huge horizontal beams supporting a large hayloft over two log cribs that serve as foundations. From the ground up, cantilever barns have flat stone footers set at each corner of the log cribs. The stones level the cribs and keep the logs off the ground to prevent moisture and insect damage. Horizontal logs that have been either hewn square or left in the round are notched and placed on top of each other, and stacked with four to eight courses per side for each crib. Crib sizes range from thirteen to about twenty-four feet on a side. The largest barn, built in 1880 by Thomas DeArnold Wilson McMahan in Sevier County, had eighteen-by-twenty-four-foot cribs. Cribs are open to the air, with unchinked logs, and have wooden doors or gates cut into them. Most cribs were once used for animal stalls or corncribs. Floors are dirt. Six of the barns had a threshing floor, reflecting the time when wheat was an important crop in the Ridge and Valley Province. Long, thick cantilever beams are the diagnostic feature of the cantilever barn. Beams are from thirty to forty feet long and from ten to eighteen inches thick.[99]

In a survey of cantilever barns in East Tennessee, Marian Moffett

This cantilever barn has double cribs serving as foundations for the large hayloft. Bradley County, Tennessee. (Source: Tennessee Historical Commission, 1984)

and Lawrence Wodehouse discovered 316 log barns that they classified as cantilever barns. Of the 316 barns, 291 had double crib foundation units. I have slightly modified their terminology, and based on their count we have 161 two crib double cantilever barns; 112 two crib single cantilever barns; 18 two crib barns with single *and* double cantilevers (Moffett and Wodehouse mysteriously called this one a "half-double cantilever"); 12 single crib double cantilever barns; 8 four crib cantilever barns, and 5 unclassified.[100] The dominant type is the two crib double cantilever barn, which accounts for 161 of the 316 barns in the Moffett and Wodehouse survey. Diagnostic traits of the barn are the two log cribs for the foundation and two sets of cantilevers that support the overhanging loft on all four sides. Imagine two square cribs set side by side. Large wooden beams are laid across the tops of both at right angles to each other. Not only that, the beams extend well beyond the dimensions of the cribs so that the structure looks top-heavy. Add to this a framed loft, a large saddle roof that has the roof ridge following the long axis, and sawmill siding used for weatherboarding all around the outside of the loft.

The second most common cantilever barn is the two crib single cantilever type, which accounts for another 112 barns out of the 316 total on the Moffett and Wodehouse survey. This barn has the same two log cribs supporting the loft. However, the gable ends of the loft are flush with the log cribs underneath so that the cantilever feature operates only along the short axis of the structure.[101]

Notch types were overwhelmingly half-dovetail notches on 208 out of 310 log structures. Barns with V notches accounted for 58 structures, and 35 barns had square notches. Nine of the barns had a mixture of half-dovetail and V notches.[102] One of the problems in identifying barn types is that sawn lumber used as weatherboarding covers the expanded outer parts of many log barns. An enormous weatherboarded barn with a huge loft may obscure the small cantilevered double cribs that support the entire structure.

My seventeen-county Tennessee survey data, which do not include Blount and Sevier counties, show a meager 20 cantilever barns. Johnson County has 9; Bradley has 6; Meigs County has 4; and there is one in Morgan County. In spite of this small number, there is an interpretable pattern. Johnson County, a mountain county in northeasternmost Tennessee in the Watauga culture hearth subregion, has an inordinate quantity of cantilevered small outbuildings that may account for the focus of some cantilever barns there. Bradley and Meigs counties, in the Ridge and Valley Province, are on the main routeway for transportation, communication, and settlement migrations down-valley from Pennsylvania and points in between. They would have been in direct line to receive traits, such as barn concepts and patterns, that diffused from Pennsylvania. The barn in Morgan County, on the Cumberland Plateau, is an anomaly and out of the loop.

The spatial pattern of Moffett and Wodehouse's 316 cantilever barns shows an inordinate concentration in Sevier County, Tennessee, with 183 barns, and in adjacent Blount County, Tennessee, with 104 barns. These two counties alone account for 287 of the 316, or 90 percent of the barns found in their surveys. This led Moffett and Wodehouse to speculate—I believe erroneously—that cantilever barns were indigenous to this part of East Tennessee. "Indigenous," in the anthropological and cultural sense, means native to an area, implying inventions and origins. For nearly twenty years I have been unable to convince them that the true origins of the cantilever barn obviously trace to Old World European source areas.[103]

The cantilever principle of construction in general, and its application to barns in Tennessee in particular, is not really unusual. The cantilever barn is a variation of the forebay barn. I believe that the cantilever log barn originated in Europe and diffused to America with German immigrants who entered through the port of Philadelphia. Ideas of barn construction diffused down the Great Valley of the Ridge and Valley Province as immigrants of German descent, along with Scotch-Irish, English, and other groups, made their way southward, generation by generation, along the Great Wagon Road and other routes. Naturally, over time barns were modified. We know for certain that barns became progressively smaller north to south from Pennsylvania to Tennessee and into Alabama and Georgia. Large barn size is believed to have been related to the need to shelter livestock in colder northern climes. With milder southern winters and open range stock raising, large barns for livestock shelters were unnecessary.[104]

How did the cantilever barn get to Tennessee? Although we have somewhat limited landscape evidence, we can still connect the dots of cantilever barn occurrences between Tennessee, Virginia, and Pennsylvania and thence to Europe.[105] Assume for the moment that the idea of the cantilever barn has made this arduous journey from Europe to Appalachia but only a few people along the way fully developed the idea in practice. Although the idea lay dormant over some of the traceable route and over much of the temporal scale, we have the emergence and near explosion of cantilever barns flourishing in Sevier and Blount counties, Tennessee, in the 1880s and 1890s. Moffett and Wodehouse said that more than half of all the 316 cantilever barns in their survey dated to this narrow time frame.[106] To me it is astonishing but not out of the question that a resurgence of log cantilever barns could emerge with the right conditions.

Something obviously happened to encourage such growth in material folk culture, and we are still not sure exactly what it was. I can only suggest that the cantilever barn possibly grew out of (1) a need for larger barns to store hay for the larger cattle herds that developed in the post–Civil War period; (2) the emergence of skilled "knowers," that is builders or farmers and their friends and families; (3) the possibility that skilled builders may have come south after the Civil War to either settle in eastern Tennessee or ply their trade here; (4) a combined use of sawmills and sawn lumber, with log construction still used but now relegated to foundation supports. Even though an explosion of cantilever barn building focuses on two counties in Tennessee over a tiny ten-year period in the late nineteenth century, we cannot and should not state that the cantilever barn is indigenous to the area or to the time. To be sure, the cantilever idea was incubating in other areas and at other times in Appalachia; but it was an idea whose time had come in Sevier and Blount counties, Tennessee.

I can think of other traits that have had somewhat similar explosive beginnings. The so called "Appalachian dulcimer" is neither native nor indigenous to Appalachia, yet some people think it is original to the region. The dulcimer also traces to European source areas through the zither and sheitholt, early stringed instruments. The "idea" and a few examples of the dulcimer came to Appalachia via Pennsylvania and the Great Wagon Road down the Great Valley of Virginia. However, the dulcimer did not emerge as a popular instrument until after the turn of the twentieth century, and this happened in parts of eastern Kentucky. I believe, as other scholars do, that dulcimer popularity grew out of the Kentucky settlement schools such as the Hindman Settlement School at Hindman, Kentucky. Here James Edward "Uncle Ed" Thomas and his students, especially Jethro Amburgey, made dulcimers and taught other people how to make and play them, and a traditional folk trait was rekindled. Notice, I did not say "was born." The Appalachian dulcimer and the cantilever barn are both folk traits that were rekindled in different but very focused places at about the same general time. I see no connection between the two traits and their time frames, but I do sense the importance of key "knowers," builders, and teachers of the culture in focused locales. While neither the dulcimer nor the cantilever barn is indigenous to any part of Appalachia, they are profound diagnostic traits that just happen to trace to European source areas. Still, both remain wonderful symbols of Appalachian folk culture.

Small Outbuildings

Henry Glassie, a well-known folklorist, described the typical Appalachian small outbuilding this way: "It has a rectangular floor plan, consistent with German and Scotch-Irish traditions, a regular double pitch roof, and a door in one gable end. . . . Although not always present its most distinguishing feature is a projecting roof, constructed on a cantilever principle typical of Pennsylvania German construction."[107] The small outbuildings of the vernacular landscape have more variety in their functions than in their form. They are characteristically used as smokehouses, root or apple houses, springhouses, dairy or milk houses, wash houses and for other specialized storage or protection functions. While most small outbuildings have one level, those built into steep hillsides may have a stone cellar and a wooden upper level. A typical small out-

Two nearly identical small Appalachian outbuildings, the older one built of logs and the newer one built with sawn lumber, tell us that traditional folk architecture endures even when materials change. Union County, Tennessee. (Source: Tennessee Historical Commission, 1979)

building is a single crib measuring about twelve by fourteen feet, square or rectangular. Diagnostic traits are the front-facing gable and the roof overhang built with a cantilever at the gable entrance to shelter the gable-end door.

Smokehouses

The smokehouse in Appalachia was integral to a culture that placed so much emphasis on raising hogs for home consumption in a "hog meat and cornpone" subsistence economy. Smokehouses are small, typically log outbuildings that have a front-facing gable and a projecting roof that overhangs the doorway. They are used for salting and smoking meat, and at one time almost every farm had one. In 1978 John Morgan, Joy Medford, and I analyzed twenty-six log smokehouses as part of a comprehensive historic buildings survey in Grainger County, Tennessee. We were interested in the log smokehouse as a typical form of folk architecture, but we were also looking for evidence of residual traditional functions and alternative uses for the structure. Twenty-four structures, or 92 percent, had the diagnostic roof overhang at the gable entrance. The smallest smokehouse measured seven feet, ten inches by ten feet, three inches; the largest was sixteen feet, nine inches square. Smokehouses with half-dovetail notches dominated, with nineteen (73 percent); six were V-notched, and one was square-notched.[108] In spite of the large number of smokehouses in the county, we discovered only one still functioning as originally intended, for salt curing and smoking pork. When I returned to Grainger County in 2001 for additional fieldwork for this book, I found that this last functioning smokehouse had disappeared.

Smokehouses were established to salt and smoke pork. (See chapter 6 for a description of hog killing.) The most common method of preserving meat was to coat it with salt and allow the meat to absorb as much as possible. During a hog slaughter, prepared cuts of meat were placed on a salting board, in a box, or in a trough made from a hollowed-out poplar trunk. Frequently salting boards two to three feet wide were wedged between logs inside the unchinked smokehouse, and the meat was rubbed with salt there. Meat was placed in salting boxes or troughs and left to absorb more salt over the course of six to eight weeks. Next the meat was hung on poles that spanned the length of the structure five or six feet above the ground. The poles were similar to joists in a loft, but there was no ceiling. A small, smoldering fire of hickory chips was set in the center of the smokehouse floor. The smoking lasted about a week, then the meat could be eaten as needed.[109] Once the slaughtering, salting, and smoking was done, the building became the storehouse for cured hogmeat for the coming year.

Data in the seventeen-county Tennessee surveys revealed 102 log smokehouses of 461 log outbuildings, or 22.1 percent. This is a strong showing. Union County led the group with 36, and adjacent Grainger County had 26; both counties are in the Ridge and Valley Province. Johnson County, a mountain county in the Unakas in northeasternmost Tennessee, also had 25 log smokehouses. None, to my knowledge, were still functioning as smokehouses.

Springhouses

In the list of priorities for successful frontier settlement, a nearby water source came first. Certainly factors of slope, soil, access, and wooded acreage were integral to first effective settlement, but nothing was as important as drinking water. Springs have always been the primary source for water, even with creeks and rivers nearby. The spring provided clear, clean drinking water, but it meant more than that. It was as much a hearth to the outdoor elements of the homestead as the cooking hearth

Springhouses protected the water supply and often had a trough inside to hold containers of perishable dairy products. (Source: Tennessee Historical Commission, 1978)

was to the family home. This cool water hearth, enjoyable as it was on a hot summer day, needed the protection of a small outbuilding called the springhouse. It shielded the drinking water source from contamination by leaves, sediment, and animals, and it was also a frontier refrigerator where milk, butter, and perishables other than meat were kept cool in troughs of spring water.

In shape and size, springhouses are much like smokehouses; from roughly six by ten feet to as much as ten by sixteen feet. Most have a stone foundation surrounding the spring. Walls can be of logs or framed sawn wood. The roof, once made of wooden shakes, now is of corrugated metal. Springhouses have the characteristic roof overhang on the gable-end entrance that is so diagnostic to Appalachian small outbuildings. Inside, stone troughs two to three feet wide by six to eight feet long are filled with cold spring water. Springhouses also have raised stone features that serve as shelves. Perishables are placed in jars or cans in the trough and on the shelves. The springhouse floor is also stone. Most springhouses in Appalachia are constructed of framed sawn wood or even stone, but a few are built with logs. In the seventeen-county Tennessee survey analysis of log buildings, there were a total of sixteen log springhouses. Union County had eight, Johnson County had four,

Grainger County had three, and Sequatchie County had one. All were in areas of limestone soils and limestone geology, where springs outnumber wells for drinking water even today. Whenever I mention springhouses to informants in the field, I get a smile and a look of satisfaction that tells me they know the meaning of good-tasting cold spring water.

Root Cellars, Apple Houses, and Canning Houses

The small outbuilding complex on a typical Appalachian farm may also include root cellars, apple houses, and canning cellars or houses. Root crops of potatoes, apples and other fruit, and canned or dried vegetables required dry, cool storage. While their names imply specialization, apple houses and potato houses do not exclusively store apples and potatoes; they can be canning cellars as well, and vice versa. These small outbuildings are similar to smokehouses and springhouses in shape and size. They also have the projecting diagnostic roof, and like the others they have the door on the gable end. The primary difference is that they have a stone cellar built into a hillside. Some buildings have a second-floor storage room built of logs or framed sawn wood above the stone cellar. The best diagnostic characteristic of the building is the stone cellar that is only partially underground. Inside are long shelves that strain under the weight of glass jars filled with canned vegetables, fruits, and meats.

The best concentration of these small cellar-buildings is in the mountain regions of the Blue Ridge and Unakas in the northeastern Tennessee counties of Johnson, Unicoi, Washington, and Greene and in the northwestern North Carolina counties of Wilkes, Allegheny, Ashe, and Watauga. They are also fairly widespread throughout Virginia's Appalachian counties. All have stone cellars, and framed sawn wood superstructures are far more common than log ones. But it is in this Watauga hearth subregion that we find solid concentrations of these small outbuildings used as food storage cellars.

The southern mountain farmstead had other structures in its outbuilding complex: outhouses to dispose of human waste; hog houses, chicken coops, granaries, and blacksmith shops; and other enclosures and storage structures. The outhouse in Appalachia has been made the object of rural jokes and has been used as a colorful index of poverty. The federal government in its war on poverty used census data on the frequency of indoor plumbing as a poverty index. I've seen homes with indoor toilets that have plastic drain pipes ("straight pipes") running out

Root cellars, apple houses, and canning cellars often used the same sort of two-story structure, with a stone cellar below and wooden storage room above. Johnson County, Tennessee. (Source: Tennessee Historical Commission, 1990)

under the house and directly to the nearest open ditch. Unsanitary indeed, but by federal indexes such houses had advanced to a higher level in modern society. A real outhouse was a wooden structure four feet square resting above a pit dug deep in the ground. Some outhouses were built as two holers, wide enough for two people to use or perhaps set up for male and female users. I do not interview many people in the field about such private matters, but some people have told me they preferred going into the woods and using oak leaves for toilet paper to using an unpleasant outhouse.

During the early frontier era, farmers allowed hogs to forage for mast—acorns, chestnuts, and other plant foods—in open woodlands. As settlement filled the region, some farmers began to keep hogs in small, low log cribs. The unchinked crude log enclosure, hugging the muddy earth, became a simple pigsty. First split rails and then wire fencing kept hogs where they could be fattened.

If the farm family kept chickens, as many did and still do, a small elevated chicken coop was preferred. While pigs wallowed low to the ground in mud, chickens were kept in raised roosts to protect them from predators. Both buildings had low single-slanted roofs that were higher in the front than in back. Doors were on the gable sides. Chicken coops had wide, low openings high on the long axis of the structure so chickens could fly up into the coop. Chicken-wire fences came later to create adequate enclosures.

Some Appalachian farms have special rooster enclosures. Inside a fenced area are individual shelters for each rooster consisting of open barrels lying on their sides or small A-frames. The roosters are raised for fighting in illegal cockfights, where fierce betting takes place over the blood sport. It is not illegal to own and raise fighting cocks, but it is illegal to have them fight to the death while humans make bets on the outcome. Marion County, Tennessee, west of Chattanooga, is well known for illegal activities. A friend of mine who had been a high school teacher there once was questioning two students who had been absent. One boy blurted out that his friend had been "fighting his chicken." He had taken a rooster to a cockfight the day before. The teacher asked, "Well, did you make any money?" The student proudly replied: "Shore did, I won sixty dollars!"

Granaries were slightly larger than smokehouses, with diagnostic features of gable-end doors, plank floors, and an unusual height above ground. These first two characteristics meant that grain (any grain other than corn, which was kept in corncribs) could be kept dry, and the height was believed to help keep vermin out. On the opposite end of the height scale was the blacksmith shop. This ground-hugging structure, often built of logs, had its own set of diagnostic traits: height under six feet at the edge of the gables; wide openings six to eight feet long and one to two or even three feet high on the long sides of the structure but still low to the ground; gable-end door; and rectangular shape. The curious openings on the long axis are especially diagnostic because a blacksmith needed daylight to illuminate his work, fresh air to feed the bellows, and room to maneuver large metal objects that were sometimes fed into the building through these low, wide "windows." Not many original blacksmith shops remain on the landscape, but in Union County, Tennessee, a very old one accurately dates by dendrochronology to 1793.[110] The abandoned shop has huge yellow poplar logs with the familiar half-dovetail notches.

This blacksmith shop, one of the oldest documented log structures in Union County, Tennessee, dates to 1793 according to dendrochronology testing conducted on it in 2002. (Sources: Tennessee Historical Commission, 1979; Reding 2002)

Other Landscape Features

Fences

The Appalachian landscape contained other material features that carried cultural identities and values. Fields bounded by split rail fences gave character to the landscape. The old fences were true folk forms: snake or worm fence, stake-and-rider fence, post-and-rail fence, buck or reindeer fence, and stone or rock fence. All but the stone fences were of wooden split rails, and all enclosed agricultural fields. Excellent old stone field fences survive on the landscape, but I know of no old brick ones. With fences you ask, "What are you trying to fence in, and what are you trying to fence out?" Field fences protected large areas, usually cornfields, but some fences kept selected livestock fenced in on pastureland. It took about eight hundred split rails to enclose an acre. If a snake fence was used, with ten-rail panels each eight feet long, nearly eight thousand rails were needed to make a fence one mile long.[111]

The snake fence takes its name from the zigzag, snakelike pattern it makes on the landscape. It is also called worm fence, Virginia fence, or zigzag fence, but snake fence is a common popular name. The rails for this and any other rail fence must be split from the trunks of trees such

This snake fence in Grainger County, Tennessee, has six-rail panels with locking cross rails. (Source: Tennessee Historical Commission, 1978)

as chestnut, oak, cedar, or old-growth heart pine. In rural Anderson County, Tennessee, you can find split rail fences, new cedar ones, because Earl and Charles Sherwood still make them. Sam Venable, journalist-humorist and friend writes: "Earl and Charles Sherwood are rail-splitters. There is nothing easy about their job. It is the epitome of labor, borne of cascading sweat and callused hands. In the simplest of terms, rail splitting is a matter of controlled destruction. The mission is to take a tree whose cellular structure has been locked rigidly in place for decades, and cleave it asunder lengthwise." The Sherwoods do it the traditional folk way, "with wedges and mauls and axes and copious amounts of rawhide muscle power."[112]

Once the rails are split, fence building begins in earnest. The easiest fence to build is a snake fence, because it requires no posts, no post holes, no permanent foundations; and the fence is completely portable. Rail fences are built in sections called panels. A panel may consist of four, six, eight, or ten rails depending on how high you want your fence. First you decide on the starting point and place one or two large flat stones on the ground. The stones keep the rails off the ground to prevent rotting. Using a rail as your measuring stick, go that far and place one or two more stones on the ground. Set your first rail down so it overhangs the stone supports by a foot or so on each end. The next panel should be laid out at about a sixty degree angle to the first. Alternate panels are at the same angle. The ends of the first rail in the second panel rest on the ends of the first rails in the first and third panels, and so on. Continue to lay the first level of rails and stones in a zigzag pattern along the desired fence line or across the field. Lay the next series of rails by overlapping the ends of the rails in the adjacent panels. Stop when you have reached the desired height of four or six or ten rails. For added support, place two rails diagonally upright to make an X in the corners at each junction to lock the rails in the panels in place.

Snake fences not only are easy to build, they are also quick. I invited my advanced cultural geography class of four students to my home one Saturday morning so they could learn firsthand how to build a snake fence. I had bought the rare chestnut rails for forty cents apiece from a farmer in Hawkins County, Tennessee. I had promised the students a beer and a sandwich for their work in this most important learning exercise in folk culture. At 9:00 we began assembling the rails for the 200-foot fence with four-rail panels. By 9:30 the fence was up, finished, done! I could not believe how fast the job went. So, true to my promise, I asked them if they still wanted beer and lunch so early that morning; and like good graduate students they said, "Of course!"

The snake fence was once the most prevalent rail fence in Appalachia. In 1870, snake fences dominated all fencing in terms of percentages for the states of Virginia (79 percent), West Virginia (85 percent), Kentucky (87 percent), Tennessee (95 percent), North Carolina (96 percent), South Carolina (98 percent), Georgia (95 percent), and Alabama (90 percent).[113] These data came on the eve of wire fencing, especially barbed wire, so the date for the fence is crucial for illustrating the importance of the split rail snake fence to Appalachian fence types. In the 1930s a resurgence of rail fence building came at an unusual but opportune time—during the Great Depression. An Oriental fungus entered American forests and systematically began killing the American chestnut trees. Originally valued for their nuts and for their fine wood, dead and dying chestnut trees were suddenly littering the landscape. Appalachian farmers made the most of an immediate resource and split chestnut trunks into fence rails. In 1968 Henry Glassie observed a concentration of snake fences in western Virginia, especially in the Ridge and Valley Province between Harpers Ferry and Upper East Tennessee and northwestern North Carolina.[114] The snake fence was the most widely used of all rail fences because it was portable and effectively kept livestock out of cultivated fields.

Post-and-rail fences were more substantial than snake fences. Floyd County, Virginia. (Photograph by J. Rehder, 1989)

In a stake-and-rider fence an additional set of rails go on top of an existing snake fence or a stone fence. The stakes are pairs of diagonal rails set to form Xs, about the length of a rail apart. The rider is a single horizontal rail that rides in the crossed rails. The stake-and-rider was meant to add tall, sharp impediments to an existing fence to deter livestock from jumping it.

The post-and-rail fence is almost self-explanatory, but it comes in two styles. The older subtype is made entirely of rails and is sometimes called a straight-rail fence. Two posts are driven into the ground about four to six inches apart; another pair is driven in about a rail's length away, and this is repeated for the length of the fence. Rails are set between the posts, with the ends overlapping much as in the snake fence. This fence is straight, so it takes fewer rails, and it is fairly permanent, but it requires more work driving posts into hard ground. Unlike the snake fence, the post-and-rail is neither portable nor temporary. The other version of the post-and-rail fence has wide single posts driven into the ground, with holes cut into them at intervals to position the rails of each fence panel. Rails are placed horizontally in the holes with their ends overlapping. The result is a solid, relatively permanent fence. This

The fence type called the buck, reindeer, Irish, Shanghai, or Swede is rare in Appalachia. Floyd County, Virginia. (Photograph by J. Rehder, 1989)

subtype of the post-and-rail was an exceptionally good fence for blooded livestock, especially horses. Today, wooden board fences nailed to posts are the fence of choice in Kentucky's bluegrass horse country.

A very rare fence in Appalachia is called the buck, reindeer, Irish, Shanghai, or Swede fence. Pairs of crossed rails form the vertical support of the fence, while diagonal rails form its linear axis. One end of the linear rail rests in the X and the other end rests on the ground. The sets of crossed vertical rails with fitted linear rails overlap one another to form this unique fence type. I know of irritatingly few examples. The only one I have ever seen on the landscape in Appalachia is a reconstructed buck fence on the Blue Ridge Parkway in Floyd County, Virginia. Another buck fence in a photograph taken about 1900 on the North Carolina Piedmont in Lincoln County shows my grandfather, Carl Burkhardt Rehder, sporting a derby and sitting on the fence. The third and fourth examples are in early photographs taken by the famous French geographer Jean Brunhes; these fences are in the Swiss Alps and in Quebec.[115] Over time, these and other rail fences met their demise after the adoption of barbed wire, smooth wire, and board fences.

Stone fences appear expressly in limestone areas. They are abundant in southern Pennsylvania and in the Great Valley of Virginia, but espe-

The limestone rock fences of Kentucky are impressive features on the landscape. Mercer County, Kentucky. (Photograph by J. Rehder, 1978)

cially in nearby limestone basins—the Bluegrass Basin in Kentucky and the Nashville Basin in Tennessee. After a wet winter with frost heaving and erosion, farmers needed to do something with the springtime harvest of stones from their fields. New England farmers used their new crop of fieldstones to build stone walls. In parts of Appalachia, some farmers hauled stones from the field in sledlike stone boats and built stone fences too. Kentucky's Bluegrass Basin probably has more stone fences per square mile than any other area outside New England. These impressive fences are made of stacked limestone. Large flat stones are laid horizontally, much like brickwork, to a height of four or five feet, then topped with limestone slabs stacked diagonally like overlapping shingles. The result is a jagged top that discourages livestock from trying to jump the fence. The stone fences in the Bluegrass Basin are thought to have at least two cultural sources. Early mythology attributed the stone fences to slave labor on plantations in the Bluegrass Basin, but Karl Raitz and Carolyn Murray-Wooley proved that the principal stone fence builders were Scottish and Irish immigrants who brought the craft of stoneworking with them from Europe after 1840.[116] The few remaining old rock fences are diagnostic to this part of the region, and their presence lends an engaging sense of history to the folk landscape.

Paling fences surrounded vegetable gardens. Made of split white oak or chestnut, the narrow vertical boards were sharpened on top to prevent chickens from alighting on the fence before dive-bombing the garden for food. The paling fence was not universal throughout Appalachia, but it was common where people needed to protect the garden from domestic farm animals as well as wildlife. The thin wooden slats were driven into the ground, and horizontal boards were fastened to them, producing a crude picket fence.

Fields and Woodlands

Since earliest European settlement, Appalachia's farm fields have been irregularly shaped and sized. Following the English land survey system of metes and bounds, it was natural for fields to take on similar irregularity. I don't believe you can find any two fields on a given farm that are the same in shape or acreage. Few are even square. Fields are relatively small, with most between one and ten acres. The woodlands are much the same, with irregular shapes. But woodlands are larger, running into the hundreds of acres on ridges and mountains in the Ridge and Valley and Blue Ridge Provinces and into the thousands on the Plateaus. However, closer inspection reveals a pattern: not a geometric one, but a geographic pattern of land use. In lowlands—bottomlands along streams, and valleys—there are cleared fields used for cropland or pasture. Since aboriginal Mississippian times, Native Americans favored bottomlands for growing their maize-beans-squash crop trilogy. European settlers traditionally grew corn on bottomlands. Steep slopes were left as woodlands. Such patterns are easily seen in the Ridge and Valley Province, where forested ridges are separated by cleared and cultivated valley lands. Such tidy patterns do not appear in the Blue Ridge and Cumberland and Allegheny plateaus; but in general, watered flatlands are cultivated while steep topography remains in woodland.

Frontier farmers practiced a form of shifting cultivation. They began with girdling, slashing, and burning woodland to clear fields. After cultivating a clearing for several years, they moved on to another site as the soil's fertility played out. According to Jordan and Kaups, shifting cultivation practices were done this way:

> Because the fields were so temporary, forest removal became magnified in importance. Backwoods clearings were small, but the pioneers relocated them so often that considerable acreages of woodland had been destroyed by the time the secondary settlers appeared on the scene. So abundant was land, however, that the backwoods farmers always had ash-richened virgin soil to cultivate and did not have to return to oldfields. Only after the frontier era had passed did the pioneer system of shifting cultivation evolve into a less satisfactory bush-fallow sequence. . . . Such a system was found in parts of the Delaware Valley by the middle of the eighteenth century, but it persisted primarily in the mountain districts of the South.[117]

In 1978 Stephan Hill, a former graduate student of mine, investigated a pattern of land rotation, a form of shifting cultivation in East Tennessee. Hill wanted to determine the validity of John Fraser Hart's thesis that Appalachian farmers still practiced shifting cultivation. Were farmers consciously shifting from oldfields to clearing woodlands to prepare a new round of fields for cultivation? John Fraser Hart had introduced the concept to the literature in a 1977 paper, and Hill believed that Hart's interpretation was not entirely correct. He concluded that land rotation was done over generations of farmers on given farmsteads and that while the landscape reflected rotation, the process was so slow that contemporary farmers in Washington County, Tennessee, were unaware of the practice.[118]

Gristmills and Country Stores

Throughout Appalachia, water-powered gristmills, country stores, churches and graveyards, and bridges, fords, and ferries for crossing streams added to the landscape mélange. Gristmills became focal points for exchanging goods, services, and gossip. Farmers brought grain to be ground and items to trade and sell. The gristmill offered mountain folk a place to exchange ideas and socialize, and to form a sense of community even in a land where independence was so important. Later, country stores served much the same functions.

Country stores began to appear in the region after the Civil War. As

The gristmill became an early focus of rural economic and social life in parts of Appalachia. The Mabry Mill on the Blue Ridge Parkway in Floyd County, Virginia, is one of the most easily recognized mills in America. (Photograph by J Rehder, 1989)

transportation slowly began to improve, with road intersections and railroad crossings as focal points, country stores took on much the same trade and social functions as gristmills. The store was a window to the outside world. It was the post office, the trade center, and the community gossip center. When my historic buildings field crews of John Morgan, Joy Medford, Stan Guffey, Vince Ambrosia, Neil Cyganiak, and others first went into the field, I told them to do three things: inform the sheriff; inform the newspaper; and inform the folks in every country store you see of your whereabouts and what you are doing. John Morgan and I had been working the Dutch Valley area in Grainger County, and as usual we stopped at the country store called Petticoat Junction, operated by Mrs. Frye. We always let Mrs. Frye know what we were doing and where we were going; it was a little like telling your mom. Besides, this store had become our favorite field grocery for this part of the survey. A few weeks later and some miles away, some people greeted us with, "Where in the world have y'all been? Why, we've waited fer two weeks expectin' y'all to be here just any minute to talk to us about our place!" We apologized for not getting there sooner, and I had to ask: "How did you know about our survey?" The folks on the porch, now rest-

The country store is a disappearing feature in rural America even though this one in Middle Tennessee had been in business since 1852. (Photograph by J. Rehder, 1976)

ing a little easier, said: "Oh, we knew you'ns were in the Valley from all the talk goin' on up to the Petticoat Junction store."

Churches and Cemeteries

Rural churches are so numerous in Appalachia that they almost defy geographic analysis. I have seen hundreds of white-painted wooden church buildings but few log churches. In my seventeen-county Tennessee data collection only two log churches were identified: one in Morgan County and one in Overton. There is an old log church with a newer brick sanctuary attached to it in Hawkins County, but Hawkins data are not yet in this analysis. However, in Overton County on the Highland Rim area just west of the Cumberland Plateau is the well-preserved Okalona log church, built in 1890 with oak logs with half-dovetail notches. During its early history, the Okalona log church was used by three congregations who rotated Sundays.

The contemporary Appalachian religious landscape reveals thousands of independent small, one-room rectangular church houses with dimensions of about twenty-five by thirty-five feet, covered with white clapboard siding or gray tarpaper. Interiors typically have hard wooden

The old log Okalona Church in Overton County, Tennessee, reflects enduring traditional folk architecture. (Photograph by J. Rehder, 2002)

benches, a preaching stand with a Bible and anointing oil, a picture of Leonardo's *Last Supper* on the back wall, benches along the front and side walls for church leaders and musicians, and wires for curtains to divide the room by gender for foot washing rituals. Crude hand-lettered signs outside invite worshipers (mostly the rural poor) to experience spirited, loud Pentecostal and Holiness three-hour evening services.[119]

Cemeteries hold a special place in Appalachian folk culture because many graveyards were on the farm. The family graveyard up on a knoll and planted with evergreen cedars became a symbol of eternal life. It also represented independence, since few rural church buildings and associated cemeteries existed before the Civil War. To be sure, church cemeteries are widespread throughout Appalachia today; but there are still many family graveyards. In a rough-cut map analysis I conducted more than twenty-seven years ago, I found that eighty-six of Tennessee's ninety-five counties each had three or more family graveyards. A filter criterion was that no church could be within two miles of a graveyard, allowing for sufficient separation for the mapped site to be a family graveyard. Counties that did not meet the criteria were large metropolitan counties, two rural counties in the Mississippi floodplain in western Tennessee, and four unexplained others. Church cemeteries and especially commercial cemeteries were late to arrive on the landscape, but once there they became the primary burying grounds.

Unusual grave coverings are grave houses and stone slab grave covers. The grave house is a small wooden structure that shelters a grave site or plot. The roof may have once been made of white oak shakes, but now it is corrugated metal. The sides are an open weave of slats like a lattice.[120] Stone slab grave coverings, also called prismatic grave structures, slab graves, box graves, sandstone graves, peaked graves, cattle rocks, and comb graves, appear relatively frequently on the Cumberland Plateau and Highland Rim Country in Middle Tennessee.[121] Two sandstone slabs six to eight feet long are pitched to form a tentlike A-frame over a grave, and shorter stone slabs cover the "gable" ends. In Morgan County, Tennessee, on the eastern end of the Cumberland Plateau, some grave coverings are limestone blocks that are "bricked" over the graves, looking as if a limestone chimney had fallen to the ground. Covering graves with grave houses, stone slabs, or other weighty rocks prevented wild animals from scavenging graves dug in shallow mountain soils.

Folk bridges for crossing streams included foot logs and swinging cable bridges, among others. Fords, toll bridges, and ferries continued in private operation until state and county governments accepted the responsibility of road and bridge maintenance in the 1930s and thereafter.[122] Log footbridges and suspended cable bridges for foot traffic are still seen in Appalachia, especially where they link a home or farm site to a state or county road. Ferries are all but gone, since bridges have replaced them. I know of no operating ferries in East or Middle Tennessee or in Appalachian Kentucky, Virginia, or North Carolina. I suspect that in the remaining parts of Appalachia, ferry crossings have been replaced by bridges as well. The relatively small size of most Appalachian streams made fords more feasible, and in some really isolated areas fords can still be found on small back roads. But they too are becoming scarce as county road crews build and maintain culverts at small stream crossings.

Throughout this chapter I have focused on the agrarian Appalachian landscape because here "vernacular" speaks best of folk cultures in an independent agrarian setting. However, in the late nineteenth and early twentieth centuries, logging and coal mining revealed other rural landscapes. In chapter 5 I discuss logging camps as well as coal mining camps, mine shafts, and villagelike company towns of cheap board-and-batten frame houses coated with the ever-present coal dust that presented a distinctive landscape expression.

One of the most basic stream crossings is a log footbridge. (Photograph by J. Rehder, 1973)

The Decline of Log Buildings in Appalachia

When I first began to explore Southern Appalachia in the late 1960s, Fred Kniffen had warned me that the buildings were going fast and that someone, perhaps me, would need to survey and record the existing relics before they all disappeared. I was a twenty-five-year-old assistant professor at Tennessee, and I took Kniffen's advice with a grain of salt, believing that Appalachia was an outdoor museum, so isolated and protected from progress that it surely would remain a gold mine of log structures. And for a decade, 1968–78, I was right. However, since the late 1970s Kniffen's prophecy has unfortunately come true. The buildings are disappearing at an alarming rate. For example, of the thirty-eight log buildings built before 1860 that were surveyed in 1978 in Grainger County, Tennessee, only sixteen remained in 2002. This represents a 58 percent loss. In adjacent Union County, Tennessee, in 1979 there were thirty-four pre-1860 log buildings; by 2002 they had declined to twelve—a 65 percent loss.[123] These two counties were so rural that they did not have a single stoplight, even though they are adjacent to Knox County, an urban county with over 300,000 people. Though they were relatively well preserved living museums, over the past thirty years they have lost buildings to neglect and the ravages of weather, to vandalism, and to destructive log collectors who have bought up unused log buildings and rebuilt them into chalets in other states.

What are some reasons for the decline? The first and most obvious is that as fewer people live in and maintain genuine folk log houses, abandonment leads to neglect. Usually the roof is the first to go, and once that happens, the rest of the structure rapidly deteriorates. Even if the roof is still intact, the sill logs may begin to sag as they are damaged by termites and moisture. Once a foundation rots, it will not be many years before the structure falls in on itself. Though such deterioration is expected, abandoned buildings invite vandalism, and this becomes a second cause for the disappearance of log buildings in Appalachia. We know it happens, and there is not much we can do about it.

Then too, when some people find a structure is no longer functional, they decide they must remove it; but being conservative, they try to find a use for the materials. In 1978 my survey team and I photographed and mapped twenty-six log smokehouses in Grainger County, Tennessee. The next spring we returned to a particular farm and discovered that the smokehouse was missing. I asked the elderly owner, "What happened to the smokehouse?" He replied, "We burned it." I must have looked puzzled, because he explained, "We needed firewood this winter, so we burned the wood [from the smokehouse] in the fireplace." A logical use of materials from a useless structure tragically meant the loss of another irreplaceable landscape feature. We knew that burning the logs kept that elderly couple warm, but it meant we would never see that log smokehouse again except in a photograph.

The frailty of the folk and their folk structures is something to be worried about. At the time the Great Smoky Mountains National Park was being created in the 1930s, Will Walker and his wife Nancy were elderly but still living on their farm on the Middle Prong of Little River. It was said that in their later years the couple heated their modest log house with rails from a nearby fence that Will had split years before. Too frail to saw them up, the couple fed the rails into the fireplace one at a time through an open doorway.[124]

An insidious force has been ravaging the landscape over the past twenty years. Commercial companies prowl the ridges, valleys, and hollers of Appalachia in search of log buildings to buy. They offer someone a small amount of money for a useless, abandoned log structure, then they place numbered metal tags on the logs, dismantle the building, load it on a flatbed truck, and take the logs to a resort community where a "new" chalet is created out of authentic Appalachian logs. It is a form of raping the landscape that destroys a small but significant part of history. In the spring of 1978, near Thorn Hill in Grainger County, Tennessee, we had just surveyed and photographed two magnificent log houses—a saddlebag double pen house and an I-house. A few weeks later we were passing by the property and found a large red flatbed truck parked in front of the remains of the saddlebag. The farm's owners came up to us with sheepish looks and told us that it was a good thing we had surveyed and photographed those buildings earlier, because they were nearly gone now. They said they didn't need the houses but they sure needed the money. Logically, the owners saw a way to remove abandoned dwellings and to make some money from them before vandals or the forces of nature beat them to it. It made sense to the owners to sell the logs from these houses, but to me and my field team it was a senseless act that furthered the deterioration of the landscape.

Interpreting Ethnic Landscape Contributions

Throughout this chapter, I have pointed out controversy surrounding sources of origin for landscape traits. Scholar-interpreters have analyzed, synthesized, and sought explanations for the presence of vernacular landscape traits. At some point, conclusions must connect these unraveling pieces of the landscape tapestry. The English contribution is impressive when one considers the total influence. The metes and bounds survey system imprinted property with patterns of unthinkable complexity. House types basically follow the English pen tradition, and for those with external chimneys, the English influence is very strong. The controversial corner notching question in log construction traces first to Scandinavian origins and later to German reintroductions. The Swedes and Finns were first to introduce corner notching to North America; the Germans reintroduced it and taught these techniques to Scotch-Irish and other European immigrants. Germans also introduced the German three-room house and several barns and small outbuildings with "Alpine" projecting roofs. With the majority of Appalachia's population considered to be of Scotch-Irish heritage, what were the landscape elements attributed exclusively to them? The dispersed settlement and the spreading arrangement of outbuildings find identity with northern Ireland and Scotland. Likewise, the rectangular floor plan has similar origins and identity with the Scotch-Irish.

For some aspects of the folk landscape, I believe our analysis is more open to question. Instead of seeking specific origins for individual vernacular landscape traits, we should be interpreting the landscape for specific times and locations from the point of view of acculturation. That is, we should recognize that some diagnostic traits attributed to Appalachian origins were indeed borrowed from other immigrant European cultures. Furthermore, it is not hard to believe that the pioneering settlers of Appalachia could have invented traits of their own such as the four crib and transverse crib barn types. Within a cultural context, the concepts of common sense and necessity as the mother of invention, as much as tradition, may answer many of the nagging questions that continue to be asked of the folk landscape in Appalachia.

5

Ways of Making a Living

There are many ways of making a living in Appalachia. In the not so distant past, subsistence agriculture, supplemented by hunting, fishing, and gathering, was the primary means of survival. In the late nineteenth and early twentieth centuries, logging, closely followed by coal mining, became the new way to make a living. But these were obtrusive commercial enterprises, often externally founded, funded, and managed by cultures not always of Appalachian stock. Logging and coal mining jobs drew folks away from their home subsistence bases to live in company towns, logging camps, and coal camps. Moonshine arrived with early European settlers in the eighteenth century and continued uninterrupted as another cash crop from corn; but marijuana patches and methamphetamine labs dominate the illegal ways of making a living today. We will explore the topics of subsistence agriculture; hunting and gathering as a folk tradition; fishing for food, not sport; logging; coal mining; and moonshining as some of the economics of Appalachian life. We need not examine every conceivable occupation exemplified by the Appalachian rural folk working in city factories, as service personnel, and in the myriad other ways of bringing home a paycheck. There are just too many jobs out there that do not have much of a folk tradition.

Subsistence Agriculture

Appalachian frontier survival relied on hunting and gathering, but once the land was settled, life depended on subsistence agriculture. The farmer-settlers and their families grew a multitude of food crops and raised livestock. Though there was supplemental trading, bartering, and selling items for cash, most Appalachian folk in the eighteenth and nineteenth centuries relied on what their own farms could produce. This was especially true for folk in the Blue Ridge and other rugged eastern mountain areas and in the sparsely settled Cumberland and Allegheny

One way of making a living is the difficult task of splitting cedar logs for fences. Anderson County, Tennessee. (Photograph by J. Rehder, 2003)

plateaus to the west. However, in the older and somewhat more progressive regions in the Shenandoah Valley and Great Valley in the Ridge and Valley Province, nineteenth-century agriculture became commercialized. By midcentury, wheat was already the dominant commercial field crop, and the Shenandoah Valley and Great Valley became the commercial breadbasket for East Coast cities. The region played the same essential role in wheat production that the Great Plains played decades later.[1] The question is not whether farming was completely subsistence, semisubsistence, or almost fully commercial, but what folkways Appalachian people used to make a living from agriculture.

Clearing Land

After selecting a piece of land based on soil, terrain, and water, a pioneer family chose two or more patches to clear for cropland. While clearing the house site, they also cleared a garden plot. Over time, this small plot adjacent to the house supplied vegetables and other foods. The primary field was for growing corn. To clear land efficiently, farmers girdled oaks, hickories, yellow poplars, and other large trees by cutting a ring of bark

"A mule pulls a straighter line than a tractor," said this farmer in Grainger County, Tennessee. (Photograph by J. Rehder, 1978)

completely around each tree with an ax. Removing a ring of sap wood made the tree die slowly, though it remained standing. This left acre-size stands of lifeless, leafless trees in ghostlike forms called deadenings. Sunlight could now reach the forest floor during the growing season, and the farmer planted his first crop among the dead trees. As time, help, and need allowed, pioneer farmers cut down trees in the deadening and hauled them to a building site to be used for barns, outbuildings, additions or replacements to the cabin, fence materials, and firewood. Unneeded limbs and brush were piled up and burned. Pulling stumps was especially difficult, and some farmers simply worked around them until they rotted away. Others used oxen, horses, or mules and even black powder or dynamite for stump removal. Once the field was clear of trees and stumps, frontier land preparation might begin with plows, hoes, and shovels. Using draft animals hitched to a plow made of wood or later of steel, farmers prepared rows for planting corn. After the seed had sprouted, the farm family used hoes and cultivators to remove weeds.

From the point of initial occupance, an early settler's subsistence farm output slowly evolved. The first year on new ground meant very lean times. Perhaps no significant crop could be planted or harvested except turnips, which provided greens very quickly and roots thereafter. After the first year, settlers relied on pork, corn, sweet potatoes, turnips, and chickens. By the third or fourth year they had a greater variety as gardens yielded more food. It would be five years or more before orchards were bearing useful fruit.[2]

Seedtime: Planting by the Signs

It's springtime, and the cornfield and vegetable garden are being plowed or hoed into rows. It is a magical, mysterious time, when decisions on what to plant and when to plant it are governed by one of the most unusual folkways in American agriculture. There is nothing more "folk" than the ritual of planting by the signs of the zodiac. Ecclesiastes 3:1–2 says, "To every thing there is a season, and a time to every purpose under the heaven: a time to be born, and a time to die; a time to plant, and a time to pluck up that which is planted." The zodiac refers to bright constellations distributed evenly along the sun's path in a belt that includes the path of the moon and the planets. The belt is eighteen degrees wide and is divided into twelve equal parts of thirty degrees each. Each of the twelve segments is called a "sign," and many of the signs are named for animals such as lion (Leo), ram (Aries), and bull (Taurus). The zodiac's signs are also identified with various parts of the human body: the head is associated with the ram (Aries), the neck with the bull (Taurus), the feet with the fish (Pisces), and so on. Each of the twelve signs has a relationship with a body part, a planet, and an earth element. These very important earthly elements rotate in sequence as fire-earth-air-water as each sign cycles through the calendar year.

Each of the twelve signs appears at least once a month; some are in effect for one, two, or even three days. Planting calendars, published as wall calendars and almanacs, show for each day of every month the relation between the prominent constellation in the sky for that date, the animal symbol, body part, planet, and element. The farmer must read the signs on the calendar and act accordingly.

This is how it works. For example, the first sign is Aries, with the ram symbol representing the head body part, the planet is Mars, and the element is fire. On this day it is good for farmers to plow, till, or even cultivate because the fire element means that conditions should be dry. You may have heard the expression "too wet to plow"; plowing requires dry conditions, so an Aries day is a good one. To plant by the signs, a farmer needs days that fall on water days with signs of Cancer, Scorpio, and Pisces. He can also set bedding plants out in either water or earth signs. Plant beans when the signs are in Gemini, arms, and air. Never plant on Sunday because it is the "sun's day," marked by fire.

A zodiac guide by T. E. Black states these routines for the fourth sign—Cancer:

> A Water Sign. No. 1 top sign for all flowers, planting and transplanting all crops that bear above the ground. When I say bear above the ground I mean cotton, corn, cane, tobacco, peas, beans, pepper, watermelons, squash, cucumbers, okra, wheat, rye, oats and etc. This sign also [is] No. 1 for all root crops. Biddies hatched in this sign for laying hens, pigs born for males and brood sows. Good for all cooking, changing jobs, moving, cut hair to stimulate growth. Good for fishing, making potato beds and seed beds.[3]

In the late 1960s one of my students, the late Larry Smith, was researching the folk nature of vegetable gardens in the city of Knoxville, Tennessee. Larry was trying to discover to what degree rural folk traditions were being carried into city life through traditional vegetable gardening. He concluded that people were planting by the signs in the city as well as out in the countryside, and he found some unusual practices not often reported. One elderly man explained his method for planting watermelons. In the spring of the year under a full moon, this informant planted watermelon seeds in his garden in the city wearing only a shirt and nothing else!

The zodiac's signs are not just for planting. They apply to harvesting and many nonagricultural chores. Paint houses on dry, fire days like Aries and Leo. Bake and cook in Aries, a dry, fire time. Hunt in the sign of Taurus, the earth sign. These are but a few of dozens of rules for plowing, planting, harvesting, and living by the signs.[4]

Harvest

In the late summer and fall of the year comes harvesttime. Field crops of corn, hay, and tobacco demand the most attention. In the summer, hay is brought in first as it is cut from the field, piled into haystacks or pressed into bales, loaded onto wagons, and hauled to the barn to serve as winter feed for livestock. Early farmers in Appalachia did not work hay into their assortment of field crops; usually they allowed livestock to forage on their own. As settlement expanded, livestock foraging declined, especially in the Ridge and Valley Province, and hay became an essential field crop, especially for farmers with dairy and beef cattle.

Tobacco as a cash crop on the Appalachian farm is harvested in late summer through midfall. Mountain-grown burley tobacco planted in

The traditional agricultural practice of harvesting with a grain cradle is now just a memory. (Source: Tennessee Historical Commission, 1979)

small patches is fairly quick to harvest. As leaves begin to yellow, the entire plant is cut at the base and impaled on a sharp tobacco stick. Stalks with leaves still attached are upended and left in the field to dry. The stalk and yellowed leaves are gathered as a unit and hauled to a large, airy barn where the tobacco stalks, still impaled on sticks, are placed in the rafters of the barn to dry and cure for the next several weeks. Then the leaves are stripped, baled, and trucked to tobacco auctions that begin in November. Unlike brightleaf and flue-cured tobaccos in the eastern Carolinas and Virginia, burley tobacco is either air cured or fire cured. Most Appalachian burley tobacco is air-cured in drafty barn lofts. Whereas fire-cured burley appears in Kentucky's Pennyroyal region north of Nashville on the far western edges of Appalachia, fire curing is a rarity in the rest of Appalachia.

The corn harvest is a fairly simple process of gathering field corn that has been allowed to dry on the cornstalks in the field. Wagons pulled through the field are loaded with ears still in their husks, or shucks, and hauled to the corncrib, a small log or slat structure built especially for corn storage. Some corn will be set aside, shucked, and shelled, with the kernels collected for home use. Little is wasted. Shucks become additional livestock fodder or mattress fillers or are artistically formed into corn shuck dolls. Cobs become corncob pipes, scrapers and scourers, or

Top: A farmer hauls hay bales through the snow on a horse-drawn sled. Russell County, Virginia. (Photograph by J. Rehder, 1972) *Bottom:* Tobacco farming is still an important way of making a living in much of Appalachia. Grainger County, Tennessee. (Photograph by J. Rehder, 1978)

Farming is still a way of life in Appalachia. This 1880 transverse crib barn built with hewn logs with V notches was still storing hay in its loft a century later. White County, Tennessee. (Source: Tennessee Historical Commission, 1984)

a rude form of toilet paper. Whoever coined the phrase "rough as a cob" apparently spoke from experience.

Harvesting plants in the vegetable garden is continuous from about late June to late October. Tomatoes and peppers come along early, as do peas and some beans and okra; corn and other beans are ready for gathering a little later in the summer. Orchard fruits and grapes arrive in late summer and early fall; mountain apples are a special treat in the fall. Most summer fruits and vegetables are eaten fresh, but some are dried or canned to be stored for the lean winter months. The bountiful earth shows itself off at harvesttime in Appalachia.

Hunting, Fishing, and Gathering

Throughout human history, people have made a living from hunting, fishing, and gathering. Before the agricultural revolution, approximately ten thousand years ago, everyone in the world survived this way. In isolated parts of the world today, such as the Amazon Basin, the Kalahari Desert of southern Africa, and Indonesia, there are small culture groups who still subsist on these ancient economic endeavors. It may seem unusual to have such seemingly primitive activities in twenty-first century America, but some people in Appalachia still supplement their income by hunting, fishing, and gathering. This section addresses hunting, fishing, and gathering in the context of folk survival rather than sport.

Hunting Wild Game

The popular image of a pioneer Davy Crockett, coonskin cap askew and long rifle at the ready, shooting a bear seems almost too fictional to believe. Certainly pioneer folk made much of providing venison, bear, coon, and squirrel, along with turkey and other fowl, for the dinner table and the family larder. Bear meat was a staple for some backwoods folk. Eighteenth-century long hunters such as Daniel Boone, Stephen Holston, and Davy Crockett went off on hunting trips that lasted months or years. Deer meat was a second choice behind bear, but deerskin was essential for making clothing and moccasins.[5]

Subsistence farming often meant that hunting game was part of supplementing the pioneer diet. Today's folk seldom need wild game for food, but hunting remains a long-held tradition for menfolk in contemporary Appalachia. Horace Kephart's 1913 book *Our Southern Highlanders* describes in wonderful detail and informal language the colorful folk ways of living and hunting in the Great Smoky Mountains long before the area became a national park. On bear hunts, men and dogs tracked black bears through the mountains, up valleys, over ridges, and through dense thickets of rhododendron and mountain laurel. Bear hunts often lasted several days, so hunting parties sheltered in small wooden shacks in the mountains, also used as drover's huts. Kephart describes the scene:

> Four dogs growled at each other under a long bunk of poles and hay that spanned one side of our cabin. The fire glared out upon the middle of an unfloored and windowless room. . . . At last we were on a saddle of the divide, a mile above sea-level, in a hut built years ago for temporary lodgement of cattle-men herding on the grassy "balds" of the Smokies. A sagging clapboard roof covered its two rooms and the open space between them that we called our "entry." The State line between North Carolina and Tennessee ran through this unenclosed hallway. The Carolina room had a puncheon floor and a clapboard table, also better bunks

than its mate; but there had risen a stiff southerly gale that made the chimney smoke so abominably that we were forced to take quarters in the neighbor State.[6]

Based on chapter 4, we can recognize the drover's hut as a dogtrot. The bear hunt then became a bit more serious for Kephart:

> On came the chase, right in my direction. Presently I could distinguish the different notes: the deep bellow of old Dred, the hound-like baying of Rock and Coaly, and little Towse's feisty yelp. . . . But the dogs caught up with him. They nipped him fore and aft. Time after time he shook them off; but they were true bear dogs . . . they knew no such word as quit.
>
> I took a last squint at my rifle sights, made sure there was a cartridge in the chamber, and then felt my ears grow as I listened. Suddenly the chase swerved at a right angle and took straight up the side of Saddle-back. Either the bear would tree, or he would try to smash on through to the low rhododendron of the Devil's Court House, where dogs who followed might break their legs. I girded myself and ran, "wiggling and wingling" along the main divide, and then came the steep pull up Briar Knob. As I was grading around the summit with all the lope that was left in me, I heard a rifle crack, half mile down Saddle-back. Old "Doc" was somewhere in that vicinity. I halted to listen. Creation, what a rumpus! Then another shot. Then the warwhoop of the South, that we read about.
>
> By and by, up they came, John and Cope and "Doc," two at a time, carrying the bear on a trimmed sapling.[7]

Dogs named Ole Bullet, Blue, and Dred conjure visions of hunting dogs in the mountains. Whether they are blue tick hounds, Plott hounds, coon dogs, bear dogs, foxhounds, hog dogs such as Louisiana's Catahoola hog dog, or some other canine variety of tracker, attacker, or retriever, these critters are more than man's best friend and worth considerably more to the owner than an ordinary pet. Hoyt Vanosdale, an East Tennessee hunter and a first-rate fisherman, told me: "Huntin' dogs are for one purpose only; and that is to hunt. You cain't keep a good huntin' dog around as a house pet. Once he becomes a pet, he'll be worthless to hunt with."[8]

Trapping wild animals for fur pelts was an early activity that gradually faded over time. There are still a few hobby trappers in Appalachian woodlands today, but in the eighteenth and nineteenth centuries there were many folk who trapped as well as hunted for their livelihood. Fur trapping resurfaced during the Depression years for survival income. Generations of trappers operated on the Cumberland Plateau. One such Tennessee trapper, J. W. Hickey, began trapping in 1905 and operated the Fur Craft Shop near Mammy's Creek on old U.S. 70 in Cumberland County on the Cumberland Plateau. Specializing in red fox furs, Hickey was selling 300 to 400 a year in the 1930s. He also trapped and sold about 150 gray fox pelts along with mink, raccoon, and other furs. Business was so brisk that he was buying furs from other trappers and marketing them from roadside stands between Knoxville and Crossville, Tennessee.[9] In the late 1960s, I remember one of his stands on U.S. 70 near Ozone, Tennessee, where red fox furs hanging from rafters under a small shed swirled in the wind.

Fishing

Unlike people in the coastal regions of America, where the close association between humans and sea has built a lasting reliance on fish and shellfish, for Appalachian folk fish have not been a major food source. Fishing was not a large part of their culture partly because there were no natural lakes and only a few large fish-holding rivers, but mostly because of the time and skill it took to catch fish in small numbers rather than in bulk. Fishing was at best supplemental, with old men and boys using fishing poles and live bait.

Appalachia's freshwater fish species include largemouth bass, smallmouth bass, crappie, walleye, bluegill (bream), catfish, horny heads (central stone rollers), and three types of trout, among others. Horny heads are tiny fish that even today are considered a springtime delicacy for folk in Upper East Tennessee and southwestern Virginia.[10] The trout species, which are akin to salmon, are brook trout (speckled trout), found in the coldest waters in small streams high in the mountains, rainbow trout that inhabit the cold waters of intermediate and lower mountain streams, and brown trout that thrive in the slightly warmer waters at even lower elevations. Rainbow and brown trout were introduced to mountain environments in the 1920s, and they play more of a role in sport fishery than in subsistence. In the logging section of this chapter, I discuss the trout habitat in the mountains. Other food items included frogs and turtles—for the most part terrapins or land-based turtles. Native Americans took greater advantage of these and other food sources from the Appalachian fishery.

Equipment for fishing included poles, hooks, lines, sinkers, trot lines, jugs, spears, gigs, fish traps and weirs, and nets, including hoop nets. Most of us have fished with a pole, perhaps a cane pole, with its hook, line, sinker, and bobber float, dangling a night crawler in the water and hoping to catch something—anything. Pole fishing, while entertaining as a hobby, was not a commercial way to fish, even in a subsistence economy. Trot lines were long strands of rope or twine strung across rivers with hooks attached to short lines tied on at regular intervals. Baited and left alone for days at a time, trot lines caught enough fish to become commercially lucrative.

Jug fishing uses a single line anchored to the bottom. Above the anchor the line holds multiple baited hooks. The upper end of the line is fastened to a glass or plastic milk jug that serves as a float. In Tennessee it is legal to jug fish with up to fifty jugs per licensed fisherman as long as you print your name, address, phone number, and the date on the jug. Jug lines, like trot lines, must be tended, and only nongame fish can be kept.

Borrowing from the Native Americans, Europeans in Appalachia used fish weirs and built fish traps on a number of broad, shallow rivers. On the Holston River in Tennessee, for example, there are the remains of three V-shaped fish weirs. Jim Cobb, a federal archaeologist and a former student of mine, described the devices in 1978:

> The lower Holston River fish traps were permanently placed structures consisting of two primary components: (1) a V-shaped rock formation that served as a dam [weir] for diverting and constricting the natural flow of a section of the river; and (2) a wooden device connected to the apex of the V-shaped dam through which the section of river was forcibly channelled. The rock wall dam [weir]—currently the most durable and visible feature of the fish trap—was usually constructed of locally available river bed load rock. The wooden device [trap] . . . was constructed of logs which formed a foundation to which an inclined framework of evenly spaced poles or slats was attached. The spacing of the poles or slats allowed water to flow through . . . while retaining all fish that were too large for the slat spacing.[11]

Fish trap fishing, which had been a European folk tradition for over 180 years here on the Holston River, ended in 1939 because weirs were a hazard to navigation as well as being illegal. Fish traps certainly had documented antecedents in protohistoric and prehistoric aboriginal traps here and elsewhere in the South.[12]

The use of hoop nets was relatively rare except in the few deepwater rivers in nineteenth-century Appalachia. Wooden hoops arranged by descending diameter were connected by two to three feet of twine netting until the narrowest point at the terminus, where bait was attached. Inside the main net, cone-shaped nets called throats funneled fish to the bait and then prevented them from coming back out. Hoops ranged from three or four feet in diameter down to eight inches and were made of white oak splints. Nets of knotted twine were fashioned in the same knot patterns as used for saltwater fishing nets. Mr. Jeff Mackie from Hancock County, Tennessee, in his eighties, was one of the last hoop net fishermen in east Tennessee north of Knoxville. He said that the best way to fish on the Clinch River was on dark nights and in muddy water. Catfish were the prime species sought, but any fish were keepers except game fish, which it is illegal to take with methods other than rod and reel.[13]

Using plant poisons to kill or stun fish is mainly known in the tropical world, but a few Appalachian folk also used the crushed seed of the mullein plant to stun or "sting" fish in Virginia's streams. Cultural geographer Gene Wilhelm Jr. writes about the folk practice and describes sites in the Blue Ridge where it occurred:

> The earliest known dates for the use of the common mullein as a piscicide in the Blue Ridge Mountains were approximately 1740 at the James River Gap, 1770 in the Wildcat-Saddle Creek area [Shenandoah National Park], and 1780 in Basin Creek, North Carolina. In the nineteenth century the use of fish poison by European settlers spread from Virginia across the middle and deep south to the Ozarks of Arkansas and Missouri and to the hilly country of Oklahoma and Texas. Cultural agents responsible for the diffusion of the trait were members of the hybrid culture of the Upland South.[14]

Gathering Wild Foods

To supplement their diets, Appalachian folk gathered wild foods: walnuts, hickory nuts, berries, wild greens, and ramps, among other things. I cover these food topics in chapter 6 and discuss ramps in chapter 8 under festivals, but the point here is that an abundance of plants were available to pioneer settlers. And contemporary Appalachian inhabitants still relish the thought of "living off the land" even for a few hours on the weekend. Today's gatherer might collect black walnuts for a special cake or perhaps pick blueberries, blackberries, or wild strawberries for their novelty and special taste. Appalachian folk do not expect to subsist on wild foods. Gathering medicinal herbs is another matter, however, and a source of pride and income.

Root Digging

Root digging is chiefly associated with wild medicinal plants, which are gathered, sorted, and dried. Most collected biota is sold for profit. Since aboriginal times, more than one thousand plants have been used for medicines in Anglo-America. Native American contributions to folk medicine were important as Native Americans and Europeans made contact. In the 1960s Edward Price, a cultural geographer, compiled data on 148 medicinal plants that had commercial value or identity in the twentieth century. More than half had been used by Native Americans or were a part of their cultures.[15]

In 1886, Kentucky writer James Lane Allen described the subsistence nature of gathering plants in mountain life:

> Formerly, also, digging "sang," as they call ginseng was a general occupation. For this, of course, China was a great market. It has nearly all been dug out now except in the wildest parts of the country, where entire families may still be seen "out sangin'." They took it into the towns in bags, selling it at a dollar and ten cents—perhaps a dollar and a half—a pound. This was mainly the labor of the women and the children, who went to work barefooted, amid briers and chestnut burrs, copperheads and rattlesnakes. . . . It was a sad day for the people when the "sang" grew scarce. . . . Not long since, too, during a season of scarcity in corn, a local store-keeper told the people of a county to go out and gather all the mandrake or "May-apple" root they could find. At first only the women and children went to work, the men holding back with ridicule. By-and-by they also took part, and that year some fifteen tons were gathered, at three cents a pound, and the whole county thus got its seed corn. Wild ginger was another root formerly much dug; also to less extent "golden-seal" and "bloodroot." The sale of feathers from a few precarious geese helps to eke out subsistence.[16]

The ecology of mountain life presupposes that people were knowledgeable about medicinal plants and that gathering wild plants and herbs was a natural part of the culture. Gene Wilhelm found that early mountain families had a sequence of herb and plant gathering in the spring and fall that carried them through the year but not much further.[17] It was only in the late nineteenth century that more medicinal plants, especially those made into patent medicines, were gathered to create a surplus, to be dried and sold for profit at some later time.

Plant materials gathered for folk and commercial medicines included roots, bark, wood, sap, and leaves. Roots came from blackberry, ginger, ginseng, ironweed, jimsonweed, pokeberry, raspberry, sassafras, snakeroot, willow, and goldenseal (yellow root). Bark came from dogwood, hickory, red oak, sweet gum, willow, and wild cherry trees. Other plant parts for folk use came from beeches, cockleburs, dandelions, dog fennel, dogwood berries, mullein, pennyroyal, poke berries, ragweed, sedge grass, and other species.[18] While dozens of plants are collected by Appalachian folk and others, here is a representative sample: ginseng, goldenseal, mayapple or mandrake, bloodroot, black cohosh, mullein, and sassafras root.[19]

Ginseng

Perhaps the best-known Appalachian plant is American ginseng (*Panax quinquefolium*), a low herbaceous plant that ranges from eastern Canada southward through the Appalachian region to Alabama and westward into the Ozarks of Missouri and Arkansas. Ginseng is a sensitive perennial whose habitat is shaded north-facing slopes in well-watered deciduous woodlands with moist, rich soils. The plant also grows in forest-sheltered damp coves. Ginseng is found under beech, maple, oak, hickory, basswood, and tulip poplar trees.[20] The perplexing plants grow either in small groupings or as solitary plants, and they are easily confused with similar looking plants. Sustaining ginseng in its wild state has been difficult, because wild ginseng does not survive in disturbed forests, and pressure from overcollecting has depleted plant populations in otherwise favorable habitats.

Ginseng grows from six inches to two feet tall. The medicinal and the most valued part of the plant is the spindle-shaped fleshy, aromatic root. It may be branched, and the roots most favored in the Chinese market resemble the shape of a human, with head, torso, arms, and legs. Three or four leaves on ginseng extend from a single whorl at the top of the stem; they are palmately compound, with finely toothed edges. The flowers are greenish white, and the fruits are two or three seeded berries grouped in the center above the leaf cluster.

Ginseng propagates from seeds that ripen in the fall but do not sprout until the second spring. The plant may be dormant as much as an entire year. In the first year, ginseng will have one compound leaf with three leaflets. Second- and third-year plants have two compound leaves. As the plant matures, it produces a maximum of five compound leaves. Ginseng can reproduce only after three years, so it is important for the collector to carefully recognize the plants that are reserved for future reproduction and to bypass obviously immature plants.[21] Collectors dig wild "'sang" in small secret patches and encourage rather than cultivate wild plants by helping with seed dispersal. In the fall when seeds ripen, clever collectors gather seeds and replant them in the area around their favorite patches.[22]

Ginseng root digging, called "'sangin'," is done in the fall; it is propitious that 'sangin' season coincides with squirrel-hunting season. Some collectors go it alone; others take along family: related menfolk including grandfathers, sons, grandsons, uncles, and cousins descend on darkly wooded, damp, north-facing slopes in search of the rare American ginseng. If they aren't hunting squirrels, collectors also search for goldenseal, which grows in the same habitat as ginseng.

Appalachian connections to ginseng have endured for more than two centuries. The European discovery of American ginseng traces to a Jesuit priest, Father Joseph François Lafitau, who found the plant near Montreal in 1716. Subsequent contacts between the Canadian priest and fellow Jesuits in China initiated an American ginseng trade to the Orient. For nearly two centuries, Appalachian folk have collected ginseng roots, dried them, and sold them to wholesale merchants who supply ginseng to Asian markets, chiefly in Hong Kong and Malaysia.[23] Medicinal uses of ginseng in Asia focus on it as a tonic that boosts energy and on its believed aphrodisiac qualities. Until the 1990s very few Appalachian folk took ginseng themselves; those that did used it for arthritis, headache, and other pains. The popular Appalachian view of ginseng was that this strange little plant was bought by gullible Chinese as an elixir for sexual problems. It was worth gathering, but only for the money.

Ginseng collecting brings good money; it always has. Prices from collector to wholesale buyer over the past four years have been between $250 and $600 a pound for dried wild American ginseng roots. In the 1980s the price was $100 to $180 a pound, and in the late 1960s ginseng was selling for $25 a pound. Ed Price cites a value of $12 to $15 a pound in the late 1950s, and James Lane Allen in 1886 gave ginseng's value as $1.10 to $1.50 a pound in Kentucky.[24] Ginseng now retails for $35 to $40 an ounce.

When I first came to live in Tennessee in 1967, I had read a little about

Dried American ginseng roots were worth $25 a pound in 1968. They now bring over $600 a pound. (Photograph by J. Rehder, 1968)

ginseng but had neither seen it nor spoken to a collector. While I was doing fieldwork on the Walden's Ridge portion of the Cumberland Plateau, I struck up a conversation with a fellow at a small gas station garage. He told me that in his spare time he collected ginseng roots and was selling them to wholesalers for $25 a pound. He reluctantly allowed me to photograph his stash, a small cardboard box full of dried roots. My informant said he had never in his life tasted ginseng, much less eaten it. When I asked what ginseng was good for, he grinned and said "money." He laughed and explained that the 'sang went to "Chiney" where people ate the stuff to improve their sex life. He said he had no use for the herb because his sex life was just fine, thank you. We had a good laugh, and I went on my way.

Ginseng collectors are a quiet, secretive bunch, careful to protect their favorite 'sang patches and goldenseal locations. They are untrusting souls, not unlike marijuana growers in the mountains today, except that ginseng is usually legal; ganja is not. Stan Guffey, a university biologist and my former research assistant, knows more about ginseng than any living person his age. While Guffey researched ginseng for his doctoral dissertation, he became a tobacco-chewing, bearded, long-haired wild man of the mountains. It was as if he had gone under cover with his investigation, and in many ways he had. It was within such a secret society that Guffey moved and worked freely. Eventually he earned the trust of 'sang collectors and for a time became a root collector himself.[25]

Today American ginseng is heavily regulated by the U.S. Fish and Wildlife Service and state environmental and conservation agencies. Permits must be bought to gather, sell, or buy both wild and cultivated forms of ginseng; some permits cost $250 a year. For the 2001–2 season, North Carolina issued fifty-two permits to ginseng gatherers, cultivators, and dealers. The geographic distribution naturally focused on the counties in the Blue Ridge Mountains, with five permits in Jackson County, four each in Avery, Buncombe, and Watauga counties, and three each in Cherokee, Macon, Madison, and Haywood counties. The remaining twenty-three permits went to one person in each of twenty counties and one each in the states of Tennessee, Georgia, and Vermont.[26] The harvest season is from August 15 to December 31, and the selling and buying season is from August 15 to the following March 31. Those caught obtaining ginseng illegally can be fined $500 and must forfeit all illegal ginseng in their possession.[27]

With the value of dried ginseng approaching $600 a pound, poachers in Appalachia have been expanding their range and digging illegally in some places. In the Great Smoky Mountains National Park and other federal parklands such as the Shenandoah National Park, authorities have foiled poachers by tagging ginseng roots with an injected orange dye and with tiny silicon granules attached the root. Legitimate dealers will not accept dyed roots. The silicon tags function like microchips, with retrievable information about the tag date and the location of the plant in the park. The invisible tags are like bar codes and are having considerable success in slowing if not completely stopping ginseng poaching on federal parklands.[28]

The wholesalers who buy from collectors are a diverse group. During the 1960s when Ed Price was investigating root digging, the wholesale houses and collecting points were in Appalachian towns such as North Carolina's Asheville, Lenoir, Roan Mountain, North Wilkesboro, and Boone; Bristol, Tennessee; Marion, Virginia; Pikeville, Kentucky; and Princeton, West Virginia. A major company in the 1960s was R. T. Greer, which had warehouses in Marion, Virginia, and Pikeville, Kentucky. Another branch of the Greer family had a warehouse in Lenoir, North Carolina. The Blue Ridge Drug Company was based in West Jefferson, North Carolina. The S. B. Penick Company, an international pharmaceuticals dealer, had a large warehouse in Asheville, North Carolina.[29] In 1984 Lisa Roberts, one of my graduate students, reported that two major wholesale companies active in the area were Wilcox Drug Company in Boone, North Carolina, and Pikeville, Kentucky, and the Coeburn Produce Company in Coeburn, Virginia.[30] Today's buyer companies include Wilcox Natural Products (formerly Wilcox Drug Company) in Boone and Wilson's Fur and Ginseng in Russell Springs, Kentucky. Some companies also grow and buy herbs, such as Janet Hodge's Hawk Mountain Trading in Smithville, West Virginia, and North Carolina Ginseng and Goldenseal in Marshall, North Carolina.

Since the 1990s, the popularity of alternative medicine in America has increased to enormous proportions. Once-quaint herbal preparations are now sold in every drugstore chain, in Walmart, GNC, and myriad other places in the American marketplace. Such popularity has led to a profusion of both folk and commercial plant collecting and cultivation in Appalachia. Root digging in Southern Appalachia can no longer be considered a purely indigenous folk practice. We now find a wider range of people—retirees and hippies from outside the culture and region who have moved in on the business. Local folk still collect wild plants, but there are others, perhaps retro hippies with environmental agendas and back-to-the-land ideals, who cultivate ginseng and other herbs for a living. The "hippies," old or otherwise, represent an outside population and culture invading the sanctity of the Appalachian folk's traditional root digging and plant collecting culture.

Goldenseal (*Hydrastis canadensis*) is also known as yellow root, orange root, wild curcuma, turmeric root, and Indian dye. It is found in eastern North America from Canada to the southern Appalachians but is concentrated in the Allegheny Plateau region of West Virginia, eastern Kentucky, and parts of eastern Ohio. Habitat for goldenseal is much the same as for ginseng, on north faces of slopes in moist soils of shady woodlands and along streams. Goldenseal belongs to the buttercup family. Its roots are the source for medicine. Iroquois, Micmac, and Cherokee populations widely used goldenseal for a tonic and a dye. The plant's roots did not become popular among European settlers until after the 1860s, when goldenseal became a leading patent medicine. Preparations were used to treat stomach disorders and skin problems and as an eyewash and a tonic or blood builder. In the early 1900s, goldenseal was harvested almost to extinction when annual production was estimated to be between 200,000 and 300,000 pounds. In 1991 goldenseal was placed on the endangered species list.[31] Goldenseal roots can be cultivated, but it takes three years of growth before they are marketable. Dry roots now sell for $45 a pound wholesale and $12 an ounce retail.[32] Today's market for goldenseal is almost universal. A combination of echinacea and goldenseal is believed to ward off colds by enhancing the body's immune system.

Mayapple or mandrake (*Podophyllum peltatum*) is a woodland plant first used as a purgative by the Cherokees. The plant came into widespread use in the late 1800s. James Lane Allen spoke of its being worth three cents a pound in 1886 and told how the sale of mandrake roots funded seed money for farmers in one Kentucky county. According to Ed Price, the market for mandrake once had been as much as 500,000 pounds a year, but by the 1950s it had dropped to half of that.[33]

Bloodroot, or red puccoon (*Sanguinaria canadensis*), is an abundant woodland plant that was used by Native Americans as well as European settlers. Native Americans used it for a dye and as a medicine for bronchitis, asthma, fever, and rheumatism. Named for the root's red juice, the plant frequently appeared in folk remedies; the juice was used as an emetic and as a treatment for colds, asthma, and emphysema. The root, however, is toxic and should not be ingested. In the late 1950s Ed Price found that the wholesale price for bloodroot ranged from fifteen to fifty cents a pound. Since a pound of bloodroot might constitute as many as two hundred plants, it was hardly worth the effort to collect it at that time. Today's value is comparatively better at $10 a pound wholesale and $3 to $6 an ounce retail. Bloodroot plants can be cultivated, but wild and cultivated ones require four years' growth to reach a marketable size.[34]

Black cohosh (*Cimicifuga racemosa*) is used for menstrual cramps and for menopausal problems. It was once called squaw root. It is still an ingredient in a patent medicine made by the Lydia Pinkham company, founded in 1875. The plant has made a relatively strong comeback and can be bought over the counter at drugstores. Black cohosh has a tall stem reaching six to eight feet and long white flower spikes with numerous small white flowers arranged along the column, resembling an ear of corn. Black cohosh lowers blood pressure and has a sedative effect. Also, it has been used for asthma, pneumonia, rheumatism, and bee stings and as a heart tonic. Overdoses can lead to severe headache, dizziness, slow pulse, nausea, and vomiting. Avoid black cohosh during pregnancy.[35] Retail prices are between $1.75 and $3 an ounce.

Mullein (*Verbascum thapsus*) is a biannual plant that produces large, wide leaves on a straight, tall stem. Small yellow flowers grow on a spire; at a later stage the flowers become seedpods. There are several medical uses of mullein. The leaves can be made into a tonic that relaxes the lungs and bronchial tubes. Mullein leaf tea can be used for coughs. The flowers and an oil extracted from them are ingredients for a sedative. While mullein has had several medicinal uses, an interesting nonmedical use in the folk culture of Appalachia has been as a fish poison. In a 1974 paper Gene Wilhelm described the unusual use of mullein seeds to stun fish in stream weirs.[36] Retail prices are about $2 an ounce.

Sassafras root (*Sassafras albidum*) is prepared in a tea as a spring tonic and blood builder. Also, it is used for hormone imbalance and prostate problems. Sassafras extract is one of the flavorings in root beer. In oil form some sassafras extracts have a narcotic effect and can be poisonous in large quantities. When I was about eight years old, my great-grandfather, John Farrington, took me to the woods in eastern North Carolina to collect sassafras roots in the fall. The fifteen-foot-tall deciduous sassafras trees were easy to identify, with dark green, lobate leaves that turned bright red in the fall. The leaves, the bark, and especially the roots had a sweet, spicy fragrance. We dug the shallow feeder roots while trying not to damage the tree. We took just a few roots home, made tea out of a couple of finger-sized pieces, and stored the rest in the pantry. Everyone in the family drank a cup of this hot, heavily sweetened aromatic tea, but we were warned not to drink more than one cup a day or it might make us sick. The aroma of sassafras tea takes me back to those days with my great-grandfather in the forests of North Carolina.

Logging

For nearly two centuries, Appalachian folk cultures were subsistence economies, but in the four decades after the Civil War, Southern Appalachia's regional treasures of timber and coal became open to national interests. In 1900, approximately 75 percent of Appalachia was still forested; 10 percent was considered virgin timber.[37] In a timber boom between 1890 and 1920, entrepreneurs, largely from outside the region, exploited the vast expanses of virgin timber and old-growth forests. Sequestered in mountain and plateau environments, these lands had escaped agriculture and settlement. They were cheap, and they came in large tracts. Railroads opened the way to the woodlands, and a recovering nation hungry for lumber created a national market for wood. Moreover, the once plentiful forest lands in New England and the Great Lakes regions had already been exploited, so the next place to seek quality timber east of the Mississippi River was Southern Appalachia.

Logging took two somewhat different approaches. Beginning about 1880, the first was the selective harvesting of only the best large, acces-

Team logging used animal power in 1925 near the Frying Pan Camp of the William M. Ritter Lumber Company in Dickenson County, Virginia. (C. C. Tiller Collection, courtesy of Archives of Appalachia, East Tennessee State University)

sible trees. Deciduous species of ash, black walnut, and yellow poplar were scouted out, marked, cut, and hauled by teams of oxen to a site where a portable sawmill transformed logs into lumber. During this phase, individual farmers became part-time loggers working their own lands for supplemental income. Landowners were paid fifty cents a foot across the stump. A tree six feet in diameter was worth a mere three dollars to the landowner. A profusion of local sawmills appeared on the landscape. In western North Carolina in 1883, three mountain counties had thirty-nine sawmills between them: Ashe County had nineteen, Haywood County had thirteen, and Jackson County had seven.[38]

A more devastating approach to logging came in 1890–1920. It involved the wholesale harvesting and clear-cutting of the region's forests. This different technology, called railroad logging, included steam locomotives, portable logging camps, steam-driven loaders, and steam-powered cable skidders for "snaking" logs up to a mile to waiting flatbed cars. It was a bigger, faster operation that raped the land and caused irreparable damage to slopes, streams, aquatic life, and certainly the vegetative cover. Conspicuous places in the Great Smoky Mountains National Park that were logged in this way eighty to one hundred years ago are still recovering.

Forest land exploitation was repeatable. Lumber barons and their management teams of scouts set about acquiring large forested tracts through purchase or lease. Next came the railroads, with spurs winding their way up narrow stream valleys. The post–Civil War railroad network in the United States grew and behaved much like the post–World War II interstate highway system. From trunk lines to secondary lines to spurs, a web of railroads not only connected cities but expanded into heretofore inaccessible forest lands. The role the railroad played in resource exploitation in Southern Appalachia cannot be overemphasized. Once the railroad entered the Appalachian forest, the landscape would never be the same. Roadbed grading and bridge building pushed the rails further into the wilderness. As they advanced into the timberland, trees flowed out in an unending log stream. Land that had trees one week was clear-cut the next. Logging was a noisy, smoky, dirty, dangerous job. From a folk standpoint, making a living in logging meant that men had to leave home either permanently or temporarily to work out of logging camps. Before the Civil War Appalachian people seldom had the opportunity to leave home for economic gain, but afterward they often would.

The number of commercial logging companies in Southern Appalachia is unknown, but in 1906 the enormous logging business in eastern Kentucky had about ten thousand full-time employees.[39] Moreover, before the formation of the Great Smoky Mountains National Park in 1926, there were eighteen logging companies working the lands that later became the park. While many companies led the charge into the wilderness, I have selected two quite different companies as examples: the relatively small Little River Lumber Company in Tennessee and the enormous William M. Ritter Lumber Company, a West Virginia giant with timberlands in Virginia, West Virginia, Kentucky, North Carolina, and Tennessee.

The drainage basin for Little River lies on the west side of what is now the Great Smoky Mountains National Park in Sevier and Blount counties in East Tennessee. The basin contains streams and tributaries connected to Little River that include the West Prong, the Middle Prong, and the main stem of Little River, which is also called the East Prong. The collective tributaries tumble down the crests of the Great Smoky Mountains from Clingmans Dome, at an elevation of 6,643 feet, to Silers Bald, Mount Davis, and Thunderhead Mountain at 5,527 feet.

The early folk use of the Little River Basin was much like that for any other part of the less accessible mountain and plateau provinces in Appalachia. By *folk use,* I mean a variation of land use centered on hunting, fishing, and gathering, with limited livestock grazing, open foraging, and extremely limited agriculture other than small vegetable gardens. Because of the steepness of the mountain terrain, poor soils, heavy forest cover, very limited flat land, scarce trails, and a complete lack of road access, the mountain perimeter of the basin was virtually uninhabited. Initial settlers Will Walker and his family entered the Middle Prong area in 1859. For several decades the Walkers subsisted on parcels of land bought and claimed in this portion of the mountain's drainage basin. The Walkers' primary folk uses of the land before logging were hunting, trapping, subsistence agriculture, and livestock raising. They practiced transhumance for cattle and sheep—in springtime the Walkers and other farmers led herds up to high mountain pastures called balds for spring and summer grazing. In the fall the fattened animals returned to farm settlements in the valleys below. By 1903 Walker was keeping cattle and tending other cattle owned by down-valley farmers for a dollar a head from April to September.[40]

Few people in Southern Appalachia fished for food except where fish were plentiful in larger rivers and certain cold mountain streams. The profusion of mountain streams on this well-watered windward side of the Smokies provided a substantial fish habitat for native brook trout (speckled trout). It is said that young Roy Myers, a local resident of Townsend and later a logging company employee, could catch seventy-five to a hundred speckled trout in two or three hours before the loggers came.[41] The point was that "before the loggers came" trout fishing was very good; the results of logging were devastating. Clear-cut areas allowed sunlight to reach streams and warm the water, reducing oxygen levels. Runoff, erosion, and silting ruined the water quality for most fish species, especially the sensitive brook trout, which were decimated except in the uppermost reaches of the colder streams high in the mountains. Beginning in the 1920s, rainbow and brown trout were introduced to the region's streams. These nonnative species fed on brook trout fry and devastated native brook trout populations.

Team Logging

Initial tree selection and cutting in the early logging period began with experts' scouting the area for specific species. Chosen by size, species, and access, individual trees were cut with two-man crosscut saws, trimmed with axes, and transported out. Teams of draft animals did all the hauling before the era of railroad logging. Team logging used oxen, horses, or mules to transport logs from the cutting sites either to streamside or directly to the sawmill. Mountain streams were too shallow to float the logs, so early loggers built splash dams to create reservoirs where logs could be concentrated in impoundments. Logs were branded with their owners' marks, particularly where more than one logging company or landowner used the stream. Once the reservoir was full of logs, men opened the gate on the dam, and a mass of floodwater and logs surged downstream for up to twenty miles from dam to sawmill. These turbulent, destructive man-made floods had serious environmental consequences. Released without warning, splash dam log floods occasionally injured or killed unsuspecting people or animals near the stream. The log slide or flume was another way of transporting logs. Built using rough-sawn planks for the V-shaped bed and poles for trestles, the log slide was lubricated with water. Logs slid downhill in the trough much like a bobsled on a bobsled run, though slower. A log slide built on the West Prong of Little River was two and one-half miles long.[42]

Railroad Logging: The Little River Lumber Company

Large-scale, technically more sophisticated practices entered the Little River watershed in 1901. The Little River Lumber Company, founded by W. B. Townsend from Clearfield, Pennsylvania, came to dominate the area and its economy. Even the local town of Tuckaleechee was renamed Townsend when the sawmill became the town's center. Beginning on the West Prong, the company initially hired the Shea brothers to do low-tech contract logging. Cutting trees was the same in team and railroad logging. Loggers first cut lead notches in the tree with double-bit axes to direct its fall; then they sawed the tree down with crosscut saws that were six to nine and even eleven feet long. Once down, logs—now no longer trees—were trimmed, cut to standard lengths, usually sixteen feet, and loaded onto the log slide. The main difference between team logging and rail logging was the use of steam-powered cable skidders for moving logs to waiting steam trains that hauled railcars full of huge logs to the permanent centralized sawmill in Townsend. Within five years the small West Prong area was logged out. In 1908 the Little River Lumber Company moved up the main stem of Little River along the East Prong to the Elkmont area, where a large logging camp was established and major logging operations continued from 1909 until 1926. The Middle Prong above Tremont got its turn to be logged between 1926 and 1939; it was the last area to be logged by the Little River Lumber Company.[43]

With each move, the company loaded all movable buildings and equipment and hauled them out to the next stream valley. As they went, they pulled up the tracks as well. While logging the East Prong main stem of Little River at Elkmont between 1909 and 1925, company president W. B. Townsend had allowed excursion trains to go as far as Elkmont, where tourists could stay at the Wonderland Hotel. Beginning in late 1925, with the area logged out, Townsend saw no reason to maintain the rail access to Elkmont, especially at a financial loss to his company.[44] It was time to shut down—the rails were needed elsewhere.

The highest elevation for a Little River logging camp in the Great Smoky Mountains was at Camp 19 above Three Forks, at an elevation of 3,400 feet. In the shadow of Clingmans Dome's 6,643 foot elevation just three miles away, the steep slopes above the camp determined the limits for logging even with overhead cable skidders. From 1915 to 1917, the company's skidder no. 1 worked the Fish Camp Prong to Three Forks area at Camp 19; then it worked the extremely steep Peawood Hollow area at elevations between 3,400 and 4,400 feet. The skidder also worked Higdon Camp, Jakes Creek, Wildcat Flats, Stringtown, and Marks Cove. The best single logging place, "the richest acre," was in the Middle Prong area at Sam's Creek between Shut-in Prong and Lost Cove below Thunderhead Mountain. Numerous trees, many five and six feet in diameter, contributed to the 100,000 board feet of lumber taken from the site. A single poplar log yielded 3,000 board feet at Lynn Camp Prong above Tremont in the Middle Prong area.[45]

Logging camp life was difficult and settlement was always temporary, but the pay was good. In the 1930s, logger Hobert Hayes was paid $82 a month and lived rent-free in a house in the Tremont area.[46] Such houses were portable boxes that could be hauled in on a railroad flatcar, off-loaded, and set up in the transient logging camps high in the mountains. They looked like railroad boxcars without wheels. Other housing consisted of board-and-batten box houses in the semipermanent settlements at Elkmont and Tremont.

In forty years of operation, the Little River Lumber Company logged over 80,000 acres of the Little River watershed and produced over 750 million board feet of lumber; some estimates said a billion.[47] In 1910 the sawmill in Townsend was producing 12,000 board feet a day. The equipment, technologically sophisticated for its time, included ten locomotives, seventy-five railroad flatcars, three log loaders, five Clyde steam-powered

Logging camps were exceptionally portable, as was the Middle Big Branch Camp of the William M. Ritter Lumber Company in Dickenson County, Virginia, in 1926. (C. C. Tiller Collection, courtesy of Archives of Appalachia, East Tennessee State University)

overhead cable skidders, two steam shovels, one ditcher, and several incline machines. Little River Lumber Company introduced eighteen miles of standard-gauge railroad to the area to haul logs to the sawmill and lumber beyond. The Little River Lumber Company's railroad joined the Southern Railway at Walland, where freight hauling and passenger excursions brought other uses to the rail line. At its peak of operation, the company had more than six hundred employees, but over its forty years of operation it had hired thousands of people to log the Little River watershed. The demise of the Little River Lumber Company came quickly as timber resources were depleted. The last load of logs came out of the Great Smoky Mountains in the Middle Prong area in December 1938, and the last log was sawn at the mill in Townsend on July 5, 1939.[48]

The William M. Ritter Lumber Company

In 1890 a Pennsylvania lumberman named William McClellan Ritter founded a lumber company in Mercer County, West Virginia. Ritter, who was raised on a farm near Houghesville, Pennsylvania, asked local lumberman John Paulhamus for advice about where he should start. For a $300 finder's fee, Paulhamus took Ritter to Bluefield, West Virginia,

McClure was one of several logging towns in Dickenson County, Virginia, in 1917. This pattern of settlement also became the blueprint for coal towns. (G. O. Edwards Collection, courtesy of Archives of Appalachia, East Tennessee State University)

where he bought a one-third interest in 1,000 acres of woodland from W. J. Denman of Ellenville, New York. From this initial purchase of land at Oakvale, about fifteen miles northeast of Bluefield, Ritter's timberland holdings expanded into McDowell and Mingo counties in West Virginia, Pike County in Kentucky, and Buchanan County, Virginia. After incorporating his businesses, Ritter expanded into North Carolina and Tennessee. In North Carolina alone, Ritter had nearly 200,000 acres that included two enormous tracts of 70,000 acres each in Macon and Clay counties and in Swain County. Ultimately Ritter became president of the largest lumber company in Appalachia. Operating each facet of the business in a corporate conglomerate, Ritter maintained companies under different holdings and names. In Burke and Caldwell counties, he had a company called Wilson Creek Lumber Company. In Swain County on the North Carolina side of the present-day Great Smoky Mountains National Park and along Fontana Lake, Ritter established the Hazel Creek Lumber Company. He launched a sawmill at the company town of Proctor and a railroad spur called the Smoky Mountain Railroad up Hazel Creek. This eastern slope area was on the exact opposite side of the Smokies from the Little River Lumber Company operations in Tennessee. By 1913, Ritter's various lumber companies had cut more than 2 billion board feet of hardwood timber in Southern Appalachia.[49] In the 1920s, realizing that timber resources would inevitably decline, Ritter began to acquire coal and coke companies in many parts of Appalachia's coal country. Ritter Lumber Company continued as a separate entity until October 1, 1960, when it became a part of the Georgia-Pacific Corporation.

Today logging harvests two tree types, with two scales of operation. Logging hardwoods is done on a smaller scale, with companies buying trees from local landowners, cutting them down, then hauling logs on trucks to a sawmill that specializes in hardwood lumber. This scale of operation is not unlike that of team logging in the 1880s, except that trucks rather than horses haul the logs. Large-scale logging in Appalachia cuts softwoods, chiefly pines, for the pulpwood-paper industry. Companies like Bowaters in Tennessee and Champion Paper Company in Canton, North Carolina, operate huge paper mills supplied by thousands of acres of company-owned and leased land. Pulpwood-paper industries remain far from folk culture, but some of the small hardwood companies retain a traditional family-folk way of life in the business of forest exploitation.

Coal Mining

"Coal, black as pitch, mined for the rich / men in suits black by souls whose lungs hack" expresses just one of my own observations about coal in the region; but others have spoken of the coal mine with words much richer and deeper. "The darkness pressed in like a vise. My hand, warm before my face, was annihilated. For night-dark is nothing like the mountain's black. The deepest moonless night yields a subtle play of shadow, a hint of color the pupil can widen to distinguish. Even stars spread a sprinkling of light. But the gut of the mountain casts no shadow."[50] How did the image of an Appalachian coal mine become such a dark and forbidding place in the psyche of Appalachian life? Is it the symbol of a depressed, melancholy life of poverty? The miner and his kin in hard times find life and much of the hope that once went with it descending into obscure, opaque depths of choking dust and darkness, much like descending into a coal mine. Mines, especially deep mines, frighten me; they frighten most people, even the most resolute miner. What can we sift out of the commercial dross of coal mining that speaks to Appalachian culture?

The tipple at the Blue Heron Mine, built in 1937 and operating until 1962, was the last of the Stearns Coal and Lumber Company's coal operations in eastern Kentucky. It is now an outdoor museum exhibit operated by the National Park Service in the Big South Fork National River and Recreation Area. (Photograph by J. Rehder, 2003)

Appalachia's identity with coal mining is not just well known; it is an obsession with an overwhelming image of twentieth-century poverty that appears to outweigh all other identifying characteristics of the region. But its spatial and temporal locations are unmistakably the Cumberland and Allegheny plateaus. Coal mining in Appalachia began in earnest in the 1880s, and the landscape presence of coal mines, coal camps, and coal towns dates from that time. Some relic landscapes from 1880 to 1940 are still present on the plateaus.

The process of exploiting Appalachia's natural resources followed a predictable pattern. We already know that early attempts to develop the region sought timber resources. Next came coal, often extracted by the same entrepreneurs and their companies. A pattern emerged in which timber companies initially bought and leased large tracts of land, encouraging railroad companies to enter the area or building their own railroads, then logged the land and followed that with coal mining at these and other coal-rich sites.

Plateau lands were unattractive and inaccessible to early settlers. Although land was available, it was a poor choice for agricultural settlement because of thin, rocky soils, lack of usable surface streams or lakes, and difficult terrain. Even where access was possible, such as through the Cumberland Gap, farmers pushed beyond the undesirable plateaus to attractive limestone soils like the Bluegrass Basin in Kentucky. It was not until the late nineteenth century that logging and especially coal mining entered the plateau regions in earnest.

Coal Mines

In 1880 to 1940, coal mining in Appalachia was done deep underground. However, on today's landscapes there are two fundamental types of coal mines: deep mines (ca. 1880 to the present) and strip mines (ca. 1950 to the present). Deep mines lie underground in dark man-made caverns. There are three types of deep mines, named for the directions they take to reach the coal seam: drift mines, shaft mines, and slope mines. Drift mines, the most common deep mines in Appalachia, enter the mountain horizontally along the same axis as the coal stratum. Shaft mines are directed to the coal seams from above. Each shaft is vertical and has a cable-operated elevator with its attendant superstructure of beams and cables and wheels poised above the mine entrance. Slope mines, the least common of deep mine types, enter the coal stratum at an angle from the surface.[51]

Deep mines are mined with "room and pillar" techniques. The first step cuts two parallel tunnels, sometimes called "butt headings," that are ten to twelve feet wide and thirty feet apart; the first tunnel provides access to the entire mine operation and is used for hauling coal out; the second tunnel is an air course and passageway. As a deep mine evolves in the room and pillar strategy, an underground landscape pattern emerges that resembles a grid-patterned town. In map or plan perspective, rooms, pillars, and tunnels resemble the street patterns of city blocks. As the mine develops, excavation creates the grid.

Entering a drift mine at the surface in 1930, one would see a gaping maw with railroad tracks on a small black, rubble-strewn roadbed following the horizontal coal seam along the main haulage tunnel. Branching off from this primary tunnel were rooms separated by pillars—columns of unmined rock. Each room had a standard twenty-four-foot width that ranged from two to eight feet high, depending on the thickness of the coal seam. The rooms represented the most serious and active part of coal mining; the room was to the miner what a field was to a farmer. Smaller tunnels called "breakthroughs" or "crosscuts" separated rooms into smaller units and provided access and airways. Pillars and

Drift coal mines entered the mountain horizontally following narrow coal seams. McCreary County, Kentucky. (Photograph by J. Rehder, 2003)

timber props attempted to keep the stone roof above from collapsing on men and machinery. Typically, a coal pillar was cut with dimensions sufficient to support the weight of 2,000 feet of overburden rock, soil, and vegetation above it. But if pillars were too small or if the company decided to go back into an old mine to do "pillar robbing"—harvesting coal-rich supporting pillars—a mine collapse was an imminent danger.[52]

At the active mine face, the farthest point of penetration in a coal mine, miners performed four tasks: undercutting, boring, blasting, and loading. In undercutting, men with picks carved out a notch by cutting a deep horizontal groove at the base of the coal seam on the floor of the mine. They did this while lying on their sides—a terribly fatiguing position—in a cold, wet, dirty, dangerous place. The notch in the undercut was analogous to the lead notch in felling a large tree. The next task was to bore holes in the coal face above the undercut so that explosives could be tamped in. Blasting used black powder or dynamite to blow the face into rubble. If too little explosive was used, the rocks in the rubble were too big and heavy. Too much explosive pulverized the face into useless, dangerous dust. And the blast might cause a roof collapse or a mine explosion by igniting methane gas or coal dust, killing dozens to hundreds of miners.

The loading phase was the bread-and-butter part of mining, and the loader was seen as the craftsman in the deep-mine coal business. His tool was a number four hand-held shovel, and he loaded about twenty-four pounds of coal per shovelful into a coal car. Coal cars were pulled first by mules and later by electric railcars that hauled the extracted coal to the surface to a tipple, or coal-screening apparatus, where the coal was sorted by size and loaded into waiting standard-gauge railroad coal cars. Loaders, as some miners preferred to be called, were paid on a complicated scale, but it was usually based on the tonnage of coal that each man could load. A miner's normal production was about eight to nine tons a day, so the Tennessee Ernie Ford song "Sixteen Tons" needs a little explanation. It seems that the legendary sixteen tons was an initiation rite in which veteran miners would ease up so a rookie miner could load a special first day's load of sixteen tons that surpassed everyone else's. Thus the rookie could prove himself a man.[53]

While it was common to have from ten to several hundred miners in a mine at any given time during the period 1880 to 1940, since the 1960s modern technologies in both cutting and loading coal have reduced the number of miners in a given mine. Today's miners work with gnawing, chewing, toothed devices called "continuous mining machines" that attack the coal face with ferocious energy. Coal automatically loads onto conveyor belts that transport the black gold to the surface, where a tipple washes coal pieces, sorts them by size, and loads them onto railroad cars. In some mining areas the coal is stored in piles or in huge concrete silos before being loaded for shipment. Where does Appalachian coal go? The Norfolk and Southern Railroad (formerly the Norfolk and Western Railroad, founded in 1881), with its enormous coal-hauling capability, carries an almost constant stream of coal cars from the coalfields of southern West Virginia and eastern Kentucky to the ocean port at Norfolk, where the coal is loaded onto ships and transported all over the world to markets as distant as Japan. The port accounts for 17 percent of U.S. coal exports, but nearly 90 percent of all U.S. coal is used for domestic coal-fired electric power plants.[54]

Since the 1950s, strip mines have created another landscape in the plateau coal country. Since coal seams lie in horizontal layers in the rugged, highly dissected terrain, they appear at the surface along contours, or areas of equal elevation. Contour strip mining begins with clearing the band of vegetation along a given elevation by clear-cut logging and harvesting the usable timber. Next, heavy bulldozers, backhoes, graders, and other equipment clear the noncoal layers of soil and rocks collectively called overburden until the coal seam is exposed. This coal is mined almost immediately. At this point the contour has been "notched"—cleared of overburden and coal—and the strip mine's high wall looks like a road cut in highway construction. Mining continues as augering equipment drills out the coal that is exposed on the high wall. Another type of strip mining removes the tops of hills and mountain peaks to reach coal seams below. Mountaintop removal mining has been returning to areas that were previously contour strip mined. The concept is to strip off the tops of accessible hills to reach the few available coal seams beneath a relatively shallow overburden. The resulting landscapes have no better appearance than contour strip mines.

Strip-mined landscapes look as if they have been butchered. From the air, notched contours carved from the gentle natural slopes look like concentric rings of destruction. In profile, topped mountains look as if an angry barber gave them a bad crewcut. Once stripped, the land can never be the same. On the Cumberland Plateau, contour strip mines created in the 1960s have not yet fully recovered their vegetative cover. Such relics are called "orphan mines," with no one to claim or reclaim them.

Appalachian strip mining on a large scale began in the 1950s and reached its zenith in the early 1970s. Contour strip mining has since declined because of land and coal depletion, low coal prices, the opening of western coal reserves, and especially environmental regulations. Since 1977, environmental protection laws call for contour strip-mined land to be returned to their original contours, a nearly impossible feat. Furthermore, land reclamation laws call for the return of a stable vegetative cover of grasses and trees—hardy, fast-growing, but dangerously thorny black locusts.

Miners

Between 1880 and 1920, three types of miners worked the early Appalachian coalfields: local white farmers from the mountains; southern blacks; and immigrants, chiefly from southern and eastern Europe. The first mine laborers came from local Appalachian stock, people who had been raised on small subsistence farms. Historian Crandall A. Shifflett describes the transition:

> Under the ground, mining coal involved laboring under conditions in sharp contrast to agriculture. The transition from farm to mine involved a change from a well-lighted, well-ventilated, open workspace to a dark, cramped, and uncertain environment. Farming was an occupation ei-

> ther without a boss or with one who only loosely supervised. At the mines a man had to answer to a foreman. Working on a farm required mostly individual labor under relaxed constraints of time and production. The mine was a competitive place where time and production determined earnings, status, and social mobility.[55]

Coal mining was far more dangerous than farming. On May 19, 1902, a mine explosion killed 216 men and boys in the deep Fraterville coal mine near Briceville, Tennessee, in the worst coal mine disaster in Tennessee history.[56] On that fateful day, one thousand children were fatherless by nightfall.

Mountain whites culturally did not fit into the structured routines of industrial life and work in the mines. Absenteeism among mountain locals, especially during hunting or 'sangin' season, left coal companies looking for other sources of labor. As more mining areas opened up, reliable low-cost, expendable labor came from aggressively recruited southern black populations and Italian and eastern European immigrants. The resulting cultural diversity in this folk mix created an almost cosmopolitan Appalachia. Yet, segregated by race and culture, people in coal camps lived in small ethnic enclaves with derogatory names such as "Colored Town" or "Dago Town" or "Hunk," a derivative of "Bohemian."[57]

The people who came to be known as coal miners entered the area from elsewhere, settled only temporarily, then moved on as coal resources and economies declined. After the 1920s, the numbers of black and European immigrant miners began to decline, so that the only remaining labor pool was again the local white populations, sons of former mountain white miners.

Coal Camps and Coal Towns

Initially, coal country lands were sparsely settled. Few towns of any consequence occupied the hollows and stream valleys, much less the plateau uplands, until the logging and coal industries entered the region. Then every coal company constructed a settlement complex of structures that housed workers, stored equipment and supplies, and processed coal for shipment. The buildings on the landscape screamed to the world, This is a coal town! The coal mine was first and best recognized by a symbolic structure—the coal tipple—a towering amalgam of beams and metal siding where coal left the community. Its prominence on the landscape served as a beacon of commerce and industrial fortitude.

Coal towns evolved over three phases. The first phase, between 1880 and 1918, was a frontier settlement with rugged people and terrain, poor living conditions, and temporary housing in tents or flimsy wooden shacks in loosely arranged coal camps. From 1918 to the mid-1930s, company towns came into their own. By 1925, there were more than five hundred coal towns in Southern Appalachia.[58] From the mid-1930s to the mid-1950s, coal mining suffered economic decline in bad years and enjoyed mechanization in good years, and both reduced the need for mine laborers. During these years, coal towns also declined in population and in numbers of structures, so that a landscape once dominated by coal mining settlements came to have fewer structures.

Housing

The coal company or a subcontractor designed and built housing for workers. Dwellings followed a well-understood hierarchy from the superintendent's mansion to a large house for the storekeeper to large houses for foremen to smaller houses for white miners and then to even smaller ones of poorer quality for black and immigrant miners in segregated sections in the town. West Virginia's huge Pocahontas Coal Field had 10,000 dwellings in a hundred coal towns in 1930.[59] Much like plantation quarter houses, miners' dwellings were small, cheaply built wooden structures. Many were simple four-room box houses with dimensions of twenty-eight by twenty-eight feet, constructed of balloon framing and covered with vertical board-and-batten siding. In 1925 the U.S. Coal Commission reported that one-third of 713 coal towns in its survey had houses constructed with board-and-batten siding, "among the cheapest, if not the cheapest, type of outside finish."[60]

In 1972 I unofficially directed my first doctoral dissertation, a study of the coal mining settlements researched and written by Mack H. Gillenwater. Mack had grown up in the Pocahontas Coal Field of West Virginia and produced a fine, credible study of the settlement morphology of the coal mining landscapes there. Following the model of what I had done for sugar plantations in Louisiana, Mack examined coal mine settlements in West Virginia in much the same way.[61]

Houses in the coal camps and coal towns were not all folk types. Mack Gillenwater's typology for dwellings built between 1880 and 1930 in West Virginia's huge Pocahontas Coal Field was as follows: one-story L, pyramidal, bungalow, basic I-house, two-story four pen, two-story shotgun, saltbox, two-story L, and superintendent's house. The *one-story L* was a three-room house that, when viewed from the front, had a pair of tandem rooms on the left side and on the right a single-room ell (or L) extending from the back room. Dimensions for the two tandem rooms together were twenty-eight feet by fourteen feet. The third room, acting as a kitchen appendage, measured twelve by sixteen feet. The one-story L house, never a dominant type in the region, was built between 1883 and 1918 and was concentrated in the Bluestone and Elkhorn valleys in southeastern West Virginia. The *pyramidal* house was a single-story four-room house that measured between twenty-four and twenty-eight feet on a side. The diagnostic trait was a pyramidal roof with all four sides sloping to a point at the apex. Pyramidal houses were concentrated in only seven towns in the Pocahontas Coal Field. Recently I found a few of them remaining in southeastern Wyoming County between Itmann and Garwood on Highway 16. The *bungalow* was a two- to four-room house with a front-facing gable and a front porch. The four-room version regularly measured twenty-eight feet on a side. A very popular, easy to build structure, the bungalow became the dominant house type in the entire Pocahontas Coal Field. The *basic I-house,* a folk house type that is ubiquitous throughout much of the United States, first came to the Pocahontas Coal Field through the town of Pocahontas, Virginia. Similar to other I-houses in Appalachia and elsewhere, this basic I-house was a two-story house that was one room deep and two rooms wide. However, in the coal camps this became a smaller house that measured twenty-eight feet wide by fourteen feet deep. To enlarge the basic I-house, a variation placed appendages on the gable ends. The idea is credited to Colonel E. A. O'Toole, a mining engineer and coal operator who was commissioned to build basic I-houses with appendages in five coal towns in McDowell County, West Virginia.[62]

The *two-story four pen,* or simply four pen house, had dimensions of thirty-two by thirty-two feet, with four rooms on each floor and separate stairwells. The four pen was occupied by two families. Four pen houses were built between 1885 and 1930 in the Pocahontas Coal Field. Elsewhere in Appalachia, four pen houses were company houses in the bituminous coal region, particularly the Connellsville coal region of western Pennsylvania in Fayette and Westmorland counties.[63]

According to Gillenwater, the *two-story shotgun,* a single-family dwelling in the coalfields, is similar to the simple one-level shotgun houses of the Mississippi Valley described by Kniffen, Rehder, and Vlach.[64] The fundamental basic shotgun house in the South is a long, narrow house that is one room wide and three or more rooms long, with a front-facing gable. However, the ones Gillenwater discovered in the coal country are two stories tall, measure fourteen feet wide and twenty-eight to thirty-eight feet long, and have two rooms on each level. Most had a kitchen appendage. Built between 1900 and 1930, the house type was popular in especially steep terrain such as upper Elkhorn Creek, North Fork, and Dry Fork in McDowell County of southernmost West Virginia.[65]

The *saltbox* house, a common New England folk house type with an asymmetrical roofline, began to appear in the Pocahontas Coal Field around the turn of the century, but none were built after 1924. A very popular row house in the Elkhorn and Tug Fork valleys in McDowell County, the saltbox could be found in coal towns such as Kimbell, Gary, Wilcoe, and Thorpe. The diagnostic trait of the saltbox is its asymmetrical roof that when seen from the gable side has a short pitch to the front of the house and an elongated pitch to the rear. Saltbox houses here were two stories tall, two rooms wide, and two rooms deep. Rooms on the first floor were of equal size, but on the second floor the two rooms in front were large and the two in back were much smaller. Dimensions for the entire house were twenty feet wide and twenty-six feet deep.[66]

Gillenwater's typology continues with his *two-story L,* a house that might also be classified as an I-house with an ell. This house became the primary house for administrative personnel: mining company office workers, storekeeper, and others who were of a higher status than coal-dust-covered miners but below the superintendent. As an upper-echelon status symbol, the house had indoor plumbing, two or more fireplaces, and a floor plan of three rooms over three rooms. The first floor had a parlor, dining room, and kitchen; the second floor had three bedrooms; the bathroom was built into the back porch. This house type, normally constructed in the coalfields after 1900, was never numerous, but each coal town had at least one to five houses like this at or near the center of town.[67]

At the pinnacle of coal town morphology and hierarchy was the superintendent's house, an elaborate mansion analogous to a southern plantation mansion in status and function but not in form. Between 1880 and 1920, the chief administrators for coal companies lived on site in a coal town, where they supervised the day-to-day operations of the company and community. In a patriarchal way, these captains of industry controlled the operations and the development of the coal town's landscape. Nowhere else was this so evident as in the appearance of the superintendent's mansion. The house was in a prominent place, usually on a slope overlooking the community. Alternative sites were at the center

The superintendent's home was the largest and most pretentious house in a coal town. Stearns, Kentucky. (Photograph by J. Rehder, 2003)

of town or the main entrance or at a bridge point in the valley. The lavish mansion was enormous, with ten to twenty rooms inside a cubic structure. The steep roof had many gables, dormer windows, and multiple chimneys. Outside, decorative trim followed the styles of the period, with early homes displaying Victorian gingerbread.[68]

In 1972 neither Mack Gillenwater nor I was as clear about house type terminology as we are now. At the time, I allowed Mack to develop his own classification and typology, and for some structures his own terminology. Reflecting on it now, I see that although coal town dwellings were commercially constructed houses, the basic I-house, two-story four pen, shotgun, and saltbox certainly have folk house connections and conceptually could be traced to folk origins outside the region. Moreover, the one-story L, two-story L, pyramidal, and bungalow types all have rural and temporal identity here and elsewhere on southern landscapes. Despite the apparent confusion, the dwellings constructed in coal camps and towns were built for one purpose—to house miners cheaply. Builders may or may not have incorporated folk elements into the architecture. The important issue is the landscape presence of coal camp housing, not folk identity.[69]

In some coal camps, smaller houses were brought in on flatbed railcars, offloaded, inhabited for a few years, then reloaded on a railcar and moved to a new active mining area. Behind each house in a coal town was a coal bin and an outhouse, and in some places, if terrain permitted, plots were provided for vegetable gardens. Gardens were uncommon because of steep slopes, shaded hollows, poor soils, and a mining culture that relied almost entirely on the company store for provisions and supplies.

The company store was clearly the most important building in a coal town. The store served as provisioner, provider, paymaster, social center, and supplier of most needs of a miner and his family. Bulk foods and canned goods; miners' clothes, lamps, hats, boots, and tools; and all kinds of dry goods for home and mine were just some of the items carried in the company store. Pay came in the form of paper scrip or coinage issued by the company. Scrip could be spent only at the company store.

In field research, I am always looking for diagnostic traits and repeatable patterns on the cultural landscape. In Appalachia's coal country, the forms and functions of coal towns can be expressed in real world examples: Tams, West Virginia, and Stearns, Kentucky.

Tams, West Virginia, 1972 and 2002

In March 2002 I went to West Virginia to retrace some of Mack Gillenwater's research and to explore the area spatially and temporally. I chose one of Mack's case studies, the coal community of Tams. Fieldwork began on a March day that was sunny but cold and very windy. Traveling south from Beckley, West Virginia, I drove along the top of the plateau on Highway 16, covering about eight miles of rolling terrain before the road descended into the head of a "holler" locally called Winding Gulf and reached Tams.

The coal mine settlement of Tams began in 1909 when W. P. Tams established the Gulf Smokeless Coal Company, based on a 3,000-acre lease in the valley. By 1910 Mr. Tams had started a drift mine, built a tipple, and set up a sawmill at a valley site six hundred feet wide that would become the town center. Forests in the leased tract provided the lumber for the first houses. Within a year, the railroad had reached the settlement of 125 houses, and coal began to be shipped out. Over the next twenty years, Tams grew to 185 houses.[70]

The population at Tams reflected the ethnicity of coal mining settlements in the region. In 1915, Gillenwater tells us, the company had 294 employees: "101 Negroes, 70 white Americans, 40 Poles, 30 Austrians, 25 Slavs, 18 Hungarians, and 10 Russians."[71] W. P. Tams Jr., son of the founder, described the population density in the settlement's ethnic enclaves: "In 1920 there was an average of seven men, women, and children per house in the Negro section of seventy-five houses; five per house in the white American section of seventy-five houses; and ten per house in the foreign section of thirty-five houses, making a total of 1,250 people."[72]

The zenith population of about 2,000 was reached in 1925 when about 233 dwellings and other domestic outbuildings constituted the coal town. Tams now had two sections divided by the coal mine entrance, tipple, and railroad complex. Upstream or north of the mine were a school, a church, and 99 houses largely occupied by blacks and immigrants. South of the mine were a store, a school, a church, and 134 houses set aside for the white miners and their families.[73]

Between 1910 and 1955, the Gulf Smokeless Coal Company owned and operated the town of Tams; afterward it became the property of the Winding Gulf Coal Company. In the intervening years, especially after 1940, the buildings began to decline. Mining operations moved northward, requiring more space for mine structures and processing equipment, so houses in the Negro section were destroyed. As mines became more mechanized, fewer miners were needed, and fewer houses were allowed to remain intact. By 1971 Mack Gillenwater observed that there were approximately 150 people living in Tams.[74] In 2002 no occupied houses remained. However, a modernized tipple designed for loading coal on trains and coal trucks occupied the landscape, along with a dozen or so newer metal-sided structures. Still, few if any buildings appeared to have survived from the 1910–25 era of company town development. No one lives there now.

Highway 16 from Tams to Itmann is a winding, busy road that passes through Ury, Helen, Amigo, Stephenson, Corinne, and other tiny unincorporated places before reaching the town of Mullens. These communities are coal towns too, but they are more than surviving; some are thriving in the otherwise dismal Appalachian coal country. Houses here range from some of the original company houses with board-and-batten siding and pyramidal roofs to more modern-looking aluminum- and vinyl-sided and even brick homes in well-kept, attractive enclaves. How can this be? I am expecting Appalachian ghettos, and for the next twenty miles or so I see mainly a string of bright pearls. Where are my settlements in coal dust coveralls? The tipples, railroads, and coal trucks are still here, but I see very little poverty—not nearly enough for the image makers. Then I reach Mullens, a town of darkened, soot-stained brick buildings. Only a few stores are doing all right; most are boarded up. The sign on the derelict Wyoming Hotel peers through a thick film of coal dust. Perhaps I spoke too soon, or maybe the landscape is telling me something. This town, small yet big for the area, the one-time center of business and commerce for this isolated part of the plateau country, has seen better days. Coal is still being mined here, but by fewer miners. The coal money made here is no longer spent in towns like Mullens or Matoaka, a slightly smaller town eighteen miles south of here that also used to be a business center. Money is made by workers living here but within commuting distance of factory and service jobs in Beckley, Bluefield, or Princeton, West Virginia. The only reason for once primary but now secondary towns to exist in this coal region is to provide minimal services and a decaying infrastructure to an already changing landscape.

Stearns Coal and Lumber Company

The Stearns Company, in McCreary County in southeastern Kentucky, exemplifies the evolving pattern of resource development and settlement in another part of the Appalachian coalfields. In 1901 Justus S. Stearns, a lumber baron from Ludington, Michigan, with timber interests in Michigan, Wisconsin, the Pacific Northwest, and Florida, began to acquire 50,000 acres of timber land called the "Big Survey" on the Cumberland Plateau in southern Kentucky and northern Tennessee. In 1902 Stearns leased another 25,000 acres in Whitley County, Kentucky, and opened his first store in what later became the company town and hub appropriately named Stearns. With logging and lumber as his company's initial focus, Justus Stearns secured transportation from the Kentucky and Tennessee Railway. The first line in 1903 went from Stearns 3.5 miles west to Barthell, the company's first coal mine. The railway extended farther west and north to Yamacraw in 1906, to Oz in 1908, and eventually to White Oak Creek by 1909, covering a distance of some twenty miles of rail on company land.

Between 1903 and 1910, the company initiated its coal mining phase that included the Barthell Mine no. 1, with the first 32,000 tons of coal extracted. Other mines followed at Barthell, Worley, and Yamacraw. The Yamacraw mine had a concrete tipple, a store, a school, numerous houses in the coal camp, and a concrete bridge across the Big South Fork. By 1910, Stearns Lumber Company was renamed Stearns Coal and Lumber Company, and the company had access to about 100,000 acres of land.

In 1916 the Stearns Company owned 75,000 acres and leased another

The company store supplied everything to a coal town. Stearns, Kentucky. (Photograph by J. Rehder, 2003)

25,000, with reserves of 50 million tons of coal and 380 million board feet of timber. One sawmill processed 70,000 board feet a day. Four coal mines were in full operation. The company had 450 buildings: nine stores, along with offices, warehouses, and barns. By 1920 Stearns had five mines open and three separate coal companies: Comargo Coal, Premier Coal on Wolf Creek, and Paint Cliff Mining Company at Oz. Between 1921 and 1929, the Stearns Company purchased more timber land, increased its timber and sawmilling work, and was the first company in the United States to build an all-electric sawmill. By 1929 Stearns had 2,200 employees living and working in eighteen coal camps and was mining a million tons of coal a year.

The company town and headquarters, Stearns, Kentucky, emerged as a representative corporate headquarters. Nearly all the buildings were painted in the company colors of white trimmed in green. There were two hundred company-owned dwellings that rented for ten to sixty dollars a month. The town had a freight depot, office building, theater, hotel, pool hall, golf course, tennis courts, and baseball field, as well as a huge company store where employees could buy just about anything with company-issued scrip. The scrip consisted of metal coins worth from ten cents to five dollars, stamped with "Stearns Coal & Lumber Company—Stearns, Kentucky." Justus Stearns had stamped his name on just about everything in his town and his company—even the currency.

In 1933 Justus S. Stearns died; his son Robert Lyon Stearns took over operations, but coal production was down by 20 percent. By 1937, deep in the Great Depression, company lands had been all but depleted of commercial timber, so Stearns sold 47,000 acres to the United States government to be added to the Cumberland National Forest that later became part of the Daniel Boone National Forest.

In 1937 J. E. Butler developed the Blue Heron Mine complex, also called Mine no. 18, on the Big South Fork of the Cumberland River and built the newest and last coal tipple for the company. During World War II, coal was in greater demand, and production increased substantially. In 1949 Robert L. Stearns Jr. became president, but the postwar years marked the end of an era. The demand for coal declined when the railroads changed from coal-fired steam engines to diesel engines and the home and factory market changed from coal to oil and gas.

One by one the coal mines closed and coal camps were abandoned as the coal seams played out. A series of bad times continued as the Barthell Mine closed, and the coal camp was torn down in 1952. The Blue Heron Mine closed in 1962, and the next year the huge sawmill burned to the ground. As of 1957, the Stearns company land encompassed 200,000 acres, of which 142,000 acres were in McCreary and Wayne counties, Kentucky. More acreage was in Scott and Fentress counties, Tennessee. A brief mining resurgence came in 1968 when Stearns reopened the Justus Mine, only to sell it in 1975 to the Blue Diamond Coal Company of Knoxville, Tennessee. Finally, in 1979 Stearns Coal and Lumber Company got out of the timber and coal business entirely when it sold off 43,000 acres to the United States government as part of the Big South Fork National River and Recreation Area. The last coal car from Blue Diamond's Justus Mine, Stearn's last operating mine, left the area in 1987, and with that coal mining terminated.[75]

Blue Heron

At the height of settlement between 1937 and 1962, the Blue Heron mine and coal camp had two hundred employees, twenty-two families living in the coal camp, a school, a church, a bath house, a company store, and one of the largest, most modern coal tipples for its time. The Blue Heron Mine operated for only eighteen months but produced 250,000 tons of coal. The tipple continued grading the nearly 5 million tons of coal that came by tram and train from other mines after the mine at Blue Heron played out. Coal at the tipple was graded into sizes: 5.5-inch block coal; 3.5- to 5.5-inch egg coal; 1- to 2-inch nut and slack coal; and 0.4- to

1-inch stoker coal. After 1962, even coal no longer had a place at Blue Heron.

When the National Park Service took over, there were no houses at the Blue Heron Mine to restore; only the tipple and the tram bridge across the Big South Fork of the Cumberland River remained. With no written history, the story of Blue Heron could be told only by recollections of the people who lived and worked there. In 1989 the National Park Service and U.S. Army Corps of Engineers restored the Blue Heron Mine complex, with oral history exhibits and ghost buildings: open steel-framed structures with corrugated steel roofs.[76]

Life in the Blue Heron Coal Camp

Like the lumber camps that came before them, coal camps were assured of a short life. They were transient, coming into existence only when coal seams were discovered. People moved frequently from camp to camp as the mines closed when coal seams played out. One informant said he had lived in seven houses while working for Stearns. The first houses at Blue Heron were hauled in on railway flatcars from the Worley camp at Mine no. 3. More houses were built to accommodate about twenty-two families. "Very few of them [miner's families] lived down here or wanted to live down here because there was nothing here," explained one former resident.

Coal camp houses were small two-room structures with balloon framing and board-and-batten siding. Later, siding changed to horizontal clapboards painted white, with green or blue trim. Unlike folk housing elsewhere, a common house type was the "Jenny Lind" structure that had two or three rooms, a single potbellied stove, and a small kitchen.[77] Houses rented for fifteen to thirty dollars a month, and water and electricity were free. Outhouses were the only sanitation system. Yard trees had the first five feet of the trunk whitewashed.

Even though Stearns had its primary store in the headquarters town, every coal camp had its own store. Blue Heron's company store supplied carbide lamps, safety shoes, boots, clothes for miners and their families, rifles, shotguns, shovels, picks, candy, soft drinks, and most foods, since few mining families tended gardens or kept livestock. Miners were not farmers. All the necessities of life could be bought with scrip from the store. But true to the "Sixteen Tons" lyrics, some people owed their soul to the company store. The store had a wide door and iron bars across the windows. Inside was a post office and a large counter where the storekeeper sat. Most merchandise was displayed on shelves lining the walls, but large, heavy items were warehoused in the back of the store. One day a woman at the Blue Heron coal camp found a Stearns five-dollar scrip coin; she didn't tell anybody about it, but she discovered something about scrip: "I found out that after you got out of Stearns, that money wasn't any good, and I thought, man, they've really got you sewed in! You either spend it here or you don't spend it [at all]!"[78]

Illegal Ways of Making a Living: Moonshine, Marijuana, and Methamphetamines

The image of Appalachia has had a longtime love affair with moonshine, also popularly known as white lightning, mountain dew, rotgut, redeye, who shot John, pop skull, and other colorful names. No other Appalachian product or activity, legal or otherwise, demands so much attention and prompts so many sly grins, smiles, and outright guffaws. Moonshine is homemade corn whiskey that is illegal because taxes due on the liquor have not been paid—not then, not now, not ever! The moonshine story is more complicated than that, but traditionally the issues of revenue, taxes, revenuers (the law enforcement officers sent out to catch moonshine distillers and sellers), and everything involved in "makin', sellin', and haulin'" high-octane liquid corn have created a mythology that rivals the tales of Robin Hood. The origins of moonshine trace to the British Isles, where Scots, Scotch-Irish, Ulstermen, Irishmen, and others distilled spirits from hardy grains of oats, rye, and barley.[79] In the intervening years, as folk migrated and diffused through Appalachia, they took with them the knowledge and skills of whiskey making. All they had to do was replace the hardy European grains with native corn.

The geography of moonshine in America focuses on the southeastern United States and centers on Southern Appalachia. Statistics on stills seized and gallons of alcohol destroyed are the best indexes of moonshine concentrations. Academic researchers have found that the safest way to investigate illegal activity is to study it "after the fact"—after lawmen have seized and destroyed moonshine stills (or marijuana plants). Nationally, still seizures show a decline from 29,087 in 1925 to 11,266 in 1955 to 2,090 in 1972.[80] In the 1954–55 raiding season, federal agents seized 11,266 stills, of which 10,526 were in the Southeast, south of the Ohio River and Pennsylvania. The top five states were North Carolina (2,044), Alabama (1,840), Georgia (1,692), South Carolina (1,420), and Tennessee (1,169). In one week in East Tennessee, agents seized 25,000 gallons of the state's 32,000 gallon total. In April 1955 agents destroyed ninety-four stills in East Tennessee; in one day they bagged twenty-three.[81] These patterns reveal two things: that the area was rich in moonshine stills and that the agents were highly effective in catching the perpetrators. Two counties in East Tennessee were centers of moonshine production: Cocke County, in the mountains just up-valley and north of the Great Smoky Mountains National Park about fifty miles from Knoxville, and Marion County in the southern part of the Cumberland Plateau just west of Chattanooga. Both counties have had a long moonshine history with nearby urban markets in Knoxville and Chattanooga. Moreover, both counties, especially Cocke County, evolved into marijuana centers, and at present methamphetamine labs are beginning to appear in these and other rural areas.

The reasons for making moonshine are several. The Whiskey Rebellion in western Pennsylvania in 1794 is popularly credited with or blamed for starting the illicit moonshine industry in America. No doubt distilling practices predated the rebellion by many decades, if not a century. The rebellion, however, had much to do with the abhorrence of paying taxes on distilled spirits, and fighting the government over taxes was reason enough. Before 1791, whiskey was untaxed; it was also untaxed in most of the years between 1802 and 1862. So it was not until 1862 that making whiskey without a license permanently became a federal offense.[82] The economic reason for making whiskey, especially untaxed whiskey, was that it was profitable. Furthermore, logic says that with the bulk and weight of corn, it was preferable to transport it in liquid form. It traveled better that way than on the unreliable hooves of fattened hogs or other livestock. Another good reason is cultural tradition, a survival from earlier times. My colleague the late Loyal Durand Jr. wrote about the tradition, "How much [of it] is a reflection of a traditional pattern of life in which a small still was as much a part of mountain occupance as the spinning wheel?"[83] So the decision is made: we have corn, we have a market, we have a high-value, lightweight, low-bulk product to ship *if* we convert the corn into likker before we ship it.

Prohibition, the Eighteenth Amendment outlawing all alcoholic beverages in the United States, began January 17, 1920, and ended December 5, 1933. In the intervening years, the floodgates opened in illicit liquor making, selling, and hauling over the entire United States, with the Appalachian states and nearby southeastern states forming a culture hearth for the venture. The markets that developed for Appalachian moonshine during Prohibition became widespread and remained that way until well after 1933 and even until the 1970s. Local laws meant that

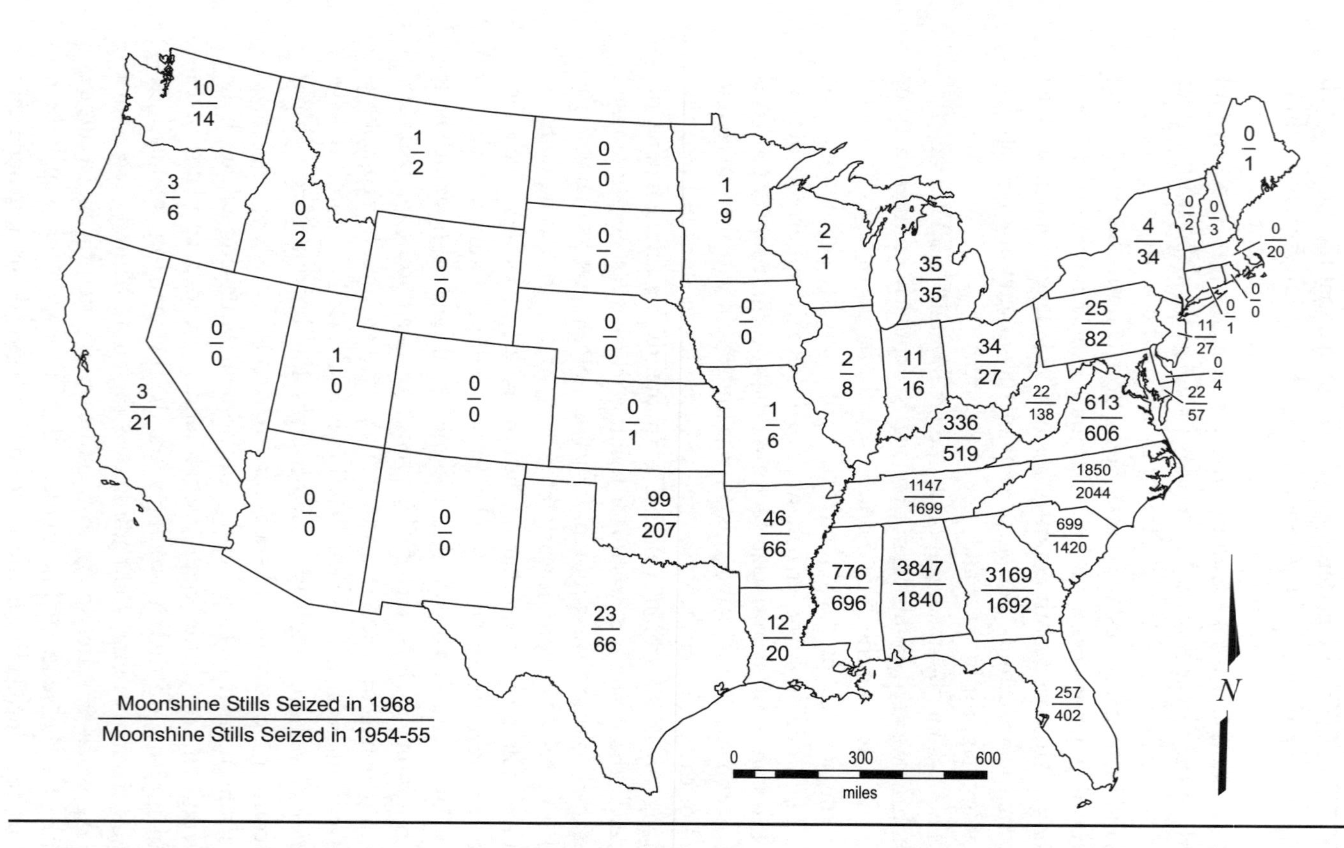

The pattern of moonshine stills seized by federal agents in 1954–55 and 1968 shows high concentrations in the South. Notice the shift of activity from the Upland South to parts of the Deep South. (Sources: Durand 1956; Nichols 1973)

residual dry counties and urban places were obvious markets for moonshine even after national Prohibition was rescinded.

Another peculiar stimulus that supports moonshine is the concept of "dry" counties. Until the 1970s there were about 446 dry counties in the southeastern United States, where prohibition was imposed by local ordinances. In Georgia alone there were 116 dry counties and only 43 wet ones.[84] Many urban and rural counties in the South remained dry until liquor by the bottle and by the drink became legally available in the 1970s. There are still dry counties in the South where the votes are now being cast in referenda to legalize liquor. Perhaps the most unusual dry county in America is Moore County, Tennessee, where the world-famous Jack Daniel's distillery is located. In and around Lynchburg, the Jack Daniel's warehouses hold millions of gallons of aging Tennessee sipping whiskey; the only problem is that no one there can legally buy a bottle or a glass. Some of the towns and counties in Kentucky's famous bourbon whiskey country around Bardstown are also dry.

The move to go from legally dry to legally wet merged entirely opposing forces in the attempt to protect counties' dry nature. As long as a county remained dry, the moonshiners were happily making and selling illegal "likker" without competition from legal wholesalers and retailers. Moreover, dry counties had happier Baptist preachers and their flocks, because righteous people felt that moonshine, though sinful, was less openly available to the general public. So moonshiners and Baptist preachers were on the same side, working hard together to prevent dry counties from becoming legally wet. Strange bedfellows indeed. Today local liquor laws are being voted out, and the Appalachian landscape changes with more liquor stores, bars, and restaurants serving up a wider variety of spirits.

Making and Hauling Moonshine

The ways of making moonshine are many, but I'll describe an old generic way with limited equipment. Making alcohol entails two fundamental processes: fermentation and distillation. The ingredients in moonshine are cornmeal, water, and malt. In later years sugar became an important addition. James Watt Raine in Kentucky in 1924 described this simple way of making moonshine:

> Corn is moistened and kept warm til it sprouts. It is then dried and carried, usually by night, to a little tub mill to be ground secretly, for grinding such corn is a federal offense. From this "sweet meal" a mash is made

> with hot water. To this some yeasty material [malt] is added and fermentation begins. For more than a week it must be kept just warm enough to ferment. . . . After it ferments, this "beer" is poured into the still, the fire lighted, and the vapors start through the copper spiral pipe. Cooled by the running water surrounding the "worm," the vapors condense into a liquid called "singlings" which drips or runs into a receptacle. After the "run" is finished, the still is emptied and the singlings poured in it to be distilled a second time into "doublings," which are thus freed from the rank oils and other impurities. While still warm, the whiskey is put into jugs and carried away for immediate sale.[85]

It's said that the best time to drink moonshine is as soon as it's cool enough!

Making moonshine required not only skill and experience but a wealth of material culture. In 1973 Edward Nichols, a former student at the University of Tennessee, explored the typology of moonshine stills and their geographic distribution. Using the vernacular of the time, Nichols described six still types and mapped them onto regions in Tennessee. Three of the still types, the silver cloud, copper pot, and Alabama black pot, are found in the Appalachian parts of Tennessee. The other three types, the box in Hickman County, the coffin in Stewart County, and the drum in southwestern Tennessee near Memphis, were beyond the Appalachian region.[86]

In 1969 Cocke County, Tennessee, led the state in moonshine production with an estimated 25,276 gallons, based on the number of stills seized. The following year the figure dropped to 16,363, indicating that the moonshine business was on its way down. Much of the change was attributed to more efficient raids by lawmen, in which seventy-eight stills were captured. Additionally, as dry counties became wet and wet ones became wetter with liquor by the drink, the decline became a trend. Ed Nichols summed it up this way: "If this trend continues, moonshining soon will vanish as an active part of American life and will be transformed from 'a legend in its own time' to a largely unrecorded segment of the American experience."[87]

Hauling moonshine has had as much of a colorful past as the story behind making the whiskey. The isolated, hidden creeks where folk made moonshine were some distance away from market cities. Hauling 'shine was a big enterprise, and the stories, movies, and songs like "Thunder Road"[88] made the venture of beating the law in souped-up old cars part of the smiling equation. Because of steep, twisting moun-

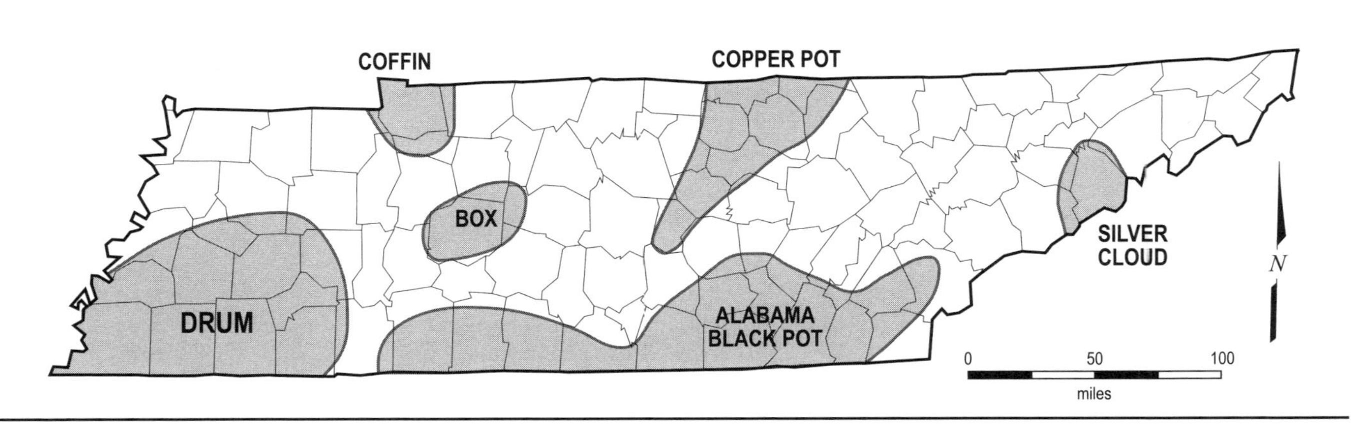

In 1972 Tennessee's moonshine stills came in different sizes, shapes, names, and geographical distributions. (Source: Nichols 1973, 53)

tain roads, specially modified cars were used as haulers. Typical haulers were either a big four-door sedan or a two-door 1940s Ford coupé with the seats removed, even the driver's seat. Our agile driver sat on a crate filled with whiskey jars. The suspension was stiffened to allow the vehicle to carry overweight loads. The engine was souped up to outrun the revenuer's federal-issue sedan. It was a cat-and-mouse game, a Robin Hood, Three Musketeers, fox-and-hounds contest. Both sides required cunning, skill, and attitude.

There is more truth than fiction behind the on-the-job skills of moonshine hauling learned by early stock car drivers and stars of NASCAR in the 1950s and 1960s. In the 1950s, when drag racing was just arriving in the Piedmont region, I went to drag races at Reedy Creek, about halfway between Winston-Salem and Lexington, North Carolina. One afternoon an old black four-door Buick sedan with a fair amount of mud on it was pulverizing every car in the field. Just about everyone drag racing in those days drove absolute stock cars, with little or no souping up. Just about everyone except the driver of this old black Buick. When the races were over, I sneaked over to take a peek at the car. It had no seats! When I first moved to Tennessee, my car needed brakes. After some inquiry I learned that I should go to a certain unmarked shop in North Knoxville where the mechanic was highly recommended. They told me, "He's real good. He puts brakes on all the local moonshiners' cars!" A high recommendation indeed, and they were right.

The serious side of moonshine, beyond its being illegal, is that it can be laced with toxic lead salts that have leached out of old automobile radiators used in the condensing step of distilling. Never have I heard anyone say, "This batch is bad, hits full of killer lead." But I have always heard, "Now this is GOOD STUFF, see the bead! I even KNOW the feller who made it." My advice to you, even if you like likker, is not to drink white lightning or moonshine or any other drink made this way. If you must have it, buy the legal stuff. It's a bit safer.

Marijuana and Methamphetamines

Of marijuana grown in the United States, 77 percent comes from five leading marijuana states: Tennessee, Kentucky, West Virginia, California, and Hawaii. Marijuana has been cultivated in clandestine plots in Appalachia since the 1960s. In 1998 the federal government designated sixty-five counties in Tennessee, Kentucky, and West Virginia an Appalachian HIDTA—high intensity drug trafficking area. The rugged rural terrain, favorable soils and climate, high unemployment, low family income, and traditionally approving culture create an environment where illegal activities and corruption flourish. Marijuana is the number one cash crop here. In 1999–2000, Tennessee's twenty-eight HIDTA counties in the mountains and on the Cumberland Plateau had 487,634 marijuana plants confiscated from 3,538 plots. Eighty-one indoor plots yielded 8,343 more plants. Tennessee's investigators arrested 1,134 people growing marijuana, and in doing so they found 250 firearms and 28 booby traps.[89] In 1999 Tennessee's illegal marijuana crop was valued at $687,226,000, compared with tobacco, Tennessee's number one legal crop, valued at $217,439,000. From Kentucky's twenty-six counties in the HIDTA, 463,373 plants were confiscated. In 1999 the value of marijuana in Kentucky was $787,088,000, exceeding all legitimate crops by $9 million. In West Virginia, 38,500 plants were confiscated from eleven counties, and the street value was $70 million, compared with legal agricultural crops worth $57 million.[90]

Methamphetamine is the new kid on the block for the manufacture, distribution, and use of illegal substances in Appalachia. Meth, now considered to be a twenty-first-century moonshine, is easily made, very dangerous, and highly addictive. The active ingredient in methamphetamine is ephedrine. Every meth recipe begins with over-the-counter cold pills containing pseudoephedrine. The pills are crushed and mixed with solvents, acids, salts, and lye and cooked to remove binders, buffers, and coatings and release the ephedrine. Heating highly volatile solvents such as acetone, ether, lantern fuel, white gasoline, alcohol, toluene, or benzene can result in horrible explosions, chemical fires, and toxic gas releases. Over the past few years, meth labs have sprouted up all over the region. In the United States in 2001, there were 10,547 meth lab incidents. For the same year, Tennessee led the South with 476 meth lab incidents—either a confiscation of illegal methamphetamines or a lab explosion.

Making a living in illegal ways seems to have a family heritage. If granddaddy made moonshine in the 1950s, then daddy grew pot in the 1970s, and now junior and other members of the family are operating meth labs. Two Tennessee counties, Cocke in the mountains of East Tennessee and Marion on the Cumberland Plateau just west of Chattanooga, were the major centers of moonshine in the 1950s. Moonshine was king. Fast forward to the 1970s and then to 2003, and the same counties have become major marijuana centers. One still finds the occasional moonshine still, but the illegal ways of making a living have progressed to marijuana patches and now to methamphetamine labs.

This journey has examined both customary and unusual ways of making a living in Appalachia. While part of it has had a folk presence, the ways of making a living reflect a rich and varied culture. We do not often experience, even vicariously, digging roots or making moonshine, or encounter logging and coal mining. Yet these are ways Appalachian people have sustained themselves and their culture as a way of life.

6

Foodways

Food, a universal necessity, represents a diverse range of diagnostic culture traits in its plant and animal varieties and in food preparation. Appalachian foodways are one of the very best ways to become acquainted with the region. Some years ago Andy Warner, a forestry expert and colleague from faraway Tasmania, visited me in Tennessee. Andy came to learn all the high-tech digital image processing that I could teach him in a one-on-one tutoring course using satellite images and remote sensing. It was Andy's first visit to the United States, and though only obliquely obliged, I was clearly excited about showing him around the area. On a hot Fourth of July, my wife and I took him on a long hike along the crest of the Great Smoky Mountains from Clingmans Dome to Silers Bald and back. Later we stopped for supper when I decided that this boy needed authentic local cuisine. Our supper consisted of fried chicken, pinto beans, turnip greens, corn bread, cat head biscuits and sawmill gravy, sweetened iced tea, and strawberry shortcake. Even the tea, a staple among British colonials, was new to him. He had never drunk iced tea presweetened until it resembled syrup—so sweet it hurt your teeth. As was the custom, the boiled pinto beans and greens were seasoned with fatback pork drippings to give them a distinctive smoky flavor. The beans came with a slice of onion; the greens had a little onion too, and if you added a shot of vinegar, it woke them up. Tabasco sauce was also an option on the beans and greens. At the end of the meal, Andy heartily thanked us for treating him to a most memorable meal. I thought the food was the most convincing introduction to Southern Appalachia that I could find.

The Hearth

The hearth has a central warmth and meaning. It is the place where the home fires burn, where a family warms itself and finds sustenance at the very heart of homelife. There is nothing so dead as a cold fireplace and

The hearth in an Appalachian home was a place of both warmth and sustenance. Tazewell County, Virginia. (Photograph by J. Rehder, 1977)

nothing so alive as a hearth with glowing coals and the aroma of food cooking over an open fire. In cultural geography we speak of culture hearths, those vital centers for groups of people who share common culture traits. Culture hearths function much as the hearth in a home does for the family. And so it is at the center of Appalachian folk life that we warm ourselves and relish a way of life by an open hearth.

Cooking, preparing food, and warming up leftovers are age-old tasks performed with a range of evolving culinary technologies. The earliest peoples in Appalachia cooked in open fire pits. Native Americans and early Europeans dug a hole in the ground perhaps three feet across and six to twelve inches deep. They cooked on sticks over the flames or placed husk-wrapped foods in the coals or on hot stones in the pit. Lacking metal pots, Native Americans boiled water by placing hot stones into water-filled skins or pottery. Cooking indoors meant that a cabin's stone or brick fireplace was both a heat source and the hearth for meal preparation. Early "stick chimneys" were made of mud and sticks. Later chimneys were limestone, sandstone, or brick. A typical fireplace had a horizontal iron bar hinged to one side of it. Pots hung on the bar with S-shaped hooks could swing in over the fire or swing out to cool and be handled. A Dutch oven, a large iron pot with a heavy lid, was placed directly in the coals and covered with a layer of coals to achieve an even temperature all around. The Dutch oven, with the emphasis on "oven," baked breads, biscuits, and corn bread and roasted meats. Heavy iron skillets or frying pans also were placed in the bed of coals. Frying pans in sizes from six to ten to fourteen inches are the most versatile of all open hearth utensils. For example, on a single ten-inch iron skillet I can prepare pancakes or French toast, brown some light bread for toast, fry bacon, sausage, or eggs, and make sawmill gravy all in one sitting without even washing the pan! After the meal, folk cleaned pots and pans with corncobs and sand.

In the late nineteenth and early twentieth centuries, open hearth cooking was replaced by the cast-iron wood-burning cookstove. Not only did the stove bring improvements to multipot cooking and oven baking, the mere existence of the appliance also meant a radical change in the architecture of the Appalachian folk house. A kitchen, either separate or attached, became an appendage to the house. There had been kitchen appendages on larger houses, especially much earlier plantation houses in the South and East,[1] but the cast-iron cookstove brought the kitchen to Appalachia. Even after other fuels and technologies entered the region in gas and electric stoves, for most families cooking traditional foods in traditional ways and warming themselves in the kitchen remained a way of life. The kitchen stove became a surrogate for the "hearth" once provided by the fireplace.

Cat Head Biscuits and Sawmill Gravy

As a child growing up with two sets of grandparents—one pair German–North Carolinian, the other English-Appalachian–Tennessee Cherokee—plus a great grandfather from Ohio who had been a longtime logger, I was blessed with solid southern foods mixed with other odd things. My mother often spoke of the cat head biscuits and sawmill gravy she ate as a child, and I too remember seeing my mother's mother from the English-Appalachian heritage create this marvelous combination of biscuits and gravy. Cat head biscuits and in particular sawmill gravy originated in Appalachian logging camps in the late nineteenth and early twentieth centuries.

But before I describe these good things, let me tell you how one of my students discovered them. In the early 1970s, I was teaching an undergraduate senior seminar for geography majors. As a field man, I knew that the students needed an experience in the field that lasted overnight or even for several days. So I took them to the University of Tennessee's geology field camp, an authentic but decrepit World War II prisoner of war camp building resembling Stalag 17, perched high on Walden's Ridge on the Cumberland Plateau. The next day, we descended into Sequatchie (Cherokee for "possum") Valley to the town of Pikeville, Tennessee. I handed each student a small white card that listed a pertinent question to investigate while in town. To one young man I handed a card that read, "What are cat head biscuits and sawmill gravy?" The students scattered, fanning out all over Pikeville. About a half an hour later, my "biscuit man" returned with a scared look on his face and tears welling up in his eyes. I asked, "Where did you go and what happened?" He replied, "Well, . . . I . . . went to the feed store and asked an old man there, 'What are cat head biscuits and sawmill gravy?' just like you told me to do. Only the old man just growled at me and said, 'Son, if you don't know, HELL, I ain't goin' to tell you!'" And with that our lad left the feed store dejected and thoroughly upset. To calm him down, I said, "Why don't you just go on down to the café. Get a little something to drink and maybe calmly run your question by them down there." After a sufficient amount of time, I decided to go check on him. He was sitting on a red upholstered stool at the counter. As I drew closer, he looked up and, with fresh and quite different tears in his eyes, pointed to his plate: "This", he said proudly, "is cat head biscuits and sawmill gravy!"

Still curious? Cat head biscuits are flour biscuits, light and golden brown; they can melt butter and melt in your mouth just as fast, and they come in the unequivocal size and shape of a cat's head. Sawmill gravy is a white gravy made from sausage drippings or bacon grease, flour, and milk. Some salt and a lot of black pepper make this thick white gravy a favorite of southern Appalachian breakfast connoisseurs—loggers, seed and feed men, café folk, and just about anyone else, especially a frightened young man thirty years ago in that small Sequatchie Valley town.

CAT HEAD BISCUITS

- 2¼ cups flour
- ⅓ tsp. baking soda
- 1 tsp. salt
- 2 tsp. baking powder
- 5 tbsp. lard
- 1 cup buttermilk

Sift and mix dry ingredients; blend in lard. Add buttermilk and mix. Take a ball of dough the size and shape of a large egg and pat it flat in your hands (about 1 or 2 inches thick). Don't use a rolling pin and biscuit cutter, or they won't be true cat heads! Bake in a woodstove oven at 350° for 10 minutes or in a modern gas or electric oven at 475° for 5 to 7 minutes or until golden brown on top.[2]

Sawmill Gravy (1)

1 lb. mild sausage	1 cup milk
2¼ tbsp. flour	½ tsp. black pepper

Fry sausage in a large iron skillet; crumble and remove browned meat and most of the drippings. Keep 2 tsp. drippings in skillet. Slowly add flour to drippings and stir until smooth. Slowly add milk and stir until smooth. Stir in black pepper and as much of the browned sausage particles as you want: all, some, or none. Serve on biscuits to four hungry folks.

A variation is to make this white gravy with the drippings after frying chicken.[3] "There may be no better gravy in the Western world than that made in a black skillet recently vacated by crisp pieces of fried chicken," proclaimed John Egerton.[4]

Sawmill Gravy (2)—"Life Everlasting"

1 tbsp. bacon grease	½ tsp. salt
3 tbsp. white cornmeal	2¼ cups milk
Dash of black pepper	

In a large iron skillet, heat the bacon grease; add cornmeal and salt. Stir, then add milk, bring to a boil, and stir to thicken. Add black pepper.[5]

This gravy is said to have come from the Little River Lumber Company logging camp at Tremont, Tennessee. According to legend, the camp cook ran out of flour, so he substituted white cornmeal. When the loggers complained that the gravy was a lot like sawdust, they started to call it "sawdust gravy," which was modified to "sawmill gravy." Whether true or not, the story is interesting, but the point is that the cultural connection between logging camp cooking and Appalachian folk food is unmistakable.

Corn: That Marvelous All-Purpose Food

Corn or maize, a Native American staple, became one of the most widely used foods in the Appalachian diet. Native Americans preferred to gather fresh corn in August and roast the ears in their shucks in the hot ashes of a cooking fire. Early European settlers borrowed this easy technique, referred to as roastin' ears, but later they began to boil fresh husked corn on the cob. Whether corn on the cob was roasted in husks or boiled, people continued to call it by its ancient name, "roastin' ears." Native American preparations of corn also included suppone, appone, samp, rockahominy, and succotash.[6] You may recognize the "pone" types that led to corn pone, the "rockahominy" that became hominy, and the well-known succotash, made of boiled corn and lima beans; all are Native American contributions.

Corn can be roasted, boiled, steamed, stewed, fried, baked, and made into perhaps the widest variety of dishes among all vegetable foods on the continent. You can eat it as roasted corn, corn on the cob, creamed corn, corn pudding, corn mush, corn bread, ash cake, griddle cake, hoecake, fritters, corn pone, crackling bread, spoon bread, corn muffins, corn dodgers (corn balls boiled in water, a bit like dumplings), hominy (big hominy), grits (little hominy), cornmeal for frying, corn bread dressing, and corn liquor, better known as white lightning or moonshine (see chapter 5).

Grits

Of all the many ways of preparing corn, grits has to be one of the more peculiar. There is no such thing as a grit. Grits are white, grainy, mushy, wet, and warm and have little if any flavor. They appear on your plate as an unattractive pale blob, so one way to eat them is to put butter, salt, and a lot of black pepper on top to give them character—or at least a little color and flavor. Adding gravy—either sawmill or redeye gravy—is an exceptional way to make grits palatable.

Native Americans boiled corn kernels and made various kinds of hominy, the ancestor of grits. In the initial preparation of hominy, the ingredients are whole dry corn kernels, water, and lye. Yes, lye, made from wood ashes. Folk collected fireplace ashes in ash hoppers or some container such as a wooden bucket or barrel. As ashes accumulated, they poured a little water on them, and eventually lye seeped out of openings in the bottom of the container. The lye was collected and stored for making lye soap and hominy. To make hominy, folk took a large iron pot, added a lot of water and a little lye, and filled it with dry, hard shelled corn. Then they boiled the mixture until the hulls slipped from the kernels. Next they dipped the corn kernels from the boiling pot and rinsed them through five or more waters to be sure the lye and the hulls were completely removed. Next they put clear, clean water in a pot and kept boiling the kernels until they became swollen and tender. The results were large corn particles that hardened again when dry but would remain soft when kept wet. People ate hominy in several ways; some ate it fresh out of the pot; others would take some out of the pot and fry it in bacon or sausage grease. In Appalachia and other parts of the South, this large-grained substance is still called big hominy. Little hominy, just another name for grits, is dried big hominy that has been pulverized into particles the size of coarse sand and stored in a dry place.

Grits go a long way whether you like them or not, and it doesn't take much time, skill, or ingredients to make a pot. When you want grits, you measure out a portion, perhaps a cup, of dry grits and put it in three cups of boiling water along with a pinch of salt. After five minutes the grits absorb water and become softer and ready to be eaten by four really hungry people.

Corn Bread

The Appalachian family's access to mill-ground cornmeal allowed people to feed themselves quite well on a bread made from corn. Native Americans made several kinds of fritterlike corn breads called pone, suppone, and appone. They looked like flapjacks, cooked on flat stones or in the ashes of an open fire; essentially they were ash cakes. European settlers adopted these forms and developed other ways to make breads from corn: cornpone, ash cake, hoecake, fritters, griddle cake, crackling bread, spoon bread, corn bread, corn muffins, and a few others. The evolution of corn breads proceeded from appone, suppone, and corn pone to ash cake to hoecake, all cooked in the ashes of an open hearth; then came heavy, flat fritters or griddle cakes made on iron griddles and light, thick corn breads baked in iron skillets and Dutch ovens. Afterward came the "hush puppy" type of corn balls, fried in oil, and corn dodgers (boiled corn ball dumplings). Then there's spoon bread, a soufflé of corn plus eggs that is so light it must be spooned onto your plate.

Ash cake, which resembles cornpone and hoecake, is a flat corn cake baked on an open hearth. Ash cakes might be cooked on a swept portion of the stones or wrapped in a cloth or cabbage leaves and placed in the ashes; some people placed the dough directly in the ashes. Hoe cake, once thought to be cooked on the blade of a farmer's hoe, was named for the hoe blade used as a spatula to scoop the flat cakes from the hearth.[7]

Old Style Cornpone

2	cups white cornmeal	½ to 2	tbsp. bacon grease
¼	tsp. salt	2	cups boiling water

Preheat oven to 425°. Grease baking tin. Sift meal into a bowl with the salt and add bacon grease. Add boiling water slowly, stirring to break up lumps. Let stand to cool and expand. Form into egg-shaped pones, place on greased baking sheet, and put in oven. Bake 20 to 30 minutes or until golden brown.[8]

The greatest change in the evolving pattern of corn bread came when iron cookware was introduced in the form of griddles, Dutch ovens, and, especially, wide and deep skillets. As people's living standards began to rise, so did corn bread. Folks began to add lard, eggs, baking powder, and soda to create a new form of risen corn bread, one that, as Joseph Dabney describes it, "had evolved into the bread that is legend: cornmeal mixed with salt, rendered fat and buttermilk, then baked in a preheated iron pan."[9]

Buttermilk Corn Bread

3	tbsp. flour	1	cup buttermilk
1¼	cups cornmeal	2	small eggs
1	tsp. salt	2	tbsp. shortening (lard)
¾	tsp. baking powder	¼	tsp. soda

Mix flour, cornmeal, salt, and baking powder in a large bowl. In another bowl mix buttermilk and eggs and beat. Melt shortening and add to the dry ingredients. Add soda, dissolved in some water. Preheat greased skillet or baking pan. Mix everything and pour into the hot pan and bake at 425° for twenty to twenty-five minutes or until golden on top and brown on the edges. Makes six servings.[10]

Hogmeat: Everything but the Oink

In much of the South, and in Appalachia in particular, hogs were a major source of meat, if not the primary one. For most of the Appalachian settlement history, hogs roamed freely through oak forests foraging on mast—acorns and other wild foods. At times, however, hogs were penned so they could be fattened on corn and other feeds. Fewer people today keep hogs for their own use, but Todd Fox, one of my undergrad-

Hog-killing time commences when cold weather settles in for the winter. DeKalb County, Alabama. (Photograph by J. Rehder, 1977)

uate geography students in 2001, gives us a contemporary and very graphic account of hog keeping, killing, and meat preparation. Fox's firsthand knowledge of the steps he and his grandfather take in preparing a hog for eating is a treasure trove of details that only one who has done it could know about. Todd Fox reports:

> My grandfather usually kept the hogs in their own lot or sty for a year, or as long as he boarded them. Others let their hogs roam freely on their land and forage for food. In other areas, elevated sties were used to house the hogs for fattening a few weeks before slaughter. Feed is another hot topic in keeping hogs. Since the feed affects how the meat tastes and how the fat cooks, personal preferences determine what the animal eats. Many of the foraging hogs eat nuts and wild berries for their main food source, while boarded hogs are fed grains, corn, and slop. My grandfa-

ther believed that sweet corn and slop made the meat sweeter and the fat easier to render.

The choice of slaughter time is pretty consistent because the weather must be cold for curing and keeping the meat. For my grandfather, this time fell close to Thanksgiving, when the weather has turned "cold to stay." Once the weather was suitable for slaughter, my grandfather and his family would tag two or three hogs for slaughter. He was very particular about choosing the hogs, using weight and age to determine quality. He preferred a young hog weighing between 175 and 250 pounds. Once the hogs were chosen, the real work began.

My family would meet at 5:00 a.m. on the day of slaughter and have a big breakfast. . . . One of the hogs would be picked out and shot between the eyes with a .22 rifle. While the hog was dazed, its jugular vein was cut with a very sharp knife and it was allowed to "bleed itself." This "bleeding" is critical, for it rids the meat of a sour taste. Once the hog is dead, its feet are tied together with a large board or a log placed longways between its legs, and it is carried to the scalding area. There the hog is dipped, or covered with scalding water to loosen the hairs. The hairs are scraped off, and the process is continued until all the hair is gone. But you must not scald the hog too much or the skin will soften and be hard to scrape. After the hair is removed, small incisions are made behind the Achilles tendons on the back legs so a stick or rack can be inserted through the holes. The hog will then be lifted and hung so it can be cleaned and all the organs kept.

After the hog is secured, two incisions are carefully made in its midsection so its organs can be extracted. The first incision is made through the skin from the throat to the anus, taking care not to cut through the abdominal wall. The second incision is made on the same path but cutting through the abdominal wall and exposing the organs. The large intestine is cut from the anus, and the organs are allowed to fall into the "gut bucket" directly below the carcass. The gallbladder is taken out so as not to spoil the other organs. The liver, heart, kidneys, and lungs are soaked in water for later use. Once all the organs are extracted and the carcass is cleaned, the head is cut off at the base of the neck and set aside for use in later products.

Once the carcass is cleaned to inspection, it is taken down and carried into the working room (the kitchen). The work is now at the point where it can be divided between groups. The men begin on the second hog while the women split into two groups, one working the carcass while the other works the organs. The organs are worked separately, starting with the valuable ones. The carcass is cut on both sides of the backbone, releasing the ribs and exposing the chops, tenderloin, and fatback. Each section is cut on the joints, releasing the hams and shoulders and leaving the middle meat around the ribs to trim. All fat is trimmed off the hams, shoulders, and middle meat. The hams and shoulders are cleaned and prepared for curing, and the middle meat is sectioned into bacon and ribs.

Once all the hogs are slaughtered and worked, the men begin helping the women with the working of the meat. The hams and shoulders are placed in saltboxes and covered. The bacon is separated from the middle meat and taken to smoke and cure along with the fatback. The organs are trimmed and worked over. The ribs and the organs are usually canned for later use in stews and recipes. The heart is boiled with the tongue and canned. The lungs (lights) are boiled with the liver and pureed for pudding and other uses. The intestines are cleaned, usually in the creek, and put up in a jar of salt water for use four or five days later. The stomach is worked much like the intestines. It is cleaned and soaked for use when the intestines are cooked. The head is the next important order of business. The eyeballs are removed, and the head is boiled until the meat and skin begin to fall off. It is then deboned and poured into a pan for cooling. This produces the headcheese or souse that we know and enjoy today. The feet and ears are thoroughly cleaned and placed in a jar of vinegar for pickling. All the rest of the loose meat and fat is used for sausage and lard rendering.

. . . Each region of the South is unique in its techniques for the slaughter and production of pork. So please do not take my information and strike up a hog slaughter conversation because you may find that the topic is serious business. . . .

These processes are traditions that have been handed down from generation to generation in my family. . . . I just hope the traditions that have existed in my family will continue through future generations. Although this is a very unlikely dream because of the accessibility of processed pork, I hope I can instill the same kind of self-sustaining abilities in my children that my grandfather and family have instilled in me.[11]

Hog meat was treated in various ways: some was eaten fresh, some was canned, but most was salted and smoked in a smokehouse. The best way to preserve pork was to place it in a salt trough or box, rub it with salt, and allow it to absorb as much of the preservative as possible in six to eight weeks. Then the meat would be hung on loft or ceiling joists in the smokehouse and a small, smoldering fire of hickory chips would waft smoke over the meat for the next week or more. This process had less to do with preservation than with giving the meat a smoky flavor.[12] The salted and smoked meat would then remain stored and often locked in the smokehouse until needed for meals. Country ham, that extremely salty smoked pork, was rinsed in fresh water to rid it of some of the salt, then fried in an iron skillet. Bacon and fatback were fried directly in the skillet. The drippings from fatback were added to just about every boiled vegetable you can name: all bean types, all cooked greens, potatoes, even corn.

Vegetable Vittles Other Than Corn

Vegetables are eaten fresh or else dried or canned for later use. A wide variety of beans dominates the Appalachian vegetable garden. There may be pinto beans, white beans, navy beans, October beans, lima beans, green beans, and snap beans in addition to a number of pea varieties. Beans are a solid starch and protein staple because most kinds can be easily dried and stored. You may have heard of "leather breeches," which are simply beans hung up on strings and dried in their hulls. Beans are good to eat whether they are fresh in the summer or dried and cooked in the winter with fatback and water, salt, and pepper. When I first moved to East Tennessee in 1967, I was struck by all the kinds of dried beans that people ate. A local favorite was "metwurst and beans": metwurst sausage, a German component, and boiled white beans with some onion on top. With a slab of corn bread and butter and sweetened iced tea, "met 'n' beans" was a low-cost, heavy, and filling meal.

Just as the aboriginal trilogy of maize-beans-squash became staple foods for the ancestral Native Americans, variant cultivated ensembles became important to Europeans. A typical twentieth-century summer mountain garden produced sweet corn, a variety of beans, peas (black-eyed peas, green peas, field peas), Irish potatoes, sweet potatoes, onions, tomatoes, cucumbers, squash, and okra. Notice I said summer garden, because some people also grew a winter garden with root crops of potatoes, carrots, and turnips. The winter garden also had cabbage and other leafy greens that were harvested before frost.

Greens, the common generic name for a half dozen or so green leafy vegetables, were dominated by turnip greens, cabbage, and the wild food poke sallet. Sometimes people grew collards but to a lesser degree; col-

Ramps and eggs, griddle cake corn bread, and slab bacon make a pretty good meal. (Photograph by J. Rehder, 1999)

lards were more common in the Deep South, farther south and east of Appalachia. The unusual one is poke sallet, because the roots of pokeweed are poisonous! The young, green leafy parts are edible, and the berries are used for folk medicine, but don't eat the roots.

Ramp, that strong, odiferous garlic-like leek, a relative of the onion, is eaten in the spring, primarily as a tonic. Fried up with eggs, ramps are palatable and relatively safe socially. See chapter 8 under festivals, where I cover the story of the ramp in its ecological and festive form.

Fruits and Nuts

Mountain fruits and nuts came from a wide variety of sources. The first tree fruit that comes to mind is mountain apples, but pears, cherries, plums, and peaches were also domesticated early. Peaches were introduced to warmer southern sites at lower elevations such as in the upcounty Piedmont areas of South Carolina, where they have flourished as a commercial crop. Wild fruits in Southern Appalachia were persimmons, pawpaws, grapes, and crab apples. Nut trees were the familiar black walnuts and hickories. Pecan, a popular southern nut tree with Texas origins, was too cold sensitive for the mountain South. Wild berry plants provided blackberries, huckleberries, blueberries, raspberries, elderberries, and wild strawberries. In 1776 William Bartram described the valley of the Little Tennessee River as "a vast expanse of green meadows and strawberry fields."[13] Domesticated strawberries, however, became a twentieth-century cultivar in warmer valley sites such as in Rhea County, Tennessee, where the Strawberry Festival has remained a major week-long event every May, although the strawberry business has declined and nearly disappeared.[14]

Mountain Sweeteners

We sprinkle the sweet white crystals over our breakfast cereal or spoon them into hot coffee and think nothing about their origins. Granulated cane sugar, either in the refined white form or in its raw brown state, was not a part of Appalachia's food system until the twentieth century. Even then, granulated sugars were imported from subtropical places like Louisiana and Florida, far from Appalachia.[15]

What did people in Appalachia do for sweeteners? At first they gathered honey from wild bee colonies found in hollow trees in the wilderness. Later they kept domestic bees in bee gum hives made from hollow trees, and still later they kept them in box hives, as they do now. Beekeeping is a source of income for some farmers, and a wide variety of honeys are produced: traditional clover honey, wildflower honey, apple and peach orchard honey, among others. But the single most important honey that is directly linked to Appalachia is sourwood honey. Up in the mountains at about 3,000 to 5,000 feet, an ecological niche supports a small flowering tree called sourwood (*Oxydendrum arboreum*). While sourwood ranges from southwestern Pennsylvania to southern Georgia, the tree reaches its maximum size on the western slopes of the Great Smoky Mountains. The late-blooming (July to August) small white flowers of the sourwood tree produce a nectar that bees collect. They bring it to the hive and make a distinctive dark, earthy, slightly acidic honey with a bite to it. No matter where beekeepers live, when the sourwood trees are in bloom, keepers haul truckloads of beehives to the mountains, where they set the bees out to collect the nectar and produce this precious golden liquid.

Mountain sweetenings often centered on syrups made from sorghum canes. Blount County, Tennessee. (Photograph by J. Rehder, 1967)

The other mountain sweetener is sorghum syrup. Cane sorghum, a sweet grass with African origins, is similar to sugarcane only in that a sweet syrup can be made from it. Unlike sugarcane, which has granulated sugar crystals as the primary product and molasses, rum, and bagasse as by-products, cane sorghum produces only syrup. There is no such thing as "sorghum molasses." The dark, sticky syrup called "molasses" is a by-product of sugarcane processing. You should never confuse the terms, but people do so all the time.

Making sorghum syrup begins with ripe cane sorghum stalks that stand ten to fifteen feet tall and have gone to seed in late summer or early fall. With seeds and leaves removed, the stalks are fed between the two rollers of a cane mill. Motive power is supplied by a horse or mule that walks in a circle pulling a long sweep pole. One end of the pole is attached to the mule, the other end to the top cog axle of the two-roller mill. As the rollers turn and crush the canes, yellow-green sorghum juice is extracted and collected in a bucket. All day, buckets of juice are collected, to be placed in vats under a tin-roofed shed nearby. Here in the midst of woodsmoke and sweet sorghum mist, the juice is boiled to evap-

Sorghum syrup is made by carefully boiling cane juice until it is reduced to syrup. Blount County, Tennessee. (Photograph by J. Rehder, 1967)

orate the water. People with skimmers and paddles work the juice along a maze of metal pathways in the vats from cooler places to hotter places, reducing the moisture and removing impurities from the green, scummy liquid. At a magical point when the consistency is just right, the syrup maker declares that it is time to begin drawing the syrup off into clean buckets, cans, or jars. The syrup is put up, canned in glass jars, and used throughout the year as a special sweetener.

Both mountain sweeteners, honey and sorghum syrup, are primarily eaten on biscuits. Poke a hole in your biscuit, pour in honey or sorghum syrup, and enjoy. However, some people also use honey to sweeten tea or coffee, put on grits, or substitute for sugar in baked goods such as apple pie and brownies. Sorghum syrup is also used in baked goods, particularly gingerbread.[16]

Meal Ensembles

The meals we eat in twenty-first-century Appalachia hardly resemble the meals we had even a decade ago. Fast food, cereals, and frozen pizzas have invaded our culture and region to the point that meal planning is almost a relic in our society. Dietary cautions about the amount of lard in our diets and the absolute necessity of eating alfalfa sprouts and tofu washed down with soy milk bother me. Where are the good old days not of eating for health's sake but eating the meals my grandparents and their grandparents ate? From 1830 to 1942, Appalachian meals were much the same. The following is a digest of several traditional meal patterns. They are the typical, not the extravagant; they are the normal, everyday fare that speaks of ethnic meals eaten by most members of the culture.

Breakfast

Sizzzzzzzzzzzlllll . . . pop . . . sizzzzzzzzzzzllllll . . . pop . . . pop. Is that bacon frying in an iron skillet? The meats of breakfast can be pork sausage, either in patties or in links, bacon, or salt-cured hickory-smoked country ham—the kind that calls for redeye gravy, not sawmill gravy. However, sawmill gravy still takes precedence. Eggs come next—fried in hot bacon or sausage grease—sunny side up, over easy, over well, or scrambled, but never as an omelet. Grits, made from corn and boiled until they are mush, ride next to the eggs. Some people put gravy—either sawmill or redeye gravy—on their grits in the same way one does with mashed potatoes. Some people mix their eggs and grits together; some come up with a grits-gravy-eggs mixture. For condiments, salt and pepper on grits is normal, though some odd people put butter and sugar or honey on their grits.

In the summer, a typical Appalachian breakfast would have a fresh slice of tomato next to the eggs or meat. This unusual trait somehow has connections to the British Isles, for folk there also offer a slice of fresh tomato on the breakfast plate. I've had this treat in many places in Great Britain, from London, to Warminster near Stonehenge, to Stratford-upon-Avon, to Derby, to Wales, usually in small bed-and-breakfast inns. An alternative to the fresh red tomato slice is fried green tomatoes, made famous by a movie of the same name. Green tomatoes dipped in cornmeal and fried in bacon grease can be a summertime feature on your breakfast plate in homes from Virginia to Alabama. Sam Hilliard, my good friend and the author of *Hog Meat and Hoecake*,[17] was visiting us in Tennessee. I decided Sam needed a good Appalachian breakfast, so I took him to Green's Restaurant on Seventeenth Street in Knoxville. (Green's is no longer there; it was swallowed up by a highway interchange.) The breakfast came, and to my delight, a large red tomato slice was on each plate of eggs, bacon, grits, and cat head biscuits. Sam, who is not known for loud exuberance, exclaimed, "Well, I'll be damned, there's a tomato on my plate!" And with that, before he took a single bite, he rushed the plate over to a sunlit window and took a photograph of his breakfast, featuring the bright red tomato slice. Everyone had a good laugh, and I'm certain the patrons and especially the folks working at Green's talked for years about that nutty professor who took a picture of his breakfast!

Breakfast bread is easy—cat head biscuits with butter and either honey or sorghum or cane syrup. A special treat is to poke a hole in a cold, hard biscuit late in the day and pour sorghum syrup, cane syrup, or molasses in the hole. My wife, who grew up eating this way, says it can be as good as ice cream. Breakfast beverages are coffee, milk, buttermilk, or water. Hot tea just doesn't happen here, even though sweetened iced tea is a mainstay at dinner and supper. Orange juice is a twentieth-century arrival and not a common element on the traditional Appalachian table.

Dinner (Elsewhere Called Lunch)

Earlier I described a dinner of metwurst and beans, but this is not a widespread pattern in Appalachia. A typical noontime dinner would have corn bread or perhaps some of those cat head biscuits left over from breakfast or a fresh batch of hot ones. Meat could be pork—a slice of ham, pork chops, roasted pork—or maybe fried chicken; however, fried chicken customarily was reserved for Sunday dinner. In summer there were fresh vegetables, but in winter dried or canned vegetables—corn, beans, greens, or stewed tomatoes—became part of dinner. If a dessert was offered, it might be fresh or dried fruit or an apple pie, or even a cake, but only on Sundays or birthdays. Beverages could be coffee, iced tea, milk, buttermilk, or water.

Supper (Elsewhere Called Dinner)

Since the noon meal was the big meal of the day, the evening supper was usually smaller and often consisted of leftovers from that midday dinner. If a new meal was to be prepared, it was similar to the noon meal, with one meat, perhaps pork chops or chicken and dumplings or squirrel and dumplings, vegetables from the larder, and the same bread-beverage-dessert pattern from dinner. An old-time favorite, though certainly not one of mine, is a light supper consisting of a tall glass of cold buttermilk and slabs of corn bread. At other times, however, families prepared grits and eggs and had an abbreviated breakfast for supper.

The folk foods of any culture can be just as important as the universal diagnostic traits of language, religion, clothing, and housing. Food, as one of my students discovered, is a cultural marker, whether it is "soul food," "Cajun," "Chinese," or "Mexican" ethnic cuisine.[18] Each in its own way says much about the culture identified with it. Sidney Saylor Farr from Stoney Fork near Pine Mountain in southeastern Kentucky moved away for a while and remembered:

> Mountain food and how it is cooked is very much a part of this sense of place. Ask any displaced Appalachian what he misses most about being away from the mountains and he will probably talk about soup beans, corn bread, sallet greens, fresh milk and butter, eggs, country ham, "and homemade biscuits every morning of the world." Some will speak of the joys of hunting wild game and birds. The women talk of missing homey things; sitting on the front porch or in the yard after all the chores are done and listening to the night sounds, watching the moon come up, or going to a church supper. They may also talk about the kitchens back home, the feeling of warmth from wood-burning stoves, the smell of coffee simmering on the back and biscuits baking in the oven, and the family gathered around the kitchen table to eat and talk.[19]

7

Folk Remedies and Belief Systems

The good Lord has put these yerbs here for man to make hisself well with. They is a yerb, could we but find it, to cure every illness.

UNNAMED EAST TENNESSEE
WILD PLANT COLLECTOR

Introduction

Treating the folk culture of self-medication and Appalachian religious beliefs in the same chapter might seem like mixing contradictory topics. But folk remedies, with their rich and mysterious ways, offer the belief, or at least the possibility, that the concoctions will work. Religious belief systems, whether organized denominations or independent groups, work from the position of faith that they too will work for the believer.

Throughout this book, I have discussed the fact of isolation in the geography of mountain life in Appalachia. And isolation from hospitals, from doctors, and from pharmacies certainly fostered a reliance on traditional folk remedies. We live in a generation where we go to a hospital emergency room for the slightest cut or sprain. In Appalachian folk culture, some people still have to make do when it comes to injury, illness, and death.

Folk Medicine

Folk medicine, folk remedies, and home remedies are all much the same thing: local people using native biota and traditional knowledge to cure their aliments, whether by superstition or by accidental science. My term "accidental science" implies that the native product is effective because it is chemically good medicine, not merely because of a superstitious belief that it might work. Certainly if a native product is both medically effective and psychologically effective, the combination can add up to a stronger "medicine" that may yield a better outcome.

Since aboriginal times, more than one thousand plants have been used for medicines in the United States and Canada. Native American contributions to folk medicine were so important that in 1812 Peter Smith wrote a pharmacology book titled *The Indian Doctor's Dispensatory.* Contacts between Native Americans and Europeans fostered significant exchanges in medical knowledge. As Native Americans introduced settlers to medicinal plants, they also instructed them on plant use.[1] This was much more efficient than if Europeans had collected unknown New World plants and had taken them back to the Old World for "testing."

There are two perspectives on medicinal plants. The first is undeniably folk because it blends folk medicine and superstition into a brew that users believe will make them well. Some remedies are hard to swallow both physically and figuratively. For example, a cure for asthma is to roll up a handful of spiderwebs and try to swallow them! To loosen phlegm from chest congestion, melt the fat of a skunk and eat two or three spoonfuls. For colds, drink the brine from a jar of sauerkraut, or drink lamb's tongue and whiskey tea.[2] These are just a few folk remedies in the region. The second perspective is the economic notion of "root digging"—collecting roots, herbs, and other native biota for resale as a part-time folk commercial venture. In chapter 5 I discussed this folk way of making a living.

Remedies

Two strategies come to mind when studying folk medicine. The first looks at the disease, ailment, or injury, then proceeds to possible remedies. The second looks at the medicine first, then shows how it is prescribed to treat symptoms. There are many ailments that may call for a folk remedy; likewise there are many remedies for each ailment or injury. Since few ailments call for a single remedy, remedies may seem to conflict. For example, the first Foxfire book lists sixteen "cures" for warts, nine for worms, eight for toothache, nine for headache, eleven for earache, sixteen for colds, twelve for burns, and nine for asthma.[3] Rather than listing all the remedies or all the ailments, I have chosen folk medicine remedies that illustrate the variety of unusual ways of "curing" ailments or injuries. I have no idea if these so-called cures work at all, but that is not the point. As long as the folk who use these methods find that they work for them, then I can only report the remedy rather than make an unstudied evaluation of its efficacy.

The ecology of mountain life presupposes that people became knowledgeable about medicinal plants and that gathering wild plants and herbs was a natural part of the culture. The plant materials used for making folk medicines included roots, bark, wood, sap, and leaves. Roots came from blackberry, ginger, ginseng, goldenseal (yellow root), ironweed, jimsonweed, pokeberry, raspberry, sassafras, snakeroot, and willow. Bark came from dogwood, hickory, red oak, sweet gum, willow, and wild cherry trees. Among other plants that found folk use were beech trees, cockleburs, dandelions, dog fennel, dogwood berries, mullein, pennyroyal, pokeberry, ragweed, and sedgegrass.[4] Besides natural biota, substances such as kerosene, coal oil, turpentine, and whiskey have been used to treat myriad ailments. The following list of remedies illustrates a wide range of substances and applications. It is not intended to be comprehensive, nor is it a proven prescription for ailments. Its sole purpose is to demonstrate the types of folk remedies used in Appalachia.

Asafetida bags. An asafetida bag is a small cloth sack filled with pungent garlic, onion, peppers, or ramps that is tied around the patient's neck. Asafetida bags were used to treat asthma, colds, and other respiratory problems. The aromatics in pine tar and other substances included in the bags worked much like today's camphor rubs. The foul-smelling bags were also believed to ward off diseases.

Blackberry juice. Blackberry juice was to be taken for diarrhea and dysentery.

Boneset (*Eupatorium perfoliatum*). The boneset plant was widely used by Native Americans and Europeans after initial contact. Its leaves were brewed into a tea for flu, colds, for reducing fever, and as a laxative. The tea had to be drunk cold; warm tea made the patient even more ill. In large doses boneset is an emetic and a purgative, causing vomiting. Boneset tea was also used for insomnia.[5]

Ginseng. For more than two centuries, ginseng (*Panax quinquefolium*) has been gathered in Appalachian forests for sale to middlemen who have supplied the root to Chinese markets. The root, a staple in Chinese medicine for over five thousand years, has been used for many things, but it is most often mentioned as an aphrodisiac, an energy booster, and an all-around tonic. The few folk Appalachian users of ginseng historically have taken it for arthritis, for colic, and for headache, with the ginseng roots tied in a cloth around the patient's head. Since the 1980s, however, commercially sold products made from American, Chinese, Korean, and Siberian ginseng (some of which are not ginseng at all) have appeared in American popular culture. You can buy ginseng in capsule, tablet, and liquid form at Wal-Mart, in mom-and-pop convenience stores, and in national drugstore chains. Not all are genuine American ginseng, but many are taken by people of all cultures. See chapter 5 for more information on ginseng habitat and collecting.

Goldenseal. Goldenseal (*Hydrastis canadensis*) is also known as yellow root, orange root, wild curcuma, turmeric root, Indian dye, and Indian paint. It is commonly found in eastern North America from Canada to the southern Appalachians. In Appalachia it is especially centered on the Allegheny Plateau region of West Virginia, Kentucky, and parts of Ohio. Goldenseal is found in much the same habitat as ginseng, on the north faces of slopes in the moist soils of shady woodlands and along streams. Goldenseal was widely used by Iroquois, Micmacs, and Cherokees as a tonic and to make a yellow dye. After the Civil War, it became a leading patent medicine. Medically, goldenseal has been used as a tonic or blood builder, for dyspepsia, as a remedy for vomiting and other stomach ailments, as a laxative, and externally as an eyewash, for itchy skin, and for sores.[6] A popular goldenseal and echinacea combination is believed to boost the immune system and ward off colds.

Honey. Honey is used for arthritis, asthma, chest congestion, and colds; for croup, drink honey and onion juice; honey and vinegar or honey and whiskey are for sore throats.

Kerosene. Strong oils or distillates like kerosene and coal oil were applied in a poultice for chest congestion, colds, and directly to the hair for head lice. For snakebite, soak the bite in coal oil until the oil turns green. Pour kerosene and sugar on infected skin. For sore throat, apply a poultice of kerosene, turpentine, and lard to the patient's neck. Never drink kerosene or coal oil.

Lady's slipper. Lady's slipper (*Cypripedium acaule*) powdered in water makes a blood builder or tonic for diarrhea.

Lard and tallow. Lard and tallow are put on burns; use mutton tallow for chest congestion; for croup, put mutton tallow and beeswax on the patient's back.

Madstone. A madstone is a calcified hair ball found in a deer's stomach. Such precious "stones" were applied to the bites of poisonous snakes and rabid animals to draw out the poison. The stone was placed on the bite, then soaked in milk or water. When the milk or water turned green, the poison was deemed to be out. The stones were so rare that the patient would go to a place where someone was known to own a madstone.

It was too risky to take the madstone to the victim; the stone might be lost or stolen. As late as the 1970s, some county courthouses in Virginia kept records on madstones.[7]

Mayapple. For constipation, mix powdered roots of mayapple (*Podophyllum peltatum*) with castor oil and roll into pills.

Milk. For snakebite, apply sweet milk and crushed cockleburs to the injury. For whooping cough, drink mare's milk.

Oak bark. For diarrhea, drink tea made from the bark of red oak (*Quercus rubra*).

Onions. Onions are prepared in a poultice and applied to the chest for colds; for croup, drink the juice of roasted onions; smear crushed onions on your head to cure headache; tie roasted onions around your neck for sore throat.

Pennyroyal. For fever, boil a cup of the leaves of pennyroyal (*Hedeoma pulegioides*) in a pint of water and drink it.

Pine needles. For colds, boil pine needles into a tea.

Poke plants. Brew the leaves of pokeweed (*Phytolacca americana*) into a tea, soak a cloth with the liquid, and apply to skin infections. For rheumatism, apply a hot roasted poke root to the painful joint; for the same ailment, one may also drink pokeberry wine and whiskey. Poke can be used as a blood builder. For toothache, drink a mixture of one part pokeberry wine and eight parts whiskey. For kidney problems, eat two pokeberries a day for two days. Eat poke leaves that have been fried in fatback grease to cure worms. Caution: Although the leaves can be made into poke sallet, other parts of the poke plant such as the berries and especially the roots are poisonous.

Poultice. A poultice is a mixture of remedies. For chest congestion, put a mixture of kerosene, turpentine, and lard on cheesecloth and apply it to the chest; for croup, mix turpentine, kerosene, and lard in a wool cloth and apply it to the chest and neck.

Ramps. For heart trouble, eat ramps (*Allium tricoccum*) and garlic raw.

Rat's vein. Parched and powdered leaves of rat's vein (*Chimaphila maculata*) can be snorted for coughs.

Red pepper. For colds, make a tea with ground red pepper.

Salt. Use salt for burns; salt and water as ear drops for earache, and ear wax; and rub chigger bites with salt and butter.

Sassafras. Sassafras (*Sassafras albidum*) is a blood builder; if taken in large quantities as strong tea, it can be a laxative.

Snakeroot. Tea made from snakeroot (*Aristolochia serpentaria*) will reduce a fever.

Soot. Soot is put on a baby's diaper rash and used for bleeding wounds and burns.

Spiderwebs. Swallow spiderwebs for asthma; apply them to cuts and wounds to stop bleeding.

Sugar. Rock candy and whisky are used for chest congestion and coughs.

Tobacco. Tobacco (*Nicotiana tabacum*) is used in a poultice on the chest for asthma, colds, and flu; apply wet tobacco to bee stings and chigger bites; eat tobacco seeds for worms.

Turpentine. Turpentine is applied directly to bleeding wounds, bee stings, and skin infections; make a poultice of turpentine, lard, and sometimes kerosene for chest congestion and croup; for headache tie on a bandage soaked in turpentine and tallow.

Urine. For earache, warm a spoonful of urine and pour some drops in the ear.

Vinegar. For colds, drink honey and vinegar.

Whiskey. Many people believe that whiskey, likker, white lightning, moonshine, or potable distilled alcohol by any other name will cure just about anything. It is used for arthritis, asthma, bleeding wounds, chest congestion, colds, colic, coughs, dysentery, gall bladder trouble, sore throat, stomachache, and toothache. For headache rub whiskey and camphor on the head. For a black widow spider bite, drink whiskey heavily from 3:00 to 7:00 p.m. In case of snakebite, drink a large quantity of whiskey; you won't get drunk, just better. These last two remedies are dangerous. Don't try them![8]

Wild cherry bark. Make tea from the green bark of the wild cherry (*Prunus serotina*) as a blood builder. For coughs, boil a cup of wild cherry bark in a pint of water until it forms a thick syrup.

Willow leaves. Leaves from the willow tree (*Salix alba*) make a tea for dysentery or to soothe poison ivy.

Yellowroot. See Goldenseal.

Superstitions in Folk Medicine

Superstitions associated with folk remedies are abundantly entertaining, even if not plausible. Some appear to have reasonable logic behind them, while others clearly spring from someone's fertile imagination. Discarded bandages used on wounds must be buried, not burned, or the wound will never heal. For asthma, drill a hole in a sourwood or black oak tree just above the patient's height. Put a lock of his or her hair in the hole, and when the patient walks by the tree, the ailment will disappear. Burn victims who have never seen their fathers can blow on their burns and draw the fire out. To cure foot cramps, turn your shoes upside down before going to bed. A sty can be removed by dragging a black cat's tail over your eye. To cure a high fever, put the patient's nail clippings in a bag and tie the bag to a live eel; the eel will carry the fever away. Never discard or burn hair clippings; bury them under a rock. This will prevent headaches. Buttermilk and lemon juice will remove freckles.[9]

Belief Systems

Every culture has a belief system. From the most primitive animism to the most complex beliefs and mystical rituals in independent and organized religions, a sense of belief in something or someone underlies the system. Organized religious churches (some belonging to denominations but some not) in Appalachia include Baptists, Catholics, Church of Christ, Disciples of Christ, Jehovah's Witnesses, Mennonites, Methodists, Moravians, Pentecostals, Presbyterians, Seventh-Day Adventists, and many others. Within each group there are divisions, and for some there are countless splinter groups. Baptists have fifty-three denominational groups; among them are Southern Baptists (fundamentalist), Southern Baptists (moderate), Independent Baptists, Primitive Baptists, Free Will Baptists, Missionary Baptists, Regular Baptists, Separate Baptists, Black Baptists, and a rare group called the Two-Seed-in-the-Spirit-Predestinarian Baptists. For example, in Tennessee there is only one church with ten members in the Two-Seed-in-the-Spirit-Predestinarian Baptists. Pentecostal groups are relatively recent compared with much older religions in Appalachia. Pentecostals also have many variations: for example, Church of God, founded in 1886; Fire Baptized Holiness (1898); Holiness (1911); and Assemblies of God (1914).[10]

Historically, representative larger religious groups in Appalachia have been Presbyterians, an early Protestant progenitor that arrived with eighteenth-century Scotch-Irish settlers; Methodists, with their widespread circuit riders; the many Baptists and their splinter groups; and twentieth-century Pentecostals, with their exuberant speaking in tongues, loud preaching, and body language that gets them described as holy rollers. Although much of Appalachia is predominantly Protestant, there are other groups of Catholic, Jewish, and Islamic adherents who make up a very small percentage of organized religions. Furthermore, throughout the history of Appalachian culture, not everyone has ex-

This church tells us that the traditional folk architecture in the old log unit endures, even though materials changed for the newer brick sanctuary. Old Union United Methodist Church, Hawkins County, Tennessee. (Source: Tennessee Historical Commission, 1987)

pressed faith in an organized religion. The early physical and cultural isolation ensured that mountain people got along quite well without a church or formal congregational worship. Rather than trying to cover every denomination, I will describe generic diagnostic traits in the region's belief systems and discuss some of the more unusual folk practices.

Generic beliefs appearing in most if not all denominations in Appalachia are monotheism, salvation, morality, and fundamentalism, including various doctrines. In addition, traditional dogmas common to most Protestant groups include the biblical story of creation, the virgin birth, the Trinity, atonement, and resurrection. Monotheism is the belief in a single deity. The belief in salvation from sin and an opportunity to go to a heaven after death is a common denominator among believers. The processes of salvation, the rituals of professing faith, along with the symbols of blessing, washing away sins, being born again, and other things in this passage from sinner to saved vary from group to group. No single belief system is right, nor are all the others wrong, but few people accept this. Each group in its ethnocentric way believes that its path to salvation and righteousness is the "correct one, the only one." Bill J. Leonard, church historian, says that "southern Christians often seem to believe more, and believe more intensely, than many groups outside the South."[11]

A common belief is that all people need salvation; but leading sinners to salvation requires considerable exertion. Evangelism and missionary work manifest themselves in door-to-door witnessing, radio evangelism, and sending missionaries to the "mission field," which can be all over the world. Ironically, Appalachia became a principal mission field for outside church missionaries in the late nineteenth and early twentieth centuries.[12]

Not all people agree on what constitutes morality. Groups differ on the social uses of alcohol, music, dancing, dress, women's use of makeup and jewelry, working on Sunday, and other things. Many are united in the things they appear to be against, such as gambling, prostitution, alcohol, and illegal drugs.

Fundamentalism is the attempt to instill unchanging doctrines or belief principles thought to be absolutely fundamental to a Christian church. Sundry denominations and independent congregations ascribe to fundamentalism while keeping their own cultural identities intact. Fundamental doctrines that have become diagnostic traits are the strongest belief in the inerrancy of the Bible and, as Leonard writes,

> the reality of the Genesis accounts of creation; Adam and Eve as the genuine "parents of the race"; the virgin birth of Christ; the substitutionary or sacrificial atonement of Christ; the bodily resurrection of Christ; the literal second coming of Christ; the reality of Christ's miracles as described in the gospels. . . .
>
> Fundamentalism is alive and well in twentieth-century Appalachia. It shapes churches, sometimes splits them, and keeps them arguing with friends."[13]

Diagnostic Traits in Appalachian Religious Practices

Appalachia's mountain religion conjures images of camp meetings, tent revivals, foot washing, snake handling, holy rollers, loud preaching, and emotional animation.[14] While carefully trying to defuse stereotypes, we cannot ignore snake handlers, holy rollers, and radio evangelists as traits that can appear in the region, though certainly not everywhere. Stereotypes become stereotypes because they already exist somewhere in the culture or because they have been invented by the imaginative but uninformed. In the former route, we recognize that colorful, unusual, and strange religious rituals take place in Appalachian cultures. Call them what you will, some are genuine and diagnostic. The term "holy roller" once was applied to people in Pentecostal/Holiness groups, but it was not a name they gave themselves, though it reflected their screaming and rolling around on the church floor during their services. Moreover, there is no single snake-handling religion. Even though it is associated with fundamentalist Pentecostal/Holiness groups, not all practice it. Appalachia has a small number of congregations who still handle poisonous snakes in ritualistic church services.

Snake Handling

Mark 16:17–18 says, "And these signs shall follow them that believe; In my name shall they cast out devils; they shall speak with new tongues; They shall take up serpents; and if they drink any deadly thing, it shall not hurt them; they shall lay hands on the sick, and they shall recover." Using this passage of scripture, a small number of known Appalachian congregations ritually pick up poisonous snakes in their bare hands to confirm the power of their faith. Some also drink poisons for the same reason. These strange and powerful acts attract the attention of scholars, journalists, and the just plain curious. Perhaps a morbid curiosity propels people to seek those who demonstrate their faith in this most unusual way. Journalists Fred Brown and Jeanne McDonald have captured an extremely rich story about serpent handlers who practice these rituals in churches called Rock House Holiness Church of God, near Parrottsville, Tennessee; the House of Prayer in the Name of Jesus Christ, at Marshall, North Carolina; the Full Gospel Tabernacle in Jesus Name, in Middlesboro, Kentucky; and the Church of the Lord Jesus, in Jolo, West Virginia.[15] Thomas Burton cites snake-handling churches past and present from Church of God, Church of God of Prophecy, Church of Jesus Christ, Church of the Lord Jesus Christ, and especially Church of God with Signs Following. Independent congregations may also practice the ritual. Most are small, scattered groups of believers who knowingly break the law. Since the 1940s, the states of Kentucky, Virginia, Tennessee, North Carolina, Georgia, Alabama, and Florida have passed laws forbidding the handling of poisonous snakes in public gatherings such as religious services.[16]

Snake handling in a church service is best described as frenzy. Believers do not simply appear at the front door of the church with serpents crawling out of their pockets. It is usually the preacher and elders who bring the snakes in wooden boxes to the church's pulpit near the altar. The boxed serpents are safely kept there for a while. But everyone knows "they's death in that there box." As the preaching and singing and getting into the spirit intensify, the preacher will open a box and take out a rattlesnake or copperhead and hold it over his head, all the while sweating and preaching. If the spirit moves them, the elders, members of their families, and other believers will come to the pulpit and pick up snakes in their hands. A preacher might hold four or five snakes at a time, letting them hang around his neck, while grasping serpents in both hands. Do the snakes bite? Yes! Are the serpents poisonous? Yes! Do people die? Yes!

From 1936 to 1991, more than seventy people died from handling poisonous snakes in church. The Brown family of Parrottsville, Tennessee, lost Melinda Duvall to snakebite in 1995 and her husband Punkin Brown in 1998. In 1991 Jimmy Ray Williams Jr. died of snakebite, eighteen years after his father had died of strychnine poisoning in a church service. Old-time pastor and snake handler Brother Charles Prince died in 1985 from a combination of snakebite and strychnine. Jimmy Williams and Buford Pack died of strychnine poisoning in 1973 at the Carson Springs Church of God in Jesus Name in Cocke County, Tennessee.[17] In Jolo, West Virginia, the Elkins family tragically lost Columbia Gay Chafin, a twenty-three-year-old mother, to snakebite in 1961. And in 1955 George Went Hensley, age seventy-five, died of a rattlesnake bite in Florida. Hensley was recognized as the founder of snake handling in Appalachia.[18]

When and how did snake handling become an Appalachian ritual? George Went Hensley, living in Owl Holler, a rough little rural ridge and valley area near Ooltewah, Tennessee, took up his first serpent in 1894 at age fourteen. By 1909 Hensley was generally recognized as the first to practice serpent handling in Appalachia in general and Tennessee in particular, although he did not claim that distinction. Eventually Hensley, part-time moonshiner and snake handler, was ordained by A. J. Tomlinson, the general overseer of the Church of God. For the next decade Hensley preached and handled serpents at churches throughout the Cleveland, Tennessee, area.[19] In about 1919 he moved to Harlan, Kentucky, where snake handling began to spread widely among rural congregations. From the late 1920s until about 1940, snake handling declined in Appalachia, but by the 1940s the movement experienced a resurgence and gained greater notoriety in the media and especially among lawmakers. The movement returned to Cleveland, Tennessee, in the Grasshopper Valley area, where Raymond Harris and Tom Harden established a new church called the Dolly Pond Church of God with Signs Following. After a member died of snakebite in 1945, Tennessee lawmakers outlawed the practice in 1947. The practice declined in the 1950s and 1960s but resurfaced once more in the 1970s.[20]

While snake handling in Tennessee, Kentucky, North Carolina, and Virginia traces to George Hensley, the practice in Alabama and Georgia independently traces to James Miller, who introduced serpent handling to Sand Mountain in northern Alabama in 1912. Miller is believed to have introduced the practice in about 1920 to Berrien and Cook counties in southern Georgia, between the towns of Tifton and Valdosta.[21]

The frequency and distribution of snake handling still remain indefinite. Estimated numbers of practicing believers are between 1,000 and 2,500. However, with church membership estimated at an average of 100 or fewer individuals per church, we can see that the movement clearly does not dominate religion in the region. The practice is far from being popular, because Protestant denominations abhor the practice, many Pentecostal groups clearly do not want to be identified with it, and in most states in Appalachia it is simply illegal. Snake-handling congregations still can be found in West Virginia, Kentucky, Tennessee, North Carolina, Georgia, and Alabama. The publicity they have received in newspapers and television documentaries and their popularity among the curious have at once stereotyped them and exaggerated their presence in the region.

Speaking in Tongues and "Holy Rollers"

Mark 16: 17 says, "they shall speak with new tongues." Speaking in tongues is also mentioned in Acts 2:1–13 and 1 Corinthians 12:10, 28. Congregations most strongly identified with these characteristics are Pentecostal and Holiness groups, but other churches occasionally will tolerate a member who spontaneously gets the spirit. The events that lead up to speaking in tongues in a Pentecostal church service evolve as the service progresses. A believer does not simply walk into church and immediately begin speaking in tongues or rolling around on the floor. It takes a little time, building up with preaching, answered with gusto-filled amens and hallelujahs, singing, and just "getting the spirit." When it hits, you know it. Many believe that the service is missing something if no one gets the spirit or speaks in tongues or rolls on the floor. It is an exuberant form of worship that "lets loose" in Appalachian churches throughout the region.

The mystery of speaking in tongues divides groups of believers. Some people think it is improper to speak in tongues; others feel they are not reaching the proper level of religious participation unless they do. Toliver Vinson, a north Georgia man, speaking in *The Foxfire Book,* explained his decision to change churches from Baptist to Holiness: "The [Baptist] church just about made up my mind for me. They said that I'd have to leave off that speaking in tongues and so forth if I stayed with them. They said their conference just wouldn't put up with it. The deacons talked with me and they said that if I'd leave that speaking in tongues off that I could stay with them. I told them no, I wouldn't do that because the Lord had given it to me and that I appreciated them and loved them and loved the church but I wasn't compromising with them or anybody else."[22]

Faith Healing

The same Bible passage that tells of snake handling and speaking in tongues continues, "they shall lay hands on the sick, and they shall recover." That passage leads to the practice of faith healing. Another mention in the Bible is in James 5:14–15. While all of us have had television evangelists and "healers" pique our curiosity about whether the act really works, there are Appalachian congregations that practice faith healing, believe it works, and above all, see no harm in trying.

Foot Washing

Foot washing is still done in mountain churches. I experienced it once in a special service. The concept is one of humbleness, unity, and faithful servitude, and it carries that same special reverence that accompanies taking communion. In churches where it is practiced, foot washing takes place in two separate rooms in the church. In small one-room churches, a curtain hanging on wire divides the room by gender. With pans of water and towels, the men wash other men's feet in one room while the women wash each other's feet in the other room. The practice comes from John 13:4–15, which describes Jesus at the time of Passover going from one disciple to another and washing each man's feet. The passage ends with, "Ye call me Master and Lord; and ye say well; for so I am. If I then, your Lord and Master, have washed your feet; ye also ought to wash one another's feet. For I have given you an example, that ye should do as I have done to you." Most southern Appalachian Baptist churches that practice foot washing see it as an ordinance along with taking communion in the Lord's Supper and full immersion baptism.[23]

Full Immersion Baptism

Baptists the world over practice full immersion. That is, in the ritual of washing away sins, the convert must be submerged completely, from head to toe. No amount of sprinkling will do the job. Appalachian Baptists are no different in the practice, but the sites and seasons of baptism are often mountain specific. "Shall we gather at the river" becomes gathering at the creek. Some churches lacking indoor baptisteries (tanks or tubs) select the warmer months from May to September for baptisms; others still do it outside in the cold. Since the water needs to be at least three feet deep, shallow creeks pose a problem, so some congregations find or dig deeper holes in the creek, or dam it, or they choose other sites at rivers, ponds, or lakes. The preacher enters the water, receives the convert, and tells her to fold her arms around her body and turn around her back to him as he stands just to one side. He holds a cloth over her nose and mouth with one hand and supports the back of her neck with his other hand. As he lowers her into the icy stream, he proclaims, "I baptize this our sister in the name of the Father, the Son, and, the Holy Spirit."[24] Among Baptist congregations, full immersion is the only acceptable baptism; for others not only does the baptism have to be full immersion, but someone who moves to another Baptist church must be baptized again.

Dinner on the Ground, Homecoming, and Decoration Day

One of the most pleasing practices of Appalachian rural churches is dinner on the ground. Though not exclusively Appalachian or even southern, it usually occurs in concert with "homecoming" and Decoration Day. The homecoming celebration, held in the fall after harvest, is a gathering of past and present members of the congregation who have returned to visit their home church. A church homecoming is much like a family reunion. The event joins important elements of mountain culture in the sense of place and the sense of family.[25] Guest preachers preach, and guest singers or musical groups celebrate with special music. Although a homecoming resembles a revival, it is not one. Revivals are serious weeklong series of sermons and special music designed to attract converts.

Decoration Day is a planned memorial day held in the spring, though not necessarily at the end of May. Its contemporary purpose is to memorialize the family members who are buried in church or family cemeteries. In the post–Civil War years, Decoration Day was the same as

Dinners on the ground at rural churches are held under dinner sheds like this one in Grainger County, Tennessee. (Photograph by J. Rehder, 2001)

Memorial Day, which memorialized fallen soldiers from that conflict.[26] On Decoration Day in Appalachia, however, church members come to clean up the cemetery, removing weeds, grass, tree limbs, dead flowers, and other debris that has accumulated over the winter. They may clean headstones. Then they decorate the gravesites with fresh real flowers or new plastic ones.[27] This sprucing up of the cemetery is followed by a church service, hymn singing, prayers, remembrances, family gatherings, and dinner on the ground.

Dinner on the grounds also occurs on the fifth Sunday of a month and on other occasions not associated with homecomings or Decoration Day. The magnitude of dinner on the ground is manifested by the dinner shed, a permanent open-sided shed thirty to sixty feet long by twelve feet or more wide, situated at the back or to the side of the church building. A double row of tables, either permanent built-in ones or folding ones set up for the occasion, is designed to hold an enormous feast of "covered dishes." A variation of dinner on the ground is a "covered dish supper" prepared for evening services. For both, people bring a magnificent variety of foods: deviled eggs, coleslaw, potato salad, ham, fried chicken, barbecue, corn, beans, field peas, sweet potatoes, hominy, bis-

Everyone enjoys the tradition of dinner on the ground. (Photograph by Bill Mynatt, 2002)

cuits, honey, corn bread, pies, cakes, and so forth. A bountiful cornucopia awaits the people who have the great fortune to participate in dinner on the ground.

Camp Meetings

Camp meetings are thought to have originated on the Kentucky frontier about 1800 as a response to the need for worship and the formation of new churches. Presbyterian preachers James McCready and Barton Stone are credited with starting them, but soon Baptists, Methodists, and Shaker groups were holding camp meetings of their own. The camp meeting, which also served as a revival, usually lasted a week and included long-winded preaching, praying, and singing along with bartering, trading, and conversation.[28] A camp meeting was more than a religious experience; it was clearly a social event, and it remains so today.

Tent Revivals

In Protestant churches throughout the South and in Appalachia, revivals are commonplace. But the tent revival is a different matter. Anyone who has seen the movie *Elmer Gantry* or has heard Neil Diamond's song "Brother Love's Traveling Salvation Show" will have a glimpse of what a tent revival has going for it. Tent revivals are rarely officially sponsored by permanent churches; so what is the difference? The purpose of any revival is to gain converts—to attract the lost, the sinner, and the unloved and persuade them to seek salvation. To support the effort, money offerings are collected in buildings as well as in tents. But a tent revival is a fly-by-night operation in which a preacher and his entourage come to a community for a short time, usually a week or less. They advertise with handbills tacked onto telephone poles, by word of mouth, by signs on the tent itself, and sometimes on radio. They erect a medium-sized tent in a vacant lot and set up wooden seats to accommodate perhaps fifty to a hundred people. The services are accompanied by hymn singing and punctuated by evangelistic preaching that is designed to save the lost, attract converts, and pay the bills.

Tent revivals have become rare in this part of the country, but recently I saw such an operation being set up in a vacant field in Bethel, Tennessee, about twenty-five miles north of Knoxville. The "Wings of Faith" revival folks led by Rev. G. W. Delotte had set up a big white tent with two portable outhouses in back and were preparing for a few days of revival services on this hot August day in 2002, my sixtieth birthday. Southern Appalachian tradition lives on.

Death Rituals

Before the arrival of contemporary funeral homes, rural mountain folk performed all the funeral duties themselves—notifying relatives and the community, preparing the body for burial, and conducting burial rites in a family graveyard. The old folkways of dealing with a death in the family can be summarized this way: A bell, usually the church bell, tolled the number of years the deceased had lived. Neighbors built a wooden coffin. Someone notified faraway relatives by a letter edged in black. The body was washed, dressed, and laid out for the wake, which was expected to last for a day. If the corpse was a man, the preacher prepared the body; if a female, the preacher's wife did so. The body needed to be dressed as soon as possible, before rigor mortis set in. Men were dressed in dark suits (often with the shirt and jacket backs split open), women in white

Slab stone grave covers protect graves in a cemetery in Overton County, Tennessee. (Photograph by J. Rehder, 2002)

dresses, and children in white clothes or blanket shrouds. At the wake (called "sittin' up with the body"), folks sat with the body overnight, feeding and comforting the relatives and softly singing hymns as the night wore on. Within a day or two, friends took the body to the dead person's church in a wagon. At the funeral everyone wore black. During the thirty-minute service, the preacher prayed and read from the Bible, the congregation sang hymns, then everyone came forward to view the body for one last time. Burial was in the family graveyard or in the church cemetery. Friends and neighbors dug and filled the grave. Flowers that had decorated the coffin at the church later covered the grave. Eventually a gravestone or a wooden marker was placed at the head of the grave.[29]

The grave is oriented east-west so that the body faces toward the rising sun. The theory is that when the rapture comes, as in the second coming of Christ, those who are buried in this position will face east upon rising, ready to be gathered up and taken to heaven.[30] In some parts of Appalachia, small wooden grave houses, built of posts and a lattice of laths with wooden or tin roofs, are used to protect graves from wild animals. On the Cumberland Plateau, where flat layers of sedimentary sandstone can be pried up with iron bars, large stone slabs are formed into tent-shaped grave houses described as "prismatic graves," "slab graves," "box graves," "peaked graves," or such.[31] Grave houses are still built today in some parts of Appalachia, especially in Kentucky.

Radio Land

Contemporary religion in Appalachia is still supplied to a homebound following through radio broadcasts. Throughout the region, radio preachers shout out their messages, mostly through tiny AM stations. It is a bright October Sunday in 2001, and I am on my way through the western edges of the Cumberland Plateau in eastern Kentucky. I have just come from Berea and am heading south, then west, through Madison, Jackson, and Laurel counties, through communities called Bighill, Sandgap, McKee, Gray Hawk, Tyner, Annville, East Bernstadt, and London. The plateau landscape is the first high ground south of the Bluegrass Basin; but this is not coal country, not yet. The minuscule farms that ride atop the rumpled ridges here raise a little tobacco and some beef cattle, but nothing else. It is a more recently settled landscape than the ancient well-bred, well-fed horse country immediately to the north. Settlement takes on an almost temporary appearance. I say "almost" because there are far more temporarily settled places south and east of here in Kentucky's true coal country. Still, there are simply not enough folk houses with much age on them. Few buildings appear to predate 1900, though I could be wrong. I see one I-house at Gray Hawk and one four pen at Annville in this fifty-two-mile traverse.

I turn on the radio searching for something to listen to. At 105.7 on the FM dial I get WTBK, a local rock station that carries Sunday religious programs and sermons covering the London-Corbin, McKee, and Barbourville listening area. A program in progress has a fellow speaking in his loudest voice: "Praise the Lord, I am happy to be here, Praise the Lord. And Praise the Lord, we will be meeting agin tonight, Praise the Lord, at seven, Praise the Lord. And Praise the Lord, we'uns hope, Praise the Lord, that you'ns can, Praise the Lord, be here. Praise the Lord."

The next 28.5 minute program is conducted by Brother Ewell Napier of the Whites Branch Holiness Church. For fourteen months Brother Ewell has suffered from a broken leg, but his voice is still strong. He has been a radio preacher for twenty-six years. He begins with a few announcements, then calls the names of people on his prayer list. Hmm— if he can find it. "Oh yes, here it is," he says as he rattles some paper at the microphone. He then recites the names of the people who have been good contributors of money to his ministry. After a couple of loud a cappella hymns (this is a one-man show), Brother Ewell begins his sermon on adultery. He explains that he feels adultery is the one sin that really needs his attention at this time. His delivery is a familiar "suck and huff" preaching style. He huffs much more than sucks, though. His huffs are the typical cough-like "hawghs" that punctuate the sermon. For example, "The good Lord, hawgh, wants you to know, hawgh, that you are welcome, hawgh. And I do too, hawgh. You can come sober or you can come drunk to my church, hawgh. And you can sit up here right beside me, hawgh, at the pulpit, hawgh." Preachers with this style of preaching call themselves "wind-sucking preachers" or wind suckers. They deliver their sermons this way because they believe that with each wind-sucking inhalation they are taking in the Holy Spirit. Such was Brother Ewell's presentation for the folks out there in radio land.

The radio begins to fade as I leave Kentucky highway 30 and join U.S. 25 and then I-75. Decades separate these roads in culture, attitudes, and basic ways of life. I experience each road in its own time. Brother Napier's sermon and old-time preaching style seem to belong exclusively to Kentucky highway 30. Only a few local people still use U.S. 25, even though it rests in large part on the bed of the old Wilderness Trail that opened central Kentucky in the late 1700s. U.S. 25 is the old road that reflects the years of commercial traffic before the coming of the interstate highway. Decrepit motels, diners, and gas stations still litter the roadside. But just yards away two concrete ribbons of three lanes each snake across the landscape with hundreds of cars and trucks speeding along their backs. The modern road, I-75, links Michigan to Florida and culturally ignores Appalachia as much as it can.

8

Folk Music, Folk Art, and Folk Festivals

Appalachia's cultural identity of independent, hardworking, isolated mountain people seems at odds with the perceived frivolity of music, art, and festivals. However, every known culture on earth has some form of art, musical sound, or ritual as an outlet for creativity. From the time of first effective settlement, Appalachian folk entertained themselves with music: a long tradition of oral history in ballads and other folk songs was widespread and common to much of the region. Commercial crafts and festivals came later. Crafting schools, promoted by outside philanthropists as a form of economic salvation for the Appalachian poor, emerged between 1893 and 1928. Most specialty festivals arrived after about 1950, and many have had their start just since the 1970s.

Folk Music

Traditional music in Appalachia has a clouded and complicated history. We will explore the rediscovery of ballads and folk songs, the musical instruments played in the region, and the connections between Appalachian folk music and religious music. From this we can observe the evolution of the old ways of self-made Appalachian entertainment and their relation to later forms of country and bluegrass music as national trends.

George Carney, America's foremost expert on the geography of folk, bluegrass, and other musical forms, said in 1998:

> Folk music is often loosely applied to cover all traditional or aurally transmitted music, that is, music that is passed on by ear and performed by memory rather than by the written or printed musical score. Folk music must be heard before it can be retransmitted. Hence, the term "aural" is used rather than "oral." In addition to its aural transmission, two other criteria are required of genuine folk music. One is that the

> origin of the melody must be unknown to the performer. A second requirement is that the melody and lyrics exist in variant forms.[1]

Carney clearly explains the first two-thirds of his statement, but what does he mean by his last two requirements? Carney wants us to think *folk* in the context of "Granny sang this song, and my daddy sang it too, but I don't know where it came from." The performer's ignorance of a song's origin perhaps makes it authentically folk, as opposed to music discovered in a professional musicologist's search for origins. Furthermore, the origin of the melody is likely to be unknown in a folk culture because the tune did not reach the culture in written form, nor was a written history likely to accompany it. The origins of most things folk are unknown, and for some that is precisely why they *are* traditionally folk. Some mythic or legendary ideas about a song's origin or antiquity may exist, but they are usually inexact, such as, " this'uns old, maybe a hundert year old." In the second requirement, folk music is aurally transmitted, and modifications are expected in the melody and lyrics of authentic folk songs because they are in flux, constantly changing. This is much like the party game "gossip" in which a story is told on one side of the room and then whispered from person to person. By the time the story reaches the last person in line, it has changed so much that everyone laughs. In contrast, once written, commercial, copyrighted songs usually have few variants.

Searching for Origins: Song Catchers

We will begin this story in 1916, at an unusual time and source of discovery for Appalachian folk music. On the surface, the music should seem obvious—it is expressed in haunting ballads and story songs using tunes as old as the hills. But the topic is much more complex and offers the temporally unexpected. Some songs trace to Old World origins, but others, thought to be old folk songs from the British Isles, have their origins in American music or African American spirituals. Perhaps I should call this section "The Origin, Nature, and Character of Appalachian Folk Music: What Cecil Sharp Found in 1916–18."

In the summer of 1916 Cecil J. Sharp, a scholarly, polite Englishman, and his assistant, Miss Maud Karpeles, arrived in America in search of old English folk songs. Sharp had spent years scouring his British homeland for old ballads, but with findings and earnings growing scarce, he turned to North America.[2] In 1915, while lecturing in Massachusetts, Sharp by chance met Olive Dame Campbell, wife of missionary and Appalachian writer John C. Campbell. The Campbells and other advisers persuaded Sharp to search in the remote, isolated southern Appalachian mountains, where they believed that old songs from the British Isles were most likely preserved.

Sharp and Karpeles were delighted and overwhelmed when they discovered well over a hundred songs in their initial forays into the mountain counties of Madison and Buncombe in North Carolina and Unicoi County, Tennessee. Between 1916 and 1918, they discovered and transcribed some five hundred different ballads and tunes in the Appalachian sections of North Carolina, Tennessee, Virginia, West Virginia, and Kentucky. Their field method was to spend about a week in a community, being introduced to people who knew old songs, then to visit other singers within walking distance. To discover folk ballads, they learned to ask for "love songs" not just old-timey songs. While a mountain woman sang a ballad, Cecil Sharp meticulously transcribed the melody note for note into a notebook and Maud Karpeles wrote down the lyrics in shorthand.[3] The team had to be quick and good, since the singer might go through the song only once.

Cecil Sharp and Maud Karpeles spent a total of forty-six weeks in Southern Appalachia in three summer field seasons from 1916 to 1918. They listened to 281 singers and obtained 1,612 variations of tunes that represented 500 different songs. Because it was impractical to print the entire collection, the published catalog had 274 entries, including 73 ballads, 134 songs, 5 hymns, 26 nursery songs, 14 jigs, and 19 play-party games. Only five of the songs are religious because Sharp intended to refrain as much as possible from collecting hymns in his search for true folk songs.[4] Had he sought all songs sung in the mountains, he could perhaps have added hundreds more to the collection.

The geography of their rediscovery of Appalachian folk songs sheds light on the region. Miss Karpeles remarked, "On the whole, the most fertile ground was either side of the big mountain range (known as the 'Great Divide') which separates the states of North Carolina and Tennessee, and this was, perhaps, to be expected, for it was in this region that the most primitive conditions prevailed."[5] Karpeles observed that the best ballad texts came from Kentucky and the finest tunes came from Virginia, but that West Virginia did not appear to be a promising field for research.

Sharp discovered that many of the songs were ballads and were most likely sung a cappella by women. Several ballads whispered more than just hints of their European origins, such as the well-known "Barbara

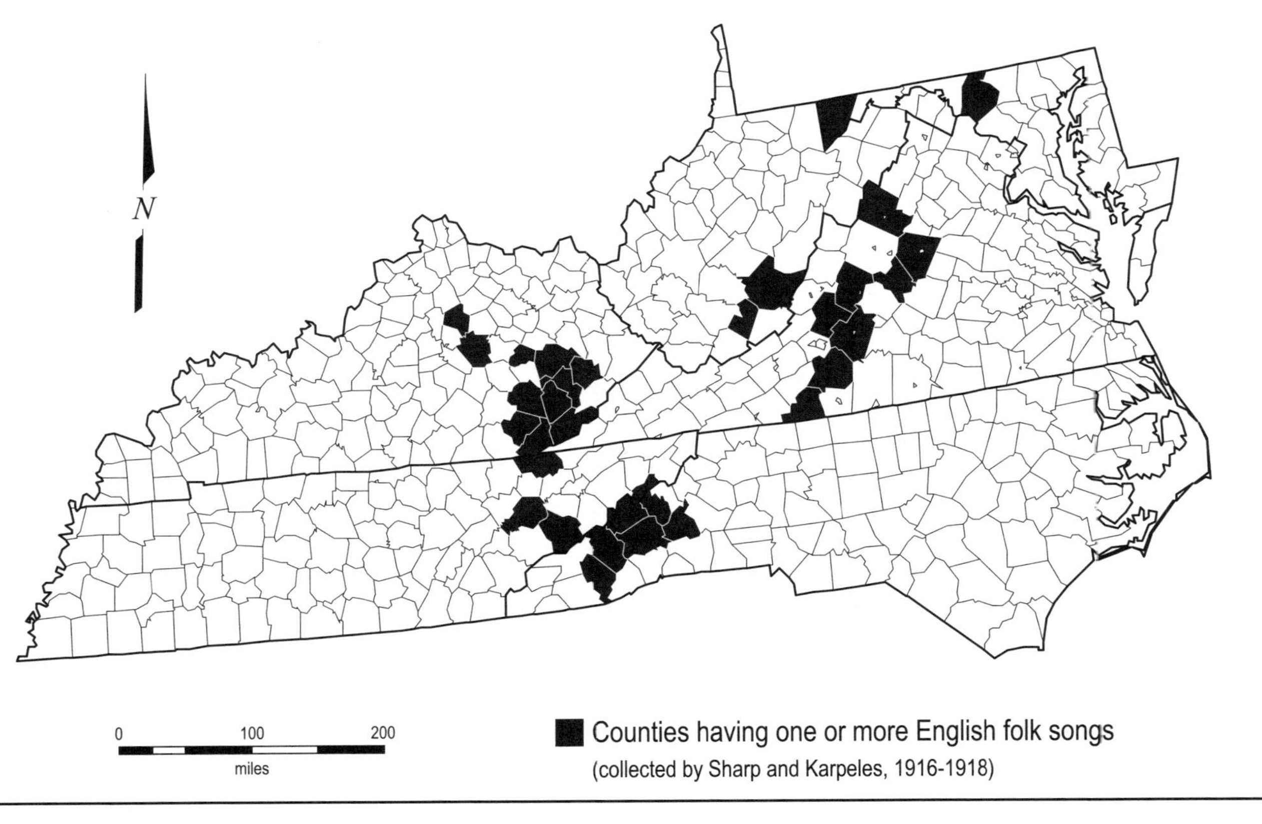

Between 1916 and 1918, Cecil Sharp and Maud Karpeles became "song catchers," collecting English folk songs in these counties in Appalachia. (Data from Sharp [1932] 1966)

Allen," "Come All You Fair and Tender Ladies," "The Gypsy Laddie," and "The Sheffield Apprentice." Others with direct links to the British Isles were "Earl Brand," "Geordie," "The Cherry Tree Carol, "Lamkin," "Lord Randal, "and "Young Hunting." Of the 73 ballads, 68 traced to Britain, and 46 of 134 folk songs were also British.[6]

A remarkable singing informant for Sharp was Jane Hicks Gentry, born in 1863 in Watauga County, North Carolina. Jane Gentry kept a boardinghouse in Hot Springs, North Carolina, and through Sharp's survey she later became an acclaimed folksinger. Mrs. Gentry was the richest repository of folk songs and was the informant-singer who supplied the most mountain songs and ballads. In August and September of 1916 and 1917, Gentry contributed forty entries to Sharp's published survey: nineteen ballads, fifteen songs, four nursery songs, one hymn, and one jig.[7]

Appalachian folk songs exhibited a number of important culture traits. Ballads were story songs performed in a narrative voice and traced directly to the British Isles. Other song types were identified as folk, jig, nursery, and play-party or game songs. Just about all were sung a cappella. Many more of the old ballads and other folk songs were sung in a minor key than those found in other geographic areas or in other derivative country and bluegrass forms. A minor key thus became a helpful diagnostic trait. Other Appalachian folk songs were sung in C, D, or G major. Ballad tempos were relatively slow; many were in 3/4 time, others in 4/4 time. Lyric themes in the ballads included, among many topics, myths, castles, love, knights of old, lords, ladies in distress, family members such as sisters, brothers, and mothers, and songs about Christmas. Story songs set in an early, mythic time and setting served up a dash of imagery. Dan Jackson, my wife's father, a descendant of the Ross clan of highland Scotland, would awaken his daughters every morning in North Carolina with this song:

Rise up, rise up my seven bold sons
And put on your armor so bright.
Let it never be said that a daughter of mine
Was married to a lord over night.

My father-in-law knew nothing of the origin of the lyrics except that his own parents traditionally recited the words upon rising before dawn on that North Carolina farm. The lines are from an early Scottish ballad called "The Douglas Tragedy," and they also appear as the Scottish ballad "Earl Brand." Cecil Sharp collected twelve Appalachian versions of "Earl Brand," four in Virginia, four in Kentucky, three in North Carolina, and one in Georgia. One of the variations from Virginia is

> Get you up, get you up, my seven sons bold
> Get on your arms so bright,
> For it never shall be said that a daughter of mine
> Shall lie with a lord all night.[8]

At this time in Appalachia, songs were rarely accompanied by any instrument but the fiddle. Cecil Sharp saw only one singer with other instrumental accompaniment, and that was a guitar in Charlottesville, Virginia. He was aware of the dulcimer through information supplied by Olive Dame Campbell, but he did not see or hear one other than in Kentucky settlement schools. The only instrumentals he wrote down in the field were some fiddle jig tunes for dances played by Reuben Hensley and Michael Wallin.[9]

Musical Instruments

Stringed instruments arrived in Appalachia in this sequence: fiddle in the eighteenth century; banjo in the mid-nineteenth century; dulcimer in the nineteenth and early twentieth centuries; guitar about 1900; bass in the 1920s; mandolin in the 1930s; and resonator slide guitar, better known by its Dobro trade name, in the 1930s. For early forms of Appalachian music the fiddle was the oldest and most widespread instrument, and it was used mostly for playing unsung jigs. Only men played fiddles, and songs were often catchy, short-bowed, jiggy numbers played at rural dances and on other festive occasions.[10] Fiddles were held low in the crook of the arm, not under the chin. Jim Russell, a farmer, fiddler, and friend who was born in 1924 in Claiborne County and lived near Norris, Tennessee, showed all these traits; he fiddled low on his arm, used short, jiggy bow strokes on hundreds of songs, and did not sing.[11] Some mountain families and communities held the fiddle in low regard because it was thought to be the instrument of the devil. It is said that A. P. Carter had to hide a fiddle from his mother because she considered it sinful to own or play one. Carter became one of the pioneers of country music in the famous Carter family, but he always played a guitar throughout his recording and broadcasting career.[12]

The banjo, called the "banjer" in folk dialect, originally came from Africa and arrived in the Americas along with African slaves in the seventeenth and eighteenth centuries. When and where white Appalachian musicians came to know the instrument and were first identified with it

Jim Russell (1924–2001) played his fiddle the traditional way, held in the crook of his arm. (Photograph by J. Rehder, 1998)

remains controversial. Cecelia Conway's 1995 book *African Banjo Echoes in Appalachia* is unquestionably the authoritative treatise on the history and folk use of the Appalachian banjo. Yet even Conway admits that the time and place of its emergence in Appalachia are unclear if not unknown.[13] Scholar Robert Winans and musicologist Alan Lomax believed that the banjo in Appalachia traced to early traveling minstrels beginning in the 1840s. Others thought the banjo came to the mountains with black laborers when railroads and coal mining entered the region after the 1870s. Still others believed that traveling medicine shows and tent performances first brought the banjo to the region, in the first quarter of the nineteenth century.[14] Once the banjo became a part of Appalachian culture, its acculturation ensured it a permanent place as an

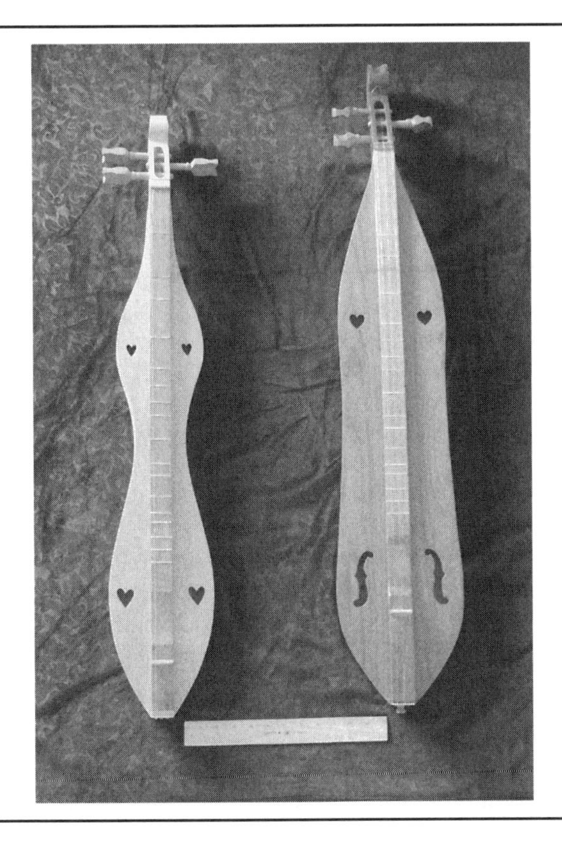

The dulcimer, a stringed instrument with European origins, has a strong cultural identity with Appalachia. Charlie Watson of Lenoir, North Carolina, built these two dulcimers: at left is a Kentucky type made in 1974 of wormy chestnut and curly maple; at right is a North Carolina type made in 1975 of black walnut. (Photograph by J. Rehder, 2002)

integral part of the character and personality of Appalachian folk music that later appeared in country and bluegrass music.

The dulcimer is another Old World instrument erroneously thought to be native to Appalachia. It is certainly Old World; it did not originate in America. The dulcimer is a three- to five-stringed instrument that is held on the lap and strummed with a plectrum made of wood or a goose quill. A single string is noted for the melody; the other strings are drones. Similar to the European fretted zither, the dulcimer was initially rare in Appalachia and was found in the nineteenth century in isolated areas in eastern Kentucky and southwestern Virginia. Most scholars consider it a relative newcomer to the mountains because its widespread diffusion came after 1900 through organized settlement schools such as the Hindman Settlement School in eastern Kentucky and the John C. Campbell Folk School in western North Carolina.[15] The earliest dulcimers, originally from Germany, likely diffused into Appalachia from Pennsylvania's Midland German hearth, where there was a three-stringed instrument called a scheitholt that was derived from the fretted zither family. During the migration periods, perhaps a few eighteenth- and nineteenth-century immigrant German farm families may have brought in either the instrument or knowledge of its construction and playing.[16]

The wider diffusion and acceptance of the Appalachian dulcimer occurred in the first third of the twentieth century when influential people such as Olive Dame Campbell and John Jacob Niles and institutions at craft centers and settlement schools promoted dulcimer crafting and playing. Carolyn White, a dulcimer player and a graduate of the Hindman Settlement School in Kentucky, told me about dulcimers and dulcimer makers she knew. James Edward "Uncle Ed" Thomas (1850–1933) was considered one of the earliest dulcimer craftsmen at Hindman. A Thomas dulcimer is on display at the Folk Art Center, at milepost 382 on the Blue Ridge Parkway near Asheville, North Carolina. In the early twentieth century, Thomas taught his skills to students and faculty such as Jethro Amburgey, who was in charge of the woodworking shops at Hindman. Later craftsmen carried the craft forward into the 1950s and 1960s, when Jethro Amburgey was ending his career in dulcimer making and Bob Mize and others were starting to make dulcimers in Upper East Tennessee, southwestern Virginia, and southeastern Kentucky. Mize has reportedly built more than two thousand dulcimers.[17]

Dulcimer shapes express regional differences. The Kentucky dulcimer has a long, narrow body but is more rounded coming into the fretboard nearest the neck at the tuning pegs. North Carolina dulcimers are slightly wider with a more elongated taper into the fretboard near the tuning pegs. Some Virginia dulcimers have deeper resonating boxes, others have double resonators, and many have all the strings tuned to D to make them sound louder.[18]

Since the 1960s, a resurgence of dulcimer making and playing has stimulated a regional pastime for all ages of folk and nonfolk alike. Almost all the craft schools teach dulcimer making; many shops sell dulcimer kits that can be assembled at home; and music classes and dulcimer groups and clubs are devoted to the Appalachian dulcimer. Craftsman Charlie Watson from Lenoir, North Carolina, built my two instruments. One dulcimer, Charlie's 120th, a Kentucky type, is made of wormy chestnut and curly maple and was finished on December 16, 1974. The other one, a North Carolina type of black walnut, was Charlie's 185th and dates to November 21, 1975. In approximately one year Charlie Watson built 65 dulcimers, which tells us something about one man's productivity. Shortly after I had purchased my first dulcimer I wanted another one, so I phoned Charlie one cold night. "Hello, is this Charlie Watson, the man who makes dulcimers?" "Yep." "Well, I'd like to buy one from you. Is that all right?" "Nope," he answered. Flabbergasted, I stammered, "Er, uh, you *are* Charlie Watson and you *do* make dulcimers, and I'd like to buy one from you. What's wrong with that?" "Cain't," he said, "cawz the only place you kin git one o' my dulcimores is at Guy's." After more laconic conversation it became clear that I could buy another from the same place where I got the first one. Charlie Watson's dulcimers were marketed by Guy's Folk Toys, in a tiny single pen cabin at a hairpin turn on U.S. 321 in Watauga County, North Carolina, and nowhere else.

American music found its way into Appalachia long before the Sharp-Karpeles surveys. It is also believed that early English and American folk and commercial tunes were widespread throughout the eastern half of the United States;[19] not until later did they gradually disappear elsewhere, to be preserved only in the isolation of Appalachia. Were it not for Cecil Sharp and Maud Karpeles, how many songs would have survived? How many old songs were lost long before Sharp arrived in Appalachia? We may never know the answers. Maud Karpeles returned to Appalachia in 1951, though, to revisit her singers and search for more old songs. She was astonished by the loss and contamination of songs, which she attributed to radio, new roads, and electricity. Still, she found ninety-one more old songs that could be classified as likely folk pieces.[20]

Religious Folk Music

Not long before the Sharp-Karpeles surveys in 1916–18, a wealth of religious folk music saturated the Appalachian region. Hymns were a general class of sacred music. Spirituals by blacks and whites largely followed lyric themes from the Old Testament. Gospel music from both black and white sources came later and used New Testament texts. Shaped note singing was a white musical form that dated at least to eighteenth-century New England and possibly traced earlier to Europe. Shaped note singing found favor in the rural South and especially Appalachia beginning in the early nineteenth century, and it continues sparingly to the present. The best early documentation on religious folk music is found in two books by George Pullen Jackson: *White Spirituals in the Southern Uplands,* published in 1933, and *White and Negro Spirituals: Their Life Span and Kinship,* published in 1943.[21]

From the time of Appalachian settlement in the eighteenth century and before many rural churches were founded, sacred songs were sung in rural homes. By the nineteenth century, group and congregational singing began to emerge and disperse as singing teachers brought spiritual music to the mountains. Moreover, "singing conventions" were organized in churches and other public areas such as courthouse squares, where people came to visit, sing old songs, and have dinner on the ground.[22] Singing techniques and rudimentary music education in these settings led to some of the traditional gospel music that later was reflected in country and bluegrass music.

Sacred hymns as folk music originated in colonial America in the mid-eighteenth century. The first American Baptist hymnbook was published in 1766. Between 1780 and 1830 a thematic pattern of lyrics in print form emerged. George Pullen Jackson richly expresses the history and lyric themes in the entertaining table of contents in *White and Negro Spirituals.* Jackson's chapter 4 covers the mid-1700s to 1830, chapter 12, the entire nineteenth century, and chapter 13, from 1900 to 1940:

> IV. Freedom in Religion Begets Freedom in Singing. Religious Folk Song Is Born 37
>
> Insurgent Baptists get to making their own songs. . . . Dissenters still use standby hymns inherited from Britain. . . . American religious turmoil reflected in homespun lyrics. Examples of hymns defending singing, use of worldly tunes, immersion; hymns for footwashing, funerals, weddings; hymns warning of hellfire, promising salvation; hymns sublimating fleshly lusts to love of God; roving preachers' tales of woe; converts' tales of conversion; songs of earthquakes, eclipses and other acts of God. . . . Song contributions of Methodists come later, are fewer but similar. They ridicule Baptist "dipping," decry the "dancing crew." Later Baptist additions to hymn and ballad stock. The come-all-ye's. . . . Bulk of the folk-hymn and ballad text is in existence with the 1830s.
>
> XII. Old-Time Religion Outmoded. Social Gospel Comes In 125
>
> Individualism, the real weakness and ideal strength of the Old-Time Religion. Radical dissent becomes institutional, loses its character. The "Christians" as an example of grouping and regrouping. Other

sects become denominations. Religious behavior becomes polite with times. Churches adapt themselves to Main Street. Changeless groups suffer eclipse. Old-Time Religion lives on in out-of-the way places.

XIII. Old Time Songs Are Pushed Aside 131
Primitive Baptists trudge along old tune trails. Urban Baptists sing a new song. Seventh-Day Adventists still somewhat faithful to their early songs. *Sacred Harp* folk sing on in style of a century ago. Methodists completely exclude folk song excepting as antiques. Presbyterians know them not. Sects born in the past hundred years discard the folky, adopt gospel hymns and/or Lowell Mason. Shaker song disappears with the Shakers.[23]

Shaped Note Singing

Shaped note singing became a fundamental trait of religious music as rural Appalachian churches and singing conventions began to proliferate. Shaped note music was a printed form in which the notes were represented by symbols: mi, diamond; fa, triangle; so, circle; and la, square. Instead of eight-note scales, shaped note music was written in four-note groups.[24] The concept of published shaped note music originated in early eighteenth-century New England and diffused over much of the eastern United States by the early 1800s. Hymnbooks with shaped notes became a mainstay and between 1815 and 1855; twenty-one hymnbooks were published in the South. The best known was *The Sacred Harp,* first published in 1844 by C. F. White and E. J. King.[25] But earlier ones included *Western Harmony,* published in Nashville in 1824, *The Columbia Harmony,* published in Wilson County, Tennessee, in 1825, *Caldwell's Union Harmony,* from Maryville, Tennessee, in 1837, and *Jackson's Knoxville Harmony,* from Madisonville, Tennessee, in 1838.[26]

Only a few shaped note tunes without the symbols still appear in Baptist hymnals. Hymns such as "Come, Thou Fount of Every Blessing" and "Come, All Christians, Be Committed" are from the 1844 *Sacred Harp.* And from Walker's 1835 *Southern Harmony* come such old favorites as "My Shepherd Will Supply My Need" and "What Wondrous Love Is This."[27]

As printed and performed shaped note music declined elsewhere, it was preserved in Appalachia; there are congregations and singing conventions where shaped note singing still survives. David Brose, folklore manager at the John C. Campbell Folk School in western North Carolina, made field recordings of shaped note singing groups in East Tennessee as well as of Cherokee, black, and white spiritual singers in western North Carolina. Brose's recordings, made between 1992 and 1994, show that diverse forms of Appalachian religious folk music are well preserved in the region. Elsewhere since about 1980, shaped note singing has had an unconventional resurgence in the United States and as far away as Japan and England.[28]

Shaped note singing, also called Sacred Harp and Old Harp singing, lives on in Appalachia. For some churches and at other venues, the shaped note experience is offered, but not every Sunday. In 1999 here were some places in East Tennessee where the singing could be heard:

June 20—Beech Grove Primitive Baptist Church between Pigeon Forge and Gatlinburg, TN on US 441—turn right on Caney Creek Road, go two miles, church is on left.
June 30—Headrick's Chapel in Wears Valley on US 321 between the Blount and Sevier County line.
August 15—Cades Cove Missionary Baptist Church, Cades Cove, Great Smoky Mountains National Park.
September 12—Wears Valley Methodist Church, Wears Valley Road, US 321 between Pigeon Forge and Townsend, TN.
October 3—Valley View Baptist Church, Wears Cove, on Valley View Road just off US 321.
October 9—Museum of Appalachia Fall Homecoming, Norris, TN off I-75 north of Knoxville.
October 10—Dollywood, Pigeon Forge, TN.
November 24—Laurel Theater 16th ave, Knoxville, TN.[29]

Few people realize that infused throughout folk, country, and bluegrass music there has been an element of sacred music. In any bluegrass performance it is customary to have one or two hymns or gospel songs. Hardly a twentieth-century recording in old folk or bluegrass is released that does not have a sacred number in the playlist. Bill Monroe, the father of bluegrass music, performed at least one sacred song in each of his concerts in his six decades of entertaining.[30]

Connections to Country Music

Another hot summer in the southern mountains; it is late July through the first of August in 1927, and Alvin Pleasant "A. P." Carter, his wife Sara, and his very pregnant sister-in-law Maybelle Carter are entering the world of recorded music for the first time. With A. P. in bib over-alls and the ladies in calico dresses, the trio arrives at Bristol, Virginia-Tennessee in a Model

A Ford. They are here to see about making recordings for Mr. Ralph Peer of the Victor Talking Machine Company. They have driven all day over the twenty-five miles of rough roads from Mace Springs, a tiny community in the shadow of Clinch Mountain near Hilton in southwestern Virginia. They carry a Martin guitar, an old Stella guitar, and an autoharp. They meet Mr. Peer in an old furniture store at 408 State Street, where they record a few of the sacred and secular songs they know.[31]

The Carters recorded four songs that first evening: "Bury Me under the Weeping Willow," "Little Old Log Cabin by the Sea," "Poor Orphan Child," and "The Storms Are on the Ocean." The next day, only Sara and Maybelle returned to the makeshift studio, where Maybelle recorded two solos, "Wandering Boy" and "Single Girl." Several months later in 1928, Ralph Peer invited the Carter family to Camden, New Jersey, where they recorded songs that launched their musical career—the songs we know so well, such as "Wildwood Flower," "Keep on the Sunny Side," and "John Hardy Was a Desperate Little Man."[32] In plaintive voices, the Carters sang the old songs from the heart with an authentic mountain twang. In "Wildwood Flower," the vocalist's wildwood "flar" rhymed with black "harr" (hair) and "arr" (hour). In one hymn, they sang of the river "Jerrdin" (Jordan) and of a heavenly "qwar" (choir). The music was real, most of it was old, and the singers were veritable folk.

It was a sure beginning, and perhaps the birth of country music. The Carters in this format of A. P., Sara, and Maybelle were to make more recordings for the Victor Company throughout the 1930s. Between 1936 and 1939, the Carters expanded their group to include Maybelle's three daughters, Anita, June, and Helen. In 1938 the Carters went to work in Del Rio, Texas, where nationwide radio broadcasts on XERA could be heard from the station's powerful transmitter across the Rio Grande in Mexico. Between 1927 and 1941, the original Carters recorded some 250 songs. In 1950 Mother Maybelle Carter, as she became known, and her three daughters became a "new" Carter family that continued to record and appear on the Grand Ole Opry until the late 1960s. In 1970 the original Carter family and bluegrass legend Bill Monroe were elected to the country music hall of fame.[33]

The Grand Ole Opry

Mountain music experienced three means of preservation and dissemination: transcription by scholars like Sharp and Karpeles; phonograph recordings by Ralph Peer and the Victor Talking Machine Company and others; and broadcasts over powerful radio stations. Beginning in 1924 in Chicago with WLS's National Barn Dance with George Hay as the announcer, radio stations evolved into powerful promoters of country music. In 1925 Hay was hired away to WSM in Nashville, Tennessee, where he initiated a live old-timey folk music program that in 1927 became the Grand Ole Opry.[34] Eventually other stations featured country music on the airwaves: some examples were XERA in Del Rio, Texas; the Renfro Valley Barn Dance from Renfro Valley, Kentucky; Louisiana Hayride from Shreveport, Louisiana; WNOX's Midday Merry-Go-Round in Knoxville, Tennessee; and one of my favorites, WCKY in Cincinnati, Ohio. Listeners across the United States could tune in to their favorite country radio station and join in the nationalizing of a heretofore isolated form of vernacular culture—mountain music.

Opry greats in the 1920s included Uncle Jimmy Thompson (1848–1931) from Smith County, Tennessee; Uncle Dave Macon (1870–1952), a banjoist and all-around entertainer from Murfreesboro, Tennessee; and Dr. Humphrey Bates from Castalian Springs, northeast of Nashville. The Delmore Brothers from northern Alabama and the Dixieliners with Sam and Kirk McGee from near Franklin, Tennessee, arrived in the mid-1930s.[35] Roy Acuff from Maynardville, Tennessee, joined the Opry in 1938 and sang his signature songs, "The Great Speckle Bird" and "Wabash Cannonball," which he had recorded in 1936. The 1940s and 1950s brought on the honky-tonk sounds of Ernest Tubb and Hank Williams and the smoother sounds of Eddy Arnold.[36] Sad and sorry lyric themes of heartbreak, lost loves, hard drinking, divorce or simply "cryin', lyin', lovin' and leavin' songs" stamped an indelible mark on the music that would continue into the 1980s. Guitarist Chet Atkins from Luttrell, Tennessee, carried an instrumentalist tradition far in time, from the 1950s to 2001, and went far beyond the expected in quality and sophistication both in his playing and as a record producer and pioneer of a smoother "Nashville Sound."[37]

Female country music stars from poor backgrounds emerged more prominently in the 1960s and 1970s. A classic country singer is the coal miner's daughter Loretta Lynn, who came from a poverty-stricken background in Butcher Hollow, Kentucky, and followed the style of Kitty Wells.[38] From rags to riches, Dolly Parton took country music from her humble beginnings in Sevier County, Tennessee, to Knoxville and then to Nashville's Grand Ole Opry and on to Hollywood in an incredible journey that is the essence of dreams. Dolly has an authentic folk heritage, and she maintains the folk legacy beneath her glitzy image, so that she is truly as down to earth as anyone you might meet. But unlike Jean

Ritchie, a Kentucky folksinger who has maintained a folk presence and a strong connection between her past and her present, Dolly Parton has eclipsed her background to present an outlandishly lavish image. Still, she has never forgotten her roots. On her 1999 compact disc "The Grass Is Blue," Dolly performs several old country and bluegrass songs. The one most dear to the depth of her culture is the ballad "The Silver Dagger," an old story song in a minor key that Cecil Sharp rediscovered in twenty versions in Southern Appalachia in 1917 and 1918.[39]

Contemporary country music has outgrown its roots. Too few young country singers and entertainers today are from rural coal camps or farm and ranch country backgrounds. Alan Jackson's satirical song "Gone Country" tells of three entertainers: an aging club singer in Las Vegas, an outdated folksinger in Greenwich Village, and a California yuppie "schooled in voice and composition." Each plans to transform himself into a country singer by wearing cowboy boots, hat, and clothes and to cash in on the country music business, down there where people are "not as backward as they used to be."[40] Songwriter Larry Cordle, originally from Cordell in eastern Kentucky, has a song titled "Murder on Music Row" that tells how contemporary greed and popularity killed the traditional sounds of country music. The lyrics describe how drums and rock guitars have replaced slide guitars and fiddles and claim that Hank Williams just wouldn't stand a chance on today's radio.[41]

Geographically, some sing closer to their roots. In eastern Kentucky, U.S. 23 is called the "country music highway" because so many performers are from the area—for example, Ricky Skaggs, Crystal Gayle, Loretta Lynn, Naomi and Wynonna Judd, Billy Ray Cyrus, Patty Loveless, Dwight Yoakam, and Gary Stewart. But the sounds contemporary country musicians make with electric amplifiers, drums, brass horns, and synthesizers slick up a form of music that requires no polishing. Take me back to song stories with simple melodies in the keys of C, D, and G, accompanied by acoustic stringed instruments.

Connections to Bluegrass Music

One name says it all—Bill Monroe, the father of bluegrass music, the progenitor of a unique style of music that instrumentally and vocally is linked to Appalachian folk and country music. Geographically, however, the music had to backflow into the folk culture of Appalachia because early on the music was commercial, largely invented and played by professional musicians. Furthermore, its founder, Bill Monroe (1911–96), came from near Rosine in the "Pennyrile" (Pennyroyal) part of west-central Kentucky, a bit west of Appalachia but not in it. Contrary to folklore, bluegrass music did not originate in Kentucky's Bluegrass Basin, nor was it named specifically for the bluegrass state. But even with a suspect origin ever so slightly beyond Appalachia spatially, bluegrass music always has been embraced culturally by Appalachia.

George Carney offers a different observation on bluegrass music origins and contends that "the culture hearth of the bluegrass sound was the mountain and piedmont sections of western North Carolina."[42] He bases the idea on the fact that so many string bands with fiddle, guitar, and especially three-finger roll banjo playing emerged in western North Carolina in the 1920s and 1930s. This occurred some fifteen years before Earl Scruggs (1924–), a North Carolina three-finger roll banjoist, joined Bill Monroe and the Bluegrass Boys in 1945. Carney cites the Snuffy Jenkins family and brothers Hoke, Vern, and Oren from the Harris community in Rutherford County, North Carolina, as having significant influence on banjo greats Ralph Stanley and Don Reno. In adjacent Cleveland County, the Scruggs clan had musician brothers Junie, Horace, and younger brother Earl.[43] Earl Scruggs taught himself the three-finger roll technique as a boy and introduced the unique banjo style to bluegrass music in 1945 when he joined Bill Monroe and the Bluegrass Boys. With the three-finger roll more strings are picked, so that the sounds are fuller and faster. The old claw-hammer and two-finger styles seem to pluck in time with the song's tempo and don't fill the spaces between notes as fast.

Bluegrass, however, is more than a style of banjo music. This "high lonesome sound," perfected by Bill Monroe, is both unique and surprisingly recent. Bill Monroe's bluegrass sound evolved when he began singing with his brother Charlie as the Monroe Brothers in the 1930s. Some people say the sound emerged first in 1940, but I agree with those who say it was in 1945 that the true full form of bluegrass sound and instrumentation arrived when Monroe added Earl Scruggs's three-finger picking style on banjo to the ensemble. In 1950 the term "bluegrass music" was applied to the sound, named after Bill Monroe's band "the Bluegrass Boys."[44] In spite of its old-timey sound, Monroe's bluegrass music predates rock and roll music only by a mere decade or less!

The diagnostic culture traits of bluegrass music are high tenor lead singer; chorus in harmony; acoustic stringed instruments of mandolin, fiddle, guitar, stand-up string bass, and, most pivotal, a Scruggs-style three-finger roll picked five-string banjo.[45] Some groups added a special acoustically augmented resonator slide guitar called a Dobro. The instrument was named for the Dopyera brothers, John and Rudy, Slovakian immigrants who invented it in 1927–28.[46] Bluegrass has been called a form of country jazz in which each instrumentalist improvises a solo in each song. Songs are fast-paced, some with frantic tempos, and sung and played in C, D, or G major.[47] On very rare occasions a song may be played in a minor key. Most bluegrass tunes are in 2/4 or 4/4 time, but some are in 3/4 time, such as the first part of Monroe's "Blue Moon of Kentucky." The songs' content is a mix of ballad and folk lyrics, depicting family members such as "Uncle Pen"; places such as "Rocky Top" and "Happy Valley"; farms and old homeplaces; coal mines, mules, and trains; or lonesome, sad, and hopeful feelings. Many songs have a story line like a ballad. In some ways the lyric themes are similar to country music, but unlike country music they usually avoid themes of alcohol, adultery, and divorce. The keys, chords, and twangs remind us that bluegrass music is not far (if at all) removed from mountain folk music.

Bluegrass bands are numerous if not popular in mainstream American culture. In the 1940s the best-known groups were Bill Monroe and the Bluegrass Boys; Lester Flatt and Earl Scruggs and the Foggy Mountain Boys; the Stanley Brothers (Ralph and Carter) and the Clinch Mountain Boys; the Bailey Brothers and the Happy Valley Boys; and Jim and Jesse and the Virginia Boys, among others. Flatt and Scruggs and the Stanley Brothers bands sounded the most like Bill Monroe's bluegrass sound. In the 1950s, bluegrass continued with the emergence of the Osborne Brothers; Mac Wiseman; Jimmy Martin and the Sunshine Boys; and Reno and Smiley and the Tennessee Cut-Ups. Between 1952 and 1962 the duo of Don Reno and Red Smiley recorded more than 250 songs for King Records. Reno's rapid-fire banjo picking and the duo's high-pitched twang endeared them to listeners of bluegrass, gospel, and country music. In the late 1950s and through the 1960s, more groups competing against rock and roll emerged, such as the Country Gentlemen, the Kentucky Colonels, and Del McCoury.[48]

Bluegrass music struck different waters in the 1960s with the folk music revival. In some ways bluegrass was treated as a musical subculture because the old performers reflected their own way of life, which was closer to the rustic listeners to country music than to the hip young urban audiences of "newgrass" folk music. Bluegrass survived the 1960s, and some unusual groups formed and recorded in the 1970s, 1980s, and 1990s: J. D. Crowe, who started as a contemporary of Sonny Osborne; Hazel Dickens and Alice Gerrard, one of the first female groups; Tony Rice; Alison Krauss; and Dobro player Jerry Douglas, among many fine musicians.[49] But the most significant contemporary culture bearer of traditional bluegrass music is Ricky Skaggs. Born in Lawrence County (home of Appalachian scholar Cratis Williams) in eastern Kentucky, Ricky Skaggs was raised on the bluegrass music of Bill Monroe, the Stanley Brothers, and Don Reno and Red Smiley. He was steeped in their music, and he has never lost it.

Bluegrass music has moved beyond its roots and regional identity. Creative names of contemporary bluegrass groups express interesting combinations: *Here Today, The Dry Branch Fire Squad,* on their *IIIrd Tyme Out* on *Blue Highway* was *All Over the Road* yet *Seldom Scene* and *Rarely Herd* as they dodged *Dreadful Snakes* to put out a *Hot Rize* fire where the *Skillet Lickers* were *Country Cookin'* some *Leftover Salmon* for a bunch of *Freight Hoppers* who were *Old and in the Way.* A little closer to home are players such as Raymond Fairchild, folk hero and five-time world champion banjo picker, from Maggie Valley, North Carolina; Dr. Mack Snoderly, a retired dentist from Clyde, North Carolina, who has been fiddle champion more than eighty-nine times; and guitarist George Webb, from Blount County, Tennessee, three who perform bluegrass professionally, semiprofessionally, and just for fun, respectively.

An example of a folk musician who became a bluegrass star is Ralph Stanley (1925–). Now in his late seventies, the legendary Ralph Stanley is one of the last remaining icons of bluegrass music. He performs two hundred concerts a year and has made over 150 record albums. Ralph and his brother Carter grew up in tiny McClure, Virginia, in Dickenson County just about twenty-five miles as the crow flies due north from the Carter family homeplace of Mace Springs near Hilton, Virginia. Ralph learned the traditional claw hammer style of banjo playing from his mother, Lucy, and he played mountain music on the banjo while he was in the army during World War II. The Stanley Brothers and the Clinch Mountain Boys formed their first band in 1946 and began playing on radio station WCYB in Bristol, Virginia. They played on the *Farm and Fun Time Show* a noontime broadcast aimed at farmers who came in from the fields to eat, rest, and listen to country and bluegrass music on the radio. Ralph had learned the Scruggs style or three-finger roll technique of banjo playing and has kept the style going for more than fifty years. The Stanleys' singing style had a high, lonesome sound, with Carter singing the melody and Ralph doing high tenor harmony. When Carter Stanley died at age forty-one in 1966, Ralph almost quit the music business, but fans persuaded him otherwise. He returned with the same style and feeling in his music by going back home to McClure in southwestern Virginia, to his roots and his traditions.[50]

A pioneer bluegrass musician and a 2002 Grammy Award winner, Dr. Ralph Stanley performs at the 2002 Museum of Appalachia Fall Homecoming. (Photograph by J. Rehder, 2002)

The range of bluegrass music—musicians and listeners—now extends worldwide, far beyond its birthplace somewhere between western North Carolina and the Pennyroyal in west-central Kentucky. The irony of the musical forms found in Appalachia is that as older folk songs are dying out, country music is now in vogue, integrating technical forms of popular and rock and roll music. But bluegrass survives—almost all of it still sounds old, and it still feels like real folk.

The Folk Music Revival

In the late 1950s and throughout the 1960s, a revival of folk music took shape with old folksingers like Woody Guthrie and Pete Seeger. Simultaneously, younger entertainers propelled the movement, such as the Kingston Trio, Peter, Paul, and Mary, Joan Baez, and Canadians Ian and

Sylvia and Gordon Lightfoot, among many others. As the folk music revival was sweeping the nation, music historian Neil Rosenberg wondered whether bluegrass was a special type of country music or a special type of folk music.[51] How did any of this music connect to Appalachia ethnically, culturally, temporally, or any other way at all?

In the 1950s and 1960s, collectors acting much like musical archaeologists sought old forms of mountain music performed in situ on front porches, at old-time fiddlers' conventions, and later at bluegrass festivals that sprouted up in Appalachia and surrounding areas. In 1955 a New Jersey suburbanite, Ralph Rinzler, a fan who was following bluegrass greats like Bill Monroe and the Stanley Brothers at open-air performances, said: "Bluegrass had not got into the folk revival. But for me it [bluegrass] was like going into another world. I was fascinated by the totally different life-style—dinner on the grounds, different speech patterns—a whole different way of life. The whole idea really astounded me—that this existed."[52]

Ralph Rinzler migrated from music fan to Greenwich Village performer to folk song collector in Appalachia to director of the Smithsonian Folklife Festivals in Washington, DC. He immersed himself in bluegrass so much that he became Bill Monroe's manager for a few months.[53] In 1960 Rinzler discovered Doc Watson (1923–) and Clarence Tom Ashley (1895–1967) at the old-time fiddlers' convention at Union Grove, North Carolina. For the next two years, Rinzler recorded Watson and Ashley on Folkways Records and promoted them in folk concert venues throughout the country. Until Rinzler discovered him in 1960, Doc Watson, a blind guitarist from Deep Gap in Watauga County, North Carolina, played backup electric guitar in local country bands. Between 1913 and 1943 Clarence Tom Ashley, from Shouns, Tennessee, near Mountain City in Upper East Tennessee, made most of his livelihood picking a banjo and singing in a traveling medicine show and busting (busking)—playing for loose change at Appalachian coal camps and sawmill camps on payday. Ashley, however, had given up musical entertaining in the 1940s until he was rediscovered by Rinzler in 1960.[54]

From this point of rediscovery, the folk music revival safely proceeded with authentic Appalachian folk tunes and folk musicians. Ralph Rinzler's friend Mike Seeger, half-brother to Pete Seeger, was so inspired by Library of Congress field recordings that he went to Southern Appalachia in the 1950s and 1960s to record more Appalachian folk music and banjo musicians. His *American Banjo Three-Finger and Scruggs Style* album was a significant banjo collection of folk musicians, mostly from

Doc Watson, a legendary folk singer, performs at the 2002 Museum of Appalachia Fall Homecoming. (Photograph by J. Rehder, 2002)

North Carolina, who played with the unique Scruggs style of three-finger roll picking.[55]

The Cultural Meaning of Appalachian Music

How do we tie up the loose threads of Appalachian music? The questions are repeatable. Is it folk? Is it traditional? Does it reflect the culture as a homemade item? The answers are a qualified yes when applied to most of the eighteenth- and nineteenth-century Appalachian ballads and tunes and to most of the country music played between 1920 and 1945. Afterward, commercial and amateur forms of bluegrass and a revival of folk music called "newgrass" in the 1960s, largely played and promoted by nonfolk people, expand the range and cloud the picture, causing an identity crisis. How folk is bluegrass, and how folk is the revived folk music of the 1960s? The acid test of the folk nature of bluegrass music is in these questions: Do the musicians learn and play songs by ear in the aural tradition? Are the instruments primarily acoustic fiddles, banjos, guitars, mandolins, Dobro guitars, and bass? Are the musicians playing songs in C, D, and G major? Are their lyrics based on rural old-time themes of home, homeplace, coal mines, mules, and other folksy topics? Another test would be authentic: old-timey, traditional characteristics such as George Hay mandated for the kind of music to be played in the early days at the Grand Ole Opry: "Let's bring it back down to earth, boys."[56]

Contemporary individuals illuminate the interpretation. A well-known mountain folk singer is Jean Ritchie, the last baby in a family of fifteen, born in 1922 in the coalfields-and-hollows country in Viper, Kentucky. Ritchie represents a bridge between an Appalachian folk upbringing and a major commercial career in folk music. Ritchie's 1955 book *Singing Family of the Cumberlands* includes forty-two songs as linkages to a legacy of folk history through song.[57] The book tells her story of growing up in a most difficult time in a difficult part of eastern Kentucky. In spite of a sea of poverty surrounding the family, the Ritchies survived, and Jean emerges from it an internationally renowned folksinger.

An important contemporary preserver of the culture is Sheila Adams, born in Sodom, North Carolina, in the mountain county of Madison. Sheila is a culture bearer who lives to perpetuate the mountain ballads that she learned from her relatives and sages in her community. Singing seems to come naturally here, and people sing as easily as they speak. Sheila's great-great aunt and teacher, Granny Dellie Chandler Norton, was her greatest inspiration and her folk culture tutor. Granny would sing a line of a ballad and ask Sheila to repeat it; then she would sing the next line and Sheila would recite both lines, and so on until she had mastered the entire piece. Adams has learned countless ballads this way and is a precious living collection of Appalachian folk music.[58]

The fiddlers' convention became the cultural mecca of Appalachian folk music pilgrimages and a focus for folk musicians and audiences. My introduction to fiddlers' conventions came in the late 1950s when I was a teenage drummer for three instrumental groups around Winston-Salem, North Carolina. One band was a country-rock group in Mount Airy, North Carolina, that had a fiddle, a flat-top guitar, an electric guitar, bass, and my drums. The fiddler and flat-top guitar player bragged that they had won or placed high in fiddlers' conventions at Union Grove and Mount Airy, North Carolina, and at the one in Galax, Virginia. Over forty years have passed, and I have forgotten the players' names, but to them a fiddlers' convention was a folk/bluegrass/country music academy from which a performer emerged with credentials. Competing in fiddlers' conventions laid claim to their cultural legacy. As a poor city boy from the projects, I envied these pickers and fiddlers because their music had the stamp of approval from their culture. Each of us—you, me, Janette Carter, Dolly Parton, Doc Watson, Ricky Skaggs, Earl Scruggs, or absolutely anyone—is a cultural mélange taken from our past, our families, and our cultural surroundings plus the persons we grow up to be. Our legacy may tell where we came from, but our present personality expresses what we have become at a contemporary time in contemporary society.

Appalachia, a known reservoir of folk culture, became a cause célèbre and justified destination for those in the search of folk music. The mountains were always perceived to be isolated pockets of archaic pioneer culture and tradition and the prime source for folk songs. Among the early collectors, the Campbells and Sharp and Karpeles in 1916–18 certainly thought that the mountains held the best chance for song survivals. Ralph Peer believed so when he went to Bristol, Tennessee, in 1927 and discovered the Carter family, yodeling Jimmie Rodgers, and other potential country singers. "Song catchers" John A. Lomax and Alan Lomax did exactly the same thing by going to the mountains from the 1930s through the 1960s seeking and recording old hymns and secular folk songs. Ralph Rinzler in 1960 thought the same way in his discovery of Doc Watson and Clarence Tom Ashley. Mike Seeger in the 1950s and 1960s also went to the region seeking banjo tunes and recorded pickers and singers. We are fortunate that these serious song seekers in their own time each had the interest, motivation, skill, and recording technology to collect and preserve this elusive, ephemeral, invisible phenomenon that is the folk music of Appalachia. In addition to preservation, others found evidence of their own heritage in Appalachian folk music. Evan Carawan, a hammered dulcimer player, explains that "music was a way to express your cultural history." Without the early collectors who acted as ethnomusicologists, and without contemporary folk singers like Sheila Adams, Jean Ritchie, Doc Watson, and others who communicated the rich traditions of the region's music, it is likely that too many of the oral traditions would have just faded away.

Folk Art and Crafts

In a made-at-home, homemade society, traditional material folk culture is the stuff that cultural geographers, anthropologists, and folk life scholars dream about. Baskets, quilts, wooden bowls and buckets, stools and chairs, horseshoes, plows, and any number of old-timey home and farm objects reflect a utilitarian way of life. I believe that southern mountain handicrafts and art should be interpreted as traditional artifacts rather than as objets d'art. Questions of authenticity haunt us even here. Are the objects folk? Were the people who made them folk, and did they make them at home, where they learned crafting skills from older folk relatives? Or were they trained at a folk school like the John C. Campbell Folk School in Brasstown, North Carolina, or Berea College in Kentucky? Perhaps the artist is not from Appalachia at all but is a transplanted New York graduate from a Manhattan art school. Fine art is one thing, but folk art must be something else.

Self-sufficiency in Folk Culture

In a frontier life of self-sufficiency, Appalachian folk had to make do. People did their own carpentry, weaving, pottery, woodworking, and many other tasks that we now call handicrafts. Log houses, outbuildings, and fences came from forests of yellow poplar, oak, and chestnut. Furniture was made from cherry, chestnut, walnut, and other hardwoods. The wood from ash trees, iron-hard yet not brittle, became handles for axes, hoes, froes, scythes, rakes, and hammers. Oak or hickory was used for wagon wheels, axles, and other durable parts of wooden farm equipment. Thin slivers of white oak were woven into chair seats and baskets. Thicker white oak shakes riven with a froe became shingles for covering roofs. Residing within each family were the traditional skills of making things not to sell but to use, things necessary to life in a technically and culturally isolated mountain environment.

The evolution of Appalachian material folk culture in handicrafts and art began with the earliest initial settlement of the region. The necessities of life, the availability of resources, and the traditions of the culture dictated material goods produced principally for home use. Things were made at home to be used at home, without much thought to producing them for an outside market. Self-sufficiency was the norm, and isolation, independence, and people's way of life dictated that most homemade things simply stayed there. This is not to say that people did not communicate and trade among themselves. Folk surely made contact with each other. They ground corn at gristmills and traded goods there. They went to the county seat to record deeds and to conduct other limited business. Socially they visited relatives and neighbors, and some people made a habit of going to church. Still, much of their culture remained at home, caught in the traditional whirlpool of an archaic culture. Not many years ago, a common assumption was that no one would sell an old quilt—it was too personal, with family memories stitched into every square inch of the fabric. A quilt was an heirloom to be kept and used but never sold. Such deliberate cultural shielding ended when strangers began to enter the region in cars and, seeing a quilt airing on a line in springtime, asked, "Is that coverlet for sale? I'll give you ten dollars for it."

Skilled people began to specialize as millers, blacksmiths, wheelwrights, gunsmiths, potters, cobblers, and weavers to serve mountain communities. As populations grew and transportation improved, even ever so slightly, this dichotomy of focused tasks led some people to recognize that they could make a living doing specialty work. Producing farm and mining equipment diversified spatially and temporally, but much of Appalachian crafting remained at home until the late nineteenth and early twentieth centuries.

The Material Folk Culture of Handcrafted Objects

Anthropologists, archaeologists, and cultural geographers are always searching for the material objects that are the hands-on substance of a culture. Material culture encompasses the visible, tangible objects that reflect the culture's folk nature. In the home, the richness of material culture is manifested by the craft objects that can be listed as material culture traits; some may be interpreted as art.

One approach to compiling a trait list of Appalachian crafts is to examine categories of crafts based on the raw materials they use: those that involve textiles, such as spinning and weaving or quilt making; those that use wood, such as furniture making and woodworking, basketmaking, and whittling or carving; and those that use clay, such as pottery making. However, for some items, such as toys and dolls, the raw materials are unimportant. Another approach to a trait list of Appalachia's material folk culture is used in the most comprehensive study of Appalachian folk art and crafts ever assembled—Allen Eaton's *Handicrafts of the Southern Highlands,* first published in 1937 by the Russell Sage Foundation. Eaton's book reflects the kinds of folk handicrafts that were present in Southern Appalachia in the 1930s: spinning and weaving; making coverlets and counterpanes; quilting and patchwork; preparing and using native dyes and herbs; making furniture and woodwork; weaving mountain baskets; whittling and carving; creating dolls, toys, and miniature furniture; crafting mountain musical instruments; producing pot-

A Grainger County, Tennessee, farmer and wood-carver greeted me one day with a basketful of hand-carved cedar pocket knives. (Photograph by J. Rehder, 1978)

tery; and other handicrafts.[59] Contemporary Appalachian folk handicrafts include just about all of those that Eaton observed in the 1930s, but here I will attempt only a brief discussion of baskets, weaving, and quilts.

Baskets

Basketry, a very old and nearly universal craft, came to the southern mountains long before European settlement. Native American Cherokees and their ancestors were Appalachia's first basketmakers and are believed to have created some of the finest baskets ever woven in the region. Baskets were used for gathering and carrying eggs, fruit, vegetables, wild berries, nuts, and many other edible and inedible things. In the absence of wooden boxes or crates, which arrived later but required sawn lumber, baskets protected small cargos over relatively short distances. Unlike the crude cloth sacks that also came later, baskets allowed the carrier to observe and preserve the condition of fragile articles like eggs and soft, ripe peaches, grapes, or plums. Baskets could also be used for storing and measuring. In the 1930s Allen Eaton saw baskets used as legal tender in country stores, where folk baskets were piled on the floor or hung from the ceiling.[60]

Basket types took their names from their shapes, their uses, and even from the family names of well-known makers. They could be round, oval, oblong, square, triangular, or shaped like a boat, melon or jug. Functionally, there were egg baskets, market baskets, workbaskets, and fruit, flower, and lunch baskets. Other types included lady baskets, knitting and sewing baskets, and tea, seed, trinket, and charcoal baskets. Families such as Rector, Purvis, and Morgan had baskets named for them because of the quality of their workmanship.[61]

The Appalachian gizzard basket is a unique type. It is also called the hip, butt, or cheek type and is found in the mountains of western North Carolina and East Tennessee. The basket has two oval or melon-shaped lobes, one on each side of the circular handle, so it resembles a pair of saddlebags. This special shape is believed to be designed to rest on the hip of a woman walking or riding horseback.[62]

The most durable and widely used material for Appalachian basketry was split white oak, *Quercus alba,* also appropriately called basket oak. Basketmakers also used splints or strips of other oak species or ash or hickory trees, peeled willow bark, honeysuckle vines, grapevines, broom sedge, or cornhusks. Native American basketmakers chose split canes, reeds, vines, and grasses, tree bark, and tree splints. Tommy Reed, a folk basketmaker I know, says he is "half Cherokee from my daddy and half Sevier County [Tennessee] hillbilly from my mother's side." Tommy, now in his late forties, has made baskets since he was a teenager. In 1999 he made two baskets for me using grapevines, honeysuckle vines, and white oak for the framework and long strips of red spruce bark for the laths. I also spoke with Emily Smith, a Cherokee basket weaver from Cherokee, North Carolina. She makes baskets from very thin white oak splints, one-sixteenth to one-eighth inch wide, and dyes them with natural dyes of bloodroot for a reddish orange color and crushed whole walnuts for a deep gray, almost black. Weaving a basket takes three or four hours, she says, but much more time is spent gathering and preparing

A Native American basketmaker exhibits his wares. (Photograph by J. Rehder, 1998)

the raw materials that go into basketmaking. She and her family may spend over a week gathering white oak, splitting it, soaking the splints, gathering the dye plants, preparing the dyes, and dying the splints. No wonder baskets cost so much. Over the years I have paid about $15 to $25 for a basket, or and sometimes a bit more; however, I have seen baskets priced at over $350.

Weaving

"Homespun" says much about a rustic expectation of Appalachia but little about the artistry of spinning and weaving. The threads of life's tapestry are reflected in the work of mountain weavers. In 2001–2 the Frank H. McClung Museum at the University of Tennessee filled a large room with coverlets and quilts for an exhibition called Textile Art from Southern Appalachia: The Quiet Work of Women. The traveling exhibition, organized by the American Textile Museum of Lowell, Massachusetts, contained forty-four coverlets and two quilts made by thirty Appalachian women weavers—twenty from southwestern Virginia, five from western North Carolina, three from East Tennessee, and one each from southern West Virginia and eastern Kentucky.

The intricately patterned bed-sized coverlets, made from homespun wool, cotton, or linen and dyed with local native plant dyes, gave the museum an unparalleled range of tapestry patterns and colors. Patterns were so complex that it was hard to believe the weavers did not have sophisticated computer graphics to generate the designs. All pieces in the exhibition dated between 1830 and 1930. About one-third were woven between 1830 and 1865 and another third between 1870 and 1900. Largely working from traditional patterns or from designs of their own making, all but one of the weavers learned her craft from family folkways. The only exception was Josie Mast from Watauga County, North Carolina, who became a professional weaver during the craft revival era in the early 1900s and was so influenced by the movement that she taught weaving classes in western North Carolina. It seemed odd that in the year 2002 I had go to a museum to see the woven art that was once so commonplace in Appalachian folk culture. The very word "homespun" still signals an ancient way of life with a granny seated at her spinning wheel and spinning the wool thread that will later be dyed and strung on a loom in the house or in a separate loom house out back. How much time did it take to weave a bed-sized coverlet? Conservatively, a month or two, and then only if the weaver spent most of her time at the loom.[63] The next time you purchase a hand-woven garment, ask yourself, How was the thread spun, woven, and transformed into the clothes on my back?

Quilts

In the tapestry of folk life nothing has become more symbolic of Appalachian culture than a homemade family heirloom—the quilt.[64] The quilt on a bed or in a family's cedar chest is not only a possession but a cultural talisman. Why do quilts carry so much symbolism, so much cultural identity? Keepsake quilts represent a family's history. The patchwork of tiny cloth swatches comes from the family's own clothes. A grandmother's handmade quilt becomes a family chronicle that tells a story. "These yellow patches were from your mother's favorite dress, and these blue swatches are from your Uncle Bill's clothes; he was killed in the Second World War, you know." A quilt requires interpretation.

People like to be remembered. John Rice Irwin, founder and curator of the Museum of Appalachia, writes:

> The men in our society have traditionally had an opportunity, or at least a hope, that they would have a chance to leave some evidence of their existence; their farm, their business, the houses they had built, the chimneys they had laid, the devices they had invented, and most of all the

A quilt can be an heirloom, a piece of art, and a cover on a cold winter night. (Photograph by J. Rehder, 2002)

> children that bore their name. But the women could hope for no such remembrances. Until recently they could have little or no hope of leaving any reminders of their passage through this world. Very often the only kind of permanent recognition a woman got was through a beautiful and expertly sewn quilt.[65]

Appalachian folk did not often practice art for art's sake, especially in making quilts. A quilt was utilitarian, made to keep the children warm. It was made of any available scraps of cloth because nothing was to be wasted and nothing much else was available. Quiltmaking took more time than equipment, but that seems to be true of most items in Appalachia's material folk culture. A small amount of raw material, a few simple tools, but much time and skill to master the craft seem to be the essence of the folk arts.

Mountain Handicrafts Revival: The Role of Crafting Schools and Guilds

For much of their history, Appalachian folk crafts were made for home use and nothing more. Few items were traded, and fewer yet were sold for cash. Crafts seemed to be a dying art. But in the late nineteenth and early twentieth centuries a revitalization and commercialization of Appalachian handicrafts began with the initial efforts of outsiders and philanthropists. In 1895 Francis L. Goodrich, a social worker with the Women's Board of Home Missions for the Presbyterian Church, founded the Allanstand Cottage Industries in Asheville, in Buncombe County, North Carolina. In the first quarter of the twentieth century Olive Dame Campbell and her husband John C. Campbell, director of the Southern Highlands Division of the Russell Sage Foundation, were actively investigating the region and supporting others' efforts to inventory folk elements of the culture. It was the Campbells who encouraged Cecil Sharp to conduct his folk music survey in 1916–18. This time, however, Mrs. Campbell enlisted Allen H. Eaton to conduct a regional survey of southern Appalachian handicrafts, a major part of the material folk culture, which appeared in 1937 as *Handicrafts of the Southern Highlands.* But it seems that until the Campbells, Sharp and Karpeles, Eaton, and the sponsorship of the Russell Sage Foundation arrived in Appalachia, before 1940 few writers other than Horace Kephart had any interest in accurately chronicling this part of the material folk culture of Southern Appalachia.

Olive Dame Campbell got the idea for the Southern Highland Handicraft Guild after a 1923 visit to northern Europe, where she observed Finnish cooperative craft schools, Danish folk schools, and others. Impressed with the European model schools, in 1925 Mrs. Campbell established the John C. Campbell Folk School in Brasstown, in Cherokee County, North Carolina. Mrs. Campbell met Allen Eaton in 1926 and persuaded him to work on the plan for a handicraft guild that became the Southern Highland Handicraft Guild in 1929.[66]

Before 1890 few organized efforts were being made to focus on handicraft education in Appalachia. But between 1893 and 1928, schools and handicraft centers began to emerge that encouraged the revival and teaching of folk crafts. In chronological order they included Berea College (1893), Allanstand Cottage Industries (1895), Biltmore Industries (1901), Hindman Settlement School (1902), Berry Schools (1902), Tallulah Falls Industrial School (1909), Pi Beta Phi Settlement School (1912),

Pine Mountain Settlement School (1913), Tryon Toy Makers and Wood Carvers (1915), and Asheville Normal and Associated Schools (1916). In the next decade, between 1920 and 1930, more schools and educational craft centers appeared, such as Rosemont Industries (1920), Pine Burr Studio (1921), the Weaving Center at Crossnore School (1922), Blue Ridge Weavers (1922), Matheny Weavers (1922), Shuttle Crafters (1923), Penland Weavers and Potters (1923), Pittman Community Center (1923), Handicraft Guild of the Diocese of Southwestern Virginia (1923), Spinning Wheel (1924), Wooten Fireside Industries (1924), Weave Shop (1924), John C. Campbell Folk School (1925), Dry Hill Community Center and Carcassonne Community Center (1925), Blue Ridge Industrial School (1926), Craft Guild of the John C. Campbell Folk School (1927), Shenandoah Community Workers (1927), and Dorland Bell School (1928).[67] The list of early crafting schools in southern Appalachia is impressive, if not overwhelming. I have chosen two schools: the John C. Campbell Folk School in North Carolina (because it is one of the more developed crafting schools in the region) and the Hindman Settlement School in Kentucky to illustrate the origin of settlement schools and their influence on Appalachian culture.

The John C. Campbell Folk School

Nestled in the mountains of southwestern North Carolina, just south of Murphy in Cherokee County, sits the 372-acre farm and campus of the John C. Campbell Folk School. Founded in 1925 by Olive Dame Campbell and Marguerite Butler, the school was created "to bring people toward two kinds of development: inner growth as creative, thoughtful individuals, and social development as tolerant, caring members of a community."[68]

The school is modeled after the Danish *folkehoskjole,* in which instruction is based on the Danish concept of "the living word" that emphasizes discussion, conversation, and hands-on experience rather than learning by reading or writing. Classes are noncompetitive and have no credits, no grades, no degrees, and no pitting of one person against another. Each year 450 classes cover 43 subjects in crafts, music, and art.

In 1923 Olive Campbell and her friend Marguerite Butler studied folk schools in Denmark, Sweden, and other parts of Europe. They discovered that the *folkehoskjole* concept in Denmark had been a major driving force in rural life that changed the Danish countryside into a creative core of spirited cultural preservation and development. Mrs. Campbell came away with a different perspective on what it meant to be folk when she wrote: "*Folk* in English has come to be associated with people of the past period—*folk*-song, *folk*-dance, while the Danish folk are the Danish people. . . . the Danish people's college cares little about imparting exact knowledge; for knowledge alone does not make man good, useful or happy. The real aim of these schools is to awaken, enliven, and enlighten."[69] Armed with a folk school model, Olive Campbell looked for a suitable location for this experimental school. She already understood that she could not impose her ideas on the mountain people—that they must form a trust in a collaborative relationship. While searching the Brasstown, North Carolina, area about one hundred miles southwest of Asheville, Miss Butler explained the concept to Fred O. Scroggs, the local storekeeper. When she returned several weeks later, more than two hundred people met her and pledged building materials and support. Scroggs donated seventy-five acres to initiate the John C. Campbell Folk School site in 1925.[70]

From these beginnings, the school developed into the one that Olive Campbell sought in the *folkehoskjole* model. Local people did indeed take up wood carving and other crafts, and the sale of handcrafted objects became a real part of the local economy. Today's students at the John C. Campbell Folk School are rarely from the local populace. Participants from all over the country pay a fee to come and learn a craft on campus in a live-learn-enjoy climate. The concept is similar to an Earthwatch or elder hostel learning experience and serves as a "summer camp" for adults from somewhere else who pay for the privilege of eating, sleeping, and learning in a folk school environment.

The Hindman Settlement School

From Knoxville, Tennessee, I have been driving just about all day to get to Hindman, Kentucky, which is just a few more miles east of Hazard in Knott County, deep in the Appalachian coalfields of eastern Kentucky. Troublesome Creek runs through the property, which consists of several small acres of bottomland where a stone school building dominates the landscape. Amid school buildings and cottages of various types and ages, we find a small log single pen house that was once the cabin of Solomon Everidge, the eighty-year-old pioneer who sought an education for the youth in his community. In 1899 Everidge walked barefoot twenty-two miles to Hazard, Kentucky, to persuade two visiting teachers, May Stone and Katherine Pettit, to come to the Forks of Troublesome Creek at Hindman. He described the two ladies who taught sewing, cooking, reading, and writing under the shade of two canvas tents there in Hazard as the "quare fotched-on women from the level land." On that day and in his own words, "Uncle Sol" Everidge pleaded:

> Women my name is Solomon Everidge. Some call me the granddaddy of Troublesome. Since I was a little shirt tail boy, hoeing corn on the hillsides, I have looked up Troublesome and down Troublesome for somebody to come in and larn us something. My childhood passed and my manhood, and now my head is ablooming for the grave, and still nobody haint come. I growed up ignorant and mean, my offsprings was wuss, my grands wusser, and what my greats will be if something haint done to stop the meanness of their maneuvers, God only knows. When I heard the tale of you two women, I walked the 22 miles acrost the ridges to search out the truth of it. I am now persuaded you are the ones I have looked for all my lifetime. Come over to Troublesome, women, and do for us what you are doing here.[71]

The Hindman Settlement School was founded in 1902 by May Stone and Katherine Pettit "to provide educational opportunities for the boys and girls of the area, keep them mindful of their heritage and to provide community service activities." The school, the first rural social settlement school in the United States, remains a nonprofit, nondenominational institution. Early in its development, Hindman's teachers arrived from prestigious northern colleges such as Vassar, Smith, Wellesley, and Mount Holyoke and brought a year or two of energized teaching and new forms of culture to mountain children. One young lady from Rhode Island, Elizabeth Watts, was recruited in 1909 by John C. Campbell and Olive Dame Campbell to teach for a year at Hindman, Miss Watts worked at Hindman as teacher and director for eighty-four years before she died there at age 102.[72]

Mountain children from all over eastern Kentucky came to Hindman to board and go to school. Carolyn White, a former student at Hindman in 1948–52, expressed her thoughts on the school: "The things I learned at Hindman took me out of the mountains in such a very positive way. There were two worlds. I learned to appreciate my mountain heritage and I learned much about the outside world. Without the Hindman Settlement School, I would have been lost."[73] Today's students no longer board at Hindman, because the regional roads and schoolbuses are much improved.

As the local public school system expanded, Hindman offered programs and services not found in public schools. The Hindman Settlement School today provides music, art, and 4-H instruction to the area's public schools through an extension program. It is a center for cultural outreach in the community and operates seven after-school programs, a Montessori preschool, a dyslexia program, an adult basic education/GED program, library and meeting facilities, and workshops on Appalachian life and culture, among other programs.[74]

In the crafts movement in the early years at Hindman, dulcimer making and playing became important new "diagnostic" elements of the folk culture in the area. James Edward "Uncle Ed" Thomas became the father of dulcimer making, and through his teaching and that of others such as Jethro Amburgey at the Hindman Settlement School, an Old World folk instrument gained notoriety. It was so popular that some people erroneously believed the dulcimer was indigenous to the region and called it the "Appalachian dulcimer." Popular though it may be, the dulcimer was an Old World instrument that was restored to the culture through the very fine and influential craftsmen at Hindman.

Folk Festivals

How does a culture express itself beyond home and family? Social groupings organized according to religious, educational, or other principles would have us believe that by compartmentalizing our activities we can understand who we are. The Appalachian folk festivals not only provide outlets for creative energy but clarify what it means to call oneself a member of Appalachian society. Most festivals are relatively new to the region. Social gatherings other than church events were rare celebrations like harvest, Christmas, birthdays, and homecomings. These were small but not insignificant precedents for today's festivals, most of which are of the twentieth century. Music festivals such as fiddlers' conventions are thought to be the oldest, dating from about eighty to one hundred years ago. Some of the official craft festivals are about that old, but much older forms of music and crafts existed in the region without organized annual public gatherings. A festival sampler would include fiddlers' conventions, folk food festivals, storytelling festivals, Cherokee powwows and other ethnic celebrations, and homecoming gatherings.

Cosby Ramp Festival

On the first Sunday in May for nearly fifty years now, the people of Cocke County, especially those in the Cosby community, celebrate the tradition of eating ramps, those thin green-leafed, white-bulbed plants found growing wild in the mountains. The ramp (*Allium tricoccum*) is a wild leek that belongs to the same genus as garlic (*A. sativum*), onions (*A. cepa*), chives (*A. schoenoprasum*), and similar pungent plants. Described as tasting like a cross between an onion and garlic, ramps lean more to the garlic side, with a slightly milder yet persistent flavor. Their musty, Oriental taste does indeed linger, sometimes for days. In earlier times ramps were eaten as a spring tonic to cleanse the blood. It is said that when schoolchildren ate ramps, they were excused from school the next day because of their repulsive breath. The ramp's rank odor also prevented close contact with someone who was sick and contagious.

Geographically, ramps are native to North America and range from eastern Canada through New England and down through the moist, cool hardwood forests of Appalachia to northern Georgia and northern Alabama. In the Great Smoky Mountains, the best ramps grow in moist, forested conditions above 3,000 feet. The plant bursts through the leaf litter or late snow in March and reaches the best eating size in late April.

The word "ramp" traces to the British Isles, where the folk name "ramsen" and the Old English *hramsa* meant wild garlic. When British settlers discovered plants in Appalachia that so resembled those in the British Isles, they called them ramps. Before European contact, Native Americans used ramps to treat colds and coughs and made a paste of the bulbs to soothe bee stings.[75]

The ramp festival at Cosby celebrates an Appalachian folk tradition of gathering, cleaning, cooking, and eating ramps. When chopped and scrambled with eggs, ramps are mild and flavorful, without too much of an aftertaste. But "Lawd have mercy, if you eat them raw, look out, even the dogs won't come up to ye." For many of the five thousand people who sit on the hillside at Kineauvista Ranch off Tennessee highway 32, a couple of miles off I-40 at exit U.S. 321 South (you'd best follow the signs), the attractions are many. The Maid of Ramps is chosen from a bevy of young ladies. Entertainment comes from local bluegrass bands, cloggers, and line-dancing clubs that perform on two permanent stages a couple of hundred yards apart, separated by a large grassy hill covered with families and fans. The Cosby Ruritan Club cooks ramps with eggs and provides raw ramp samples. The menu is simple: eggs and ramps, griddle cake cornbread, and slab bacon; pinto beans, griddle cake cornbread, and slab bacon; and barbecued chicken dinners. There is an admission charge, but just the panoramic vista of the Great Smoky Mountains from Snowbird Mountain to Mount Le Conte is worth the price.

Gatlinburg Scottish Festival and Games

The pipes are piping in the glen at Mills Park on the far side of Gatlinburg, Tennessee, on this warm May morning in 1999. Members of forty-eight Scottish clans are assembled in tents arranged in a horseshoe-shaped Scottish village. Sixteen piping bands (bagpipes and drums) compete in musical events. From the sound of things, the bands are playing all the time. Eating can be habit forming at festivals, and this one has interesting foods. Haggis, a mush made of oatmeal, suet, mystery meat, onions, and spices cooked in a sheep's stomach, tops the list of unusual food traits in Scottish culture. The main attraction at any Scottish festival is the games. To the uninitiated, Scottish games appear to be merely feats of strength; however, there are also events such as the kilted mile foot race and battle ax throwing for accuracy. Athletes compete in professional or amateur categories, mostly in throwing contests involving heavy stones, metal blocks, sheaves of hay, or long, heavy poles, as in the caber toss, the signature event in Scottish games.

Such was the setup for the eighteenth Gatlinburg Scottish Festival and Games in 1999. I spoke with Ed Miller, Edinburgh folksinger and fellow geographer (master's degree, University of Texas, Austin), who said he did not know what to make of all the kilts and bagpipes here in Tennessee. Back home in Scotland, Ed observed dryly, "you don't see much of this around." The festival allows folks to wear $500 kilts for a weekend in May. Perhaps they also attend other clan gatherings such as the enormous Grandfather Mountain Highland Games and Gathering of Scottish Clans held in July near Linville, North Carolina. Were these folks playing a grown-up version of cowboys and Indians, only this time in plaid and pipes? It may look like make-believe, but it is a genuine search for roots and a celebration of heritage. Without their kilts, bagpipes, haggis, games, dances, Mac names, and other Scottish culture traits, these folks' lives just might have been very plain indeed. Aye!

Fiddler's Grove: Old-Time Fiddlers' and Bluegrass Festival

Music has always drawn people together, to play and to entertain at home, at parties, at church, and at fiddlers' conventions. This one is the granddaddy of folk music festivals. For more than seventy-five years, the old-time fiddlers' convention has been held in Union Grove in Iredell County, about a country mile west of exit 65 on I-77 in the upper piedmont section of North Carolina. Over the Memorial Day weekend crowds assemble in the Fiddler's Grove Campground, where the stages are set for another round of mountain tradition—the fiddlers' convention.

A fiddlers' convention is a contest to see who can play the best, and often the fastest, mountain music. There are senior and junior competitions for fiddle, banjo, autoharp, mandolin, harmonica, guitar, dulcimer, hammered dulcimer, bass fiddle, and the acoustic slide guitar with the trade name Dobro. Traditional bands are organized into old-time and bluegrass divisions. In the 1997 festival, forty-one old-time bands and ten bluegrass bands competed at the senior level. In an effort to keep traditional music alive, other categories include the Certified Old-Time Fiddle Category (only fiddlers who are fifty-five or older, play strictly by ear, and have no formal violin training); Twin Fiddle Category (two fiddlers in harmony); and Heritage Tune Exhibition (designed to preserve older traditional music; here, the fiddler must give the origin of the tune and discuss its history before playing it). The official schedule begins Friday evening and ends Sunday afternoon, but fans and performers begin arriving on Thursday because only five thousand tickets are sold, first come, first served.[76]

The fiddlers' convention at Union Grove has become a cultural event and an institution in the history of folk music in Appalachia. I became aware of it in the 1950s, and in the 1960s Ralph Rinzler and other earnest song seekers rediscovered Appalachian folk music at Union Grove. For without this fiddlers' convention, the songs of Doc Watson and Clarence Tom Ashley quite likely would not have been heard around the world.

West Virginia State Folk Festival

Glenville, population two thousand, in the heart of West Virginia, is where every year since 1950 crowds gather to hear good old-time music, listen to stories, admire and buy handmade crafts, and celebrate the remembrance of traditional ways of life. Some say it is the oldest continuously running festival of its kind. The West Virginia State Folk Festival was initiated by Patrick Gainer in 1950 as an outgrowth of a folklore course he taught at Glenville State College. The festival began as an afternoon exhibition and evening program in the 1950s, but it has grown into a four-day celebration of mountain folk music in concerts and jams around town, fiddle and banjo contests, unaccompanied gospel singing, shaped note singing workshops, square dancing, storytelling, a spelling bee, foods, crafts, parades, and other attractions. There is no "in crowd" of musicians because the festival organizers have not intended to provide a stage for professional musicians but try to feature people who have learned in the oral tradition or firsthand from their family. For the pickers and singers who gather here, it is a religious pilgrimage to honor old-time music. The West Virginia State Folk Festival is free; it begins on the third Thursday in June and runs to Sunday morning.

Sequoyah Festival Powwow and Fort Loudoun Eighteenth-Century Trade Fair

This is a tale of two cultures and two festivals, in two days on one island. I am on the south end of an island in the Tennessee Valley Authority's Tellico Reservoir (formerly the Little Tennessee River) in Monroe County near Vonore in East Tennessee. It is the birthplace of Sequoyah, ancient patriarch, chief, and inventor of the Cherokee alphabet. A brick museum dedicated to Sequoyah dominates a hill to our left, but to the right is a large circle, perhaps fifty yards in diameter, surrounded by craftspeople. Most are Cherokees. A huge white canvas tepee looms on the right, but all around the circle are makeshift stalls, awning-shaded camper trailers, and booths selling anything that can be called Native American–made. Bill Red Bear, a craft colleague of my daughter's, sells hides and furs. The list of trinkets and trade items is almost endless: feathers, beads, necklaces, deer hides, a wide variety of furs, costumes, silver jewelry, drums, bows, arrows, flutes, pots, and baskets.

East of the circle are replicas of two Cherokee house types: a summer house and a winter house. Both are framed with poles. The summer house, a rectangular shelter about twenty feet by ten, has a long ridgepole, and the roof is covered with bark or slabs of wood. Three sides of the house are open; the west side of the long axis is covered with saplings woven into wattle or lath. Mark Finchum (a Cherokee reenactor and a former student of mine) explained that in addition to the covered west elevation, either one or both gable ends could have been covered, but the east long side was always left open. Traditionally, the summer house faced east, open to the early morning sunrise.

The winter house is round, made of vertical poles with a conical roof. It has a bark or wood slab roof, and all sides are covered with saplings woven into a wattle-and-daub wall, sealed with clay. A single doorway allows entry into this warm, protective shelter and admits sunlight. At the powwow Finchum and his wife spent two cool nights sleeping on the ground in the summer house, under furs and wool blankets. Their reenactment demonstrated folkways in cooking, crafting, blowgun shooting, singing, and other living conditions of the eighteenth-century Cherokees.

On the north end of the island stands a replica of Fort Loudoun, a British outpost that was occupied from 1756 to 1760. Reenactors have set up an eighteenth-century trade fair with British troops and white settlers camped in tents sleeping, cooking, and trading down by the river. Marauding Cherokee reenactors do battle with the British troops, and at other times they carry on trade. Everyone is in period costume, which means that on this hot September day in 1999 the British and the white settlers are muffled to the ears in wool or linen. I find it amusing and culturally refreshing that so many people are willing to endure the hardships of eighteenth-century life while bikini-clad girls on Jet Skis skim across the waters of Tellico Lake.

National Storytelling Festival

Saturday October 3, 1998, is prime time for crowds and tellers, but it is mighty quiet here in Jonesborough, Tennessee: no music, no loud antique engines, only the sounds of eating, shuffling feet, and a little talk can be heard in this oldest of storytelling festivals. It is meant to be this way. The organizers of the event have cleverly organized storytelling into a moneymaking extravaganza. I was expecting a more folksy atmosphere, where old people in bib "overhauls" would drift down out of the mountains to sit on the courthouse steps and spin yarns to a cluster of enthralled listeners.

Storytelling has become big business; you can get a master's degree in it at East Tennessee State University in nearby Johnson City, Tennessee. About half a dozen large white tents that each seat two hundred or more are scattered about the town. The festival is a pricey affair; $65 a day entitles you to wear a small cloth badge (your ticket) and to attend all the events. But all sorts of non-Appalachian professional storytellers appear on the program: Celtic speakers, Pacific Northwest Indians, Jews, Greeks, Sicilians, a Chinese storyteller from Los Angeles, French Canadians, some Chicagoans, and even a New England Yankee. Ethnically the event surely is culturally diverse, but too much so. Could it be that our authentic mountain storytellers, other than the most famous—Ray Hicks (1922–2003) from Beech Mountain, North Carolina—have been too shy to come to Jonesborough? "If you'ns is out thar, cum on dowen, we needs ye!"

Museum of Appalachia Tennessee Fall Homecoming

Must be, let's see, a few thousand people—make that tens of thousands—driving up to John Rice Irwin's sixty-five-acre farm, the Museum of Appalachia near Norris, Tennessee. It is a perfectly clear, warm October morning in 1998, and the place is already teeming with people on foot, in cars, and in campers. An assortment of portable antique steam- and gas-powered tractors and engines are grinding corn, riving cedar shakes, and making homemade ice cream. It seems that almost every festival and fair around these parts has the same engines and tractors popping and puffing away, meticulously fussed over by men in bib overalls. Don't these guys have regular jobs?

John Rice Irwin charges $20 a head for this sparkling Saturday on the farm, but it's worth it. The best parts of the Fall Homecoming are the music and folk demonstrations. In Irwin's words, it is "one of the nation's largest and most authentic old-time mountain, craft, and music festivals, involving hundreds of musicians and Appalachian folk." Even when nobody is around except the caretakers, the museum is a gold mine for the student and researcher of Appalachian folk culture. On other visits you can witness more of the museum's collections, but on this October weekend the barns, log houses, and outbuildings are mere backdrops for musicians and demonstrators of Appalachian folkways. Some of the musicians are professional, but many are not. The demonstrators are authentic Appalachian people. If you attend only one Appalachian folk festival each year, this should be the one.

The Tennessee Fall Homecoming began in 1980 when Berea, Kentucky, bookseller George Brosi asked John Rice Irwin for permission to bring a vanful of books from the Council of Southern Mountains bookstore to the Museum of Appalachia for a weekend sale. Brosi had a small grant to help with expenses, but Irwin, thinking turnout might be low, rounded up some local musicians to encourage attendance at the book sale.[77] Since then the gathering has grown into a folk festival of music and demonstrations as one of the top one hundred events in North America and in the top twenty October events in the southeastern United States.

Nearly a hundred (well, at last count ninety-nine) individual musicians and groups play and sing old-time mountain songs, bluegrass, old classic country, and gospel music, among other things. The instruments are acoustic, and some of the world's best pickers appear here on four stages. There are so many musicians that even with four stages and rotating musicians from one stage to the next, each group performs only three or four songs. But my, oh my, the music! Here is a sample of the folks who appeared at the 1998 Fall Homecoming: Dr. Ralph Stanley, legendary bluegrass musician and his Clinch Mountain Boys; Raymond

Fairchild, five-time world champion banjo picker from Maggie Valley, North Carolina; Dr. Mack Snoderly from Clyde, North Carolina, winner of eighty-nine fiddlers' contests; Sparky Rucker, nationally acclaimed folksinger and folklorist; Mac Wiseman, a national treasure of bluegrass and country music; and John Hartford, banjo player and composer of "Gentle on My Mind." Janette Carter, daughter of A. P. and Sara, the famous Carter family pioneers in country music, has appeared here many times. She made it plain that she would still be back at the Carter fold at the Carter homeplace to do her Saturday night old-time music program there at Hiltuns (Hilton), Virginia.

The Fall Homecoming also features some eighty-nine crafts and demonstrations of mountain activities such as rail splitting and fence making; white oak basketmaking; whittling; quilting; constructing corn shuck and apple head dolls; and making sassafras tea, lye soap, and molasses, among other things. I had such a good time it was hard to call it fieldwork.

This is my third time around at John Rice Irwin's Museum of Appalachia's Tennessee Fall Homecoming, this time the twenty-first, and I am in bluegrass heaven. It is a warm October 15, 2000, and the same steam and single-cylinder engines and old-time tractors are popping, hissing, and doing all the things they did two Octobers ago. But there is something special in the air besides woodsmoke this Sunday morning. Earl Scruggs, a legend among three-finger roll banjo pickers and quite possibly the cofounder of bluegrass music, is here to do his only public appearance this year. Scruggs is seventy-six years old now and still occasionally makes recordings, but he has not come out of Nashville for a public performance for quite some time. He is here today because PBS is producing a documentary video on country and bluegrass music and its living legends. PBS filmed Ralph Stanley's performance here on Thursday October 12. Today it is Earl Scruggs's turn on stage to play his banjer (as he calls it) with a group of all-star backup musicians—Marty Stuart on mandolin, Gary Scruggs on bass, Jerry Douglas on a resophonic guitar (a Dobro type instrument), and Glen Duncan on fiddle. Scruggs's three-finger roll style has never left him as he blazes away at "Foggy Mountain Breakdown," his signature song, made famous by the movie *Bonnie and Clyde.* He delights the audience with another of his compositions, "Earl's Breakdown," in which he detunes the banjo at critical points. The result is an original blues effect that is a Scruggs characteristic. Scruggs and the band play and sing seventeen songs such as "My Saro Jane," "Salty Dog Blues," "Sally Gooden," "In the Pines," "Little Maggie," "John Henry," and "Black Mountain Blues," along with two requisite gospel songs, "Paul and Silas" and "Precious Memories." After a fast hour and a half of solid music, Earl Scruggs with Family and Friends leaves the stage. Scruggs goes into a hot tent to sign autographs for fans for the next hour and a half. The day is long, hot, and dusty. Scruggs is ready to retire to Nashville, but I am thrilled to have seen him in person.

Now it is October 11, 2001, a little past 10:00 a.m., and hordes of people are making their way onto the grounds at the Museum of Appalachia's twenty-second Tennessee Fall Homecoming. Exactly one month ago we suffered devastating national tragedies at the World Trade Center and the Pentagon and in a Pennsylvania field; but today is bright. Music fills the air, as does the smoke from fires that make sassafras tea, sorghum syrup, and lye soap. Familiar faces, sights, and sounds of folk culture on display greet me as I prepare to photograph musical acts on the program. I have specifically come to listen to American folk music stars Doc Watson and Ralph Stanley, who "headline" stage one. I casually stop by the information tent, and across the way I hear a familiar singing voice and flat-picked guitar on stage one. It's Doc Watson! Picking along with his grandson Richard Watson (son of the late Merle Watson) and accompanied by Jack Lawrence, a longtime guitar accompanist, Doc Watson sings, picks, and comments for a forty-five-minute set. Among the songs are "Matchbox Blues," "Frankie and Johnny," "I Am a Pilgrim," and the Homer and Jethro version of "I'm Movin' On." That afternoon about five Watson does a repeat performance with wonderful enthusiastic joy. In between, Ralph Stanley and his Clinch Mountain Boys, who truly are boys in their twenties and thirties, take the stage for two thirty-minute sets. Taking advantage of some familiar retro tunes, Stanley sings three songs from the movie *O Brother, Where Art Thou?* He sings "Man of Constant Sorrow" then, a cappella, "Oh, Death" and "Angel Band." He had recorded these songs decades before the movie was ever conceived. It is a thrill to watch and listen to these living legends today; but several musicians I saw just a year ago have passed away: John Hartford (songwriter, at age 63), Bob Douglas (fiddler, at 101), and Jim Russell (local fiddler, at 76).

The year is 2002, it is October, and I am slogging through the rain, mist, and mud at John Rice Irwin's Museum of Appalachia Fall Homecoming. I have a five-day pass, and it has rained solidly for four days running. The music and folk demonstrations this time are as good as they can be given the weather conditions. This soggy week has folks wearing garbage bags for ponchos and sitting for hours in a steady rain. They are here to see 350 folkways demonstrators and musicians, including Ralph Stanley and Doc Watson. And just like all the other foolish people here, I too am sitting in the rain just twenty feet from Doc Watson and savoring every minute!

The festivals in Appalachia have become so numerous that you must be duly selective about which ones to attend. Most folk festivals and fiddlers' conventions work hard to maintain an authentic folk flavor, but some arts and crafts festivals do not. Too many festivals originate from public relations efforts by chambers of commerce or visitors' centers that see them only as a way to encourage tourism. Some people fail to recognize that authenticity in folkways can be a valuable commodity; and like so many aspects of contemporary Appalachia, real folk culture traits are harder to discover because so many are vanishing.

At a festival recently, I discovered a display of antique engines and tractors—old stuff, traditional things. Always searching, I found Alden and Unavee Fowler from Parrottsville in East Tennessee's infamous moonshine capital of Cocke County exhibiting a portable corn grinding machine built in 1912. Alden Fowler, seventyish, in his Liberty brand bib overalls, talked about the drought of '98, saying he had never seen it this dry. Yet his twenty-foot-deep well was still providing cold, clear water, and his two springs up the hill above his house were still flowing "right well." We talked about his water but didn't discuss moonshine. Unavee Fowler was making and selling old-timey sunbonnets and aprons at this mountain festival and at similar folk gatherings in this part of Appalachia. At least for now, this generation still carries on the tradition.

9

Folk Speech: Terms and Sayings

Origins

"Folk Speech Is English, Too" could be the title of this chapter, but it's already been taken.[1] A mythology surrounds the origins of the dialects heard in the southern mountains. One assumption is that since to some the speech sounds coarse, unsophisticated, and uncultured, it has no place in America and certainly could not have come from the proper speech patterns currently used in the British Isles. Whether someone thinks something is uncultured or cultured, the truth is that everything thought by, invented by, and used by humans is some part of Culture; and language commands the paramount position in every culture. Appalachian speech is not entirely vulgar, but to some listeners it seems as unkempt as a pigpen. Others who think they know better believe the dialect is pure Elizabethan English. The truth probably lies somewhere between.

The British Isles contributed the most significant components of Appalachian speech. Dialect, vocabulary, even grammar trace to the northern British Isles, meaning northern England and especially Scotland, and to northern Ireland. Linguistic contributions from southern English, German, French, and Native American sources have enriched the language.

Scholars have long sought the connections between British English and Appalachian forms. Perhaps the best-known work is Hans Kurath's *A Word Geography of the Eastern United States.*[2] A popular work, *The Story of English,* by McCrum, Cran, and McNeil, was the basis of a television series by the same title.[3] Both sources compare vocabularies to some degree, but they leave it to the reader to infer deeper connections. Other sources have treated the subject in different ways. Cecil J. Sharp and Maud Karpeles sought British ballads in Southern Appalachia. John C. Campbell examined the surnames of 1,200 families in North Carolina,

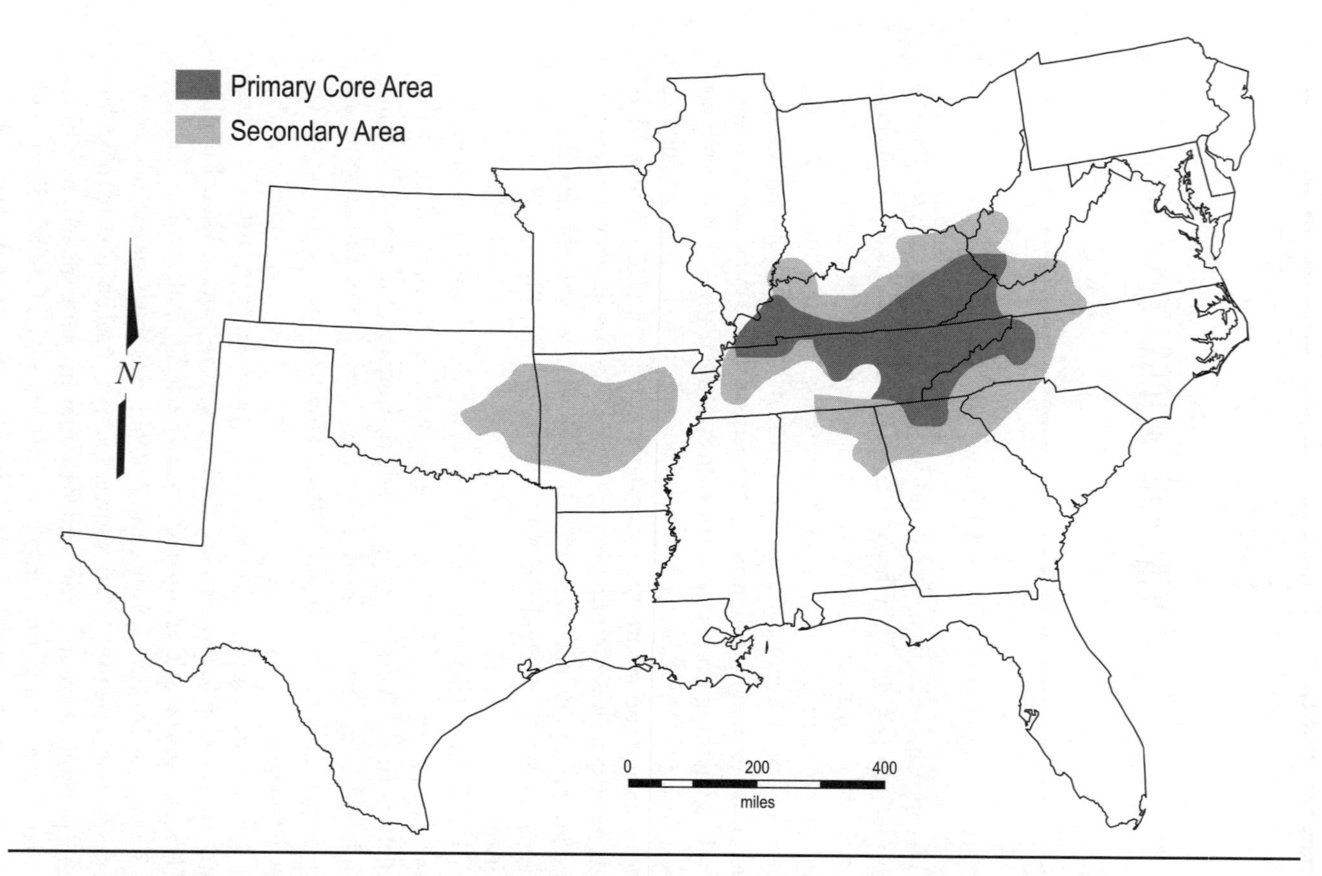

Appalachian speech in the Upland South can be found in a centralized core region and in two secondary subregions. (Source: Carver 1987, 166)

Tennessee, and Kentucky and found equal proportions of Scotch-Irish and English names.[4] Cratis Williams, a highly respected scholar and Appalachian native from Ricky Skaggs's neck of the woods in Kentucky's Lawrence County, wrote many marvelous pieces on Appalachian speech. Cratis Williams's early work also sought the cultural identity of Appalachian people.[5]

Grammatical traits are thought to be deeper, stronger evidence. In seeking the origins of Appalachian speech, Michael Montgomery made comparisons between Scotch-Irish, English, and Appalachian grammatical characteristics. The best part of Montgomery's material contains forty grammatical usages that are apparently reliable diagnostic traits. Of the forty usages, seventeen are Scotch-Irish, four are southern British, thirteen are general British, and six are uncertain.[6] (See table 9.1.)

Sayings

Appalachian speech and speech patterns elsewhere in the South are overwhelmingly rich in folk sayings. Expressions for distances and measurements appear inconsistent, even ambiguous, but are understood by the speaker and sometimes by the listener. "Several" means many, certainly a good many, perhaps hundreds. Expressions like "a right smart," "gobs of 'em," "biggun," "ever little bit," "don't you guess," "still yet," "done completely" are imprecise yet meaningful, conveying the speaker's general understanding of distance and amounts without the need for calipers, rulers, clocks, or scales. Time was once measured as the amount of daylight between dawn and dusk. Long distances were measured in days' journey, not in miles. When someone says "I'll be with you directly" he means ultimately or soon, but not right away. "Children grow up directly" means they grow up fast and steadily but not in a rush. Other measures were based on the number of hands, as in a horse's height at the withers, the length of one's own feet, the number of ax handles, lengths of chain, rods, or hollers into the woods.

A positive comment quickly following a negative one somehow defuses it and makes things right: "She ain't too good lookin', bless her heart." "He's jus' too danged stubborn fer his own good, sweet Jesus, bless 'em."

Appalachian speech is uniquely blessed with comparative sayings:

Limber as a dishrag. Dishrags are well-worn pieces of cloth, soggy and always flexible; some people and things seem just as loose.

TABLE 9.1
Appalachian Grammatical Elements and Their European Origins

A. VERBS

1. Third-person plural *s* with noun subjects but not with pronoun subjects: Folks *knows* all about it. [Scotch-Irish]
2. Regularized past *ed:* The wind *blowed* hard. He *throwed* the ball. [southern British]
3. Get to or got to (meaning begin to or commence): After the lightnin' storm, the woods *got to* burnin'. [uncertain]
4. Multiple modal verbs: She *might could* look a little better. [Scotch-Irish]
5. Liked to (meaning almost): We was late and *liked to* never got home. [general British]
6. A- prefixing: The trout was *a-swimmin'* faster than I could see. [southern British]
7. Perfective "done": Pap *done* finished plowin' afore sundown. [Scotch-Irish]
8. Preposed "used to": *Used to,* we didn't get to town much. [uncertain]
9. Used to + would, could: We don't kill hogs anymore but we *used to could.* [Scotch-Irish]
10. Need + past participle: Jack *needs taught* to mind his own business. [Scotch-Irish]

B. PRONOUNS

1. Y'all, you-all: *Y'all* come on up to see us next Sunday after church. [Scotch-Irish]
2. You'ns, we'uns: *You'ns* come down outta that tree! [Scotch-Irish]
3. You, ye: Stuck, are *ye?* Well, I'm jus' gonna hafta come up atter *ye.* [Scotch-Irish]
4. Combinations with "all": *You-all, We-all, who-all, they-all, what-all, where-all, ye-all, why-all, you-all, your-all's* [Scotch-Irish]
5. Hisn, hern, theirn, ourn, yourn: Was that milk pail *ourn* or *yourn?* [southern British]
6. Hisself, theirselves: Jack was lonely up on the mountain all by *hisself.* [general British]
7. Ethical dative: When I saw trouble acomin', I brought *me* a gun. [general British]
8. "They" existential (meaning there). *They's* only three chickens in the henhouse. [Scotch-Irish]
9. "It" existential: *It's* many people that left our holler. [general British]
10. Deletion of the subject relative pronoun: *They's* only three people could have been there. [general British]
11. Hit (meaning it). *Hit* don't matter no more. [general British]
12. Everwhat, everwhich, everwho: *Everwho* stole that truck didn't git much of a bargain. [uncertain]

C. NOUNS

1. Zero plural for a quantifier: About two mile away, you can buy three bushel o'apples fer a dollar. [general British]

(continued)

Table 9.1 *(continued)*

2. Syllabic plural after *sp, st, sk:* Wasp*es* build nest*es* inside these post*es*. [southern British]

D. Prepositions

1. Compounds: Get that polecat *out from up under* the smokehouse. His favorite place was *across down over* Waldens Ridge. [uncertain]
2. Anent (meaning opposite, near): Grandpa built his store *anent* the old home-place in 1925. [general British]
3. Fornent, fernent (meaning against or next to): The plow is *fornent* the corncrib. [Scotch-Irish]
4. Till (meaning before): She said she'd be here at seven *till* five. [Scotch-Irish]
5. Wait on (meaning wait for). I *waited on* her till six-thirty. [Scotch-Irish]
6. Again, against (meaning before, by the time that): She told me she'd be back *again* six or seven. [general British]

E. Conjunctions

1. Again, against (meaning before): I'll be ready *again* you are. [general British]
2. Whenever (meaning as soon as, at the time that): I'll fix the gate *whenever* I git around to it. [Scotch-Irish]
3. Till (meaning so that): Some chores on the farm was hard but we did 'em *till* we could eat. [uncertain]
4. "And" as an absolute: Jack took a swing, *and* him drunk as a skunk. [Scotch-Irish]
5. Nor (meaning than): She says she's better *nor* you. [general British]

F. Adverbs

1. Positive for anymore: I don't care, it's still good work *anymore*. [Scotch-Irish]
2. All the far (meaning as far as): That's *all the far* as he wanted could go. [uncertain]
3. Right as an intensifier: Hit's *right* airish this morning in the holler. [general British]
4. Yonder: The old log cabin is just over *yonder*. [general British]
5. Yan: Frogs was everywhere here and *yan*. [Scotch-Irish]

Source: Places of origin in brackets are from Michael Montgomery, "The Roots of Appalachian English: Scotch-Irish or British Southern?" *Journal of the Appalachian Studies Association* 3 (1991): 183–84, 186–88.

Tough as whitleather. Whitleather, an exceptionally tough, strong leather, was used to make harnesses. The expression refers to a person's tough attitude or an object's strength or durability.

Rough as a cob. A dry corncob is rough and scratchy, traits attributed to people and things.

Tight as Dick's hatband. Stingy. It also refers to close-fitting garments and other things. Dates to the eighteenth century.[7]

Mean as a snake. Given to angry, aggressive behavior.
Cocky as a bantam rooster. Feisty, prideful, quick to anger.
Not big as a minute. Tiny.
Nervous as a long-tailed cat in a roomful of rocking chairs. Nervous, really on edge.
Bright as a button. Shiny, pretty, or smart.
Pretty as a penny. Shiny, bright.
Lazy as sin. Shiftless, slothful, lethargic.

Other expressions include:

Frog strangler. Heavy rain, enough to drown amphibians.
Rollin' sand in your shoes. Traveling widely.
I don't care if I do. A positive reply, like "Yes, Ma'am." Momma says, "Won't you have pie?" The preacher says, "I don't care if I do."
No you ain't. Expressed as astonishment. She brags, "We'uns done went to Asheville yesterday." The reply is, "No you ain't!" This means the same as "No way, José!"
Jeet jet? Say it fast, and it means "Did you eat yet?"
We ain't short around here except where we ain't growin'. We aren't behind the times, and we can keep up with anybody.

I heard this last expression for the first time on December 21, 2001, at my barber's shop. I commented that the shop was really looking first class, with a small decorated Christmas tree, whereupon Hoyt, my barber for over thirty years, said with great pride," We ain't short around here except where we ain't growin'."

Pronunciation

Here in Southern Appalachia the dialect is totally oral. We are in "hard *r* country," where the letter *r* gets extra heavy treatment in pronunciation, as in "arn" for iron, "tar" for tire, "warsh" for wash, "warter" for water, "far" for fire, "quar" for choir, "barn" for born, "arter" for after, and "orter" for ought to. You might hear this: "Far engines run on rubber tars, and we fix 'em with balin' war and plars."

We are also in "stiff lower lip country" as opposed to the British stiff upper lip. Hold your lower lip tight against your teeth and pretend you have (or maybe you really do have) a wad of smokeless tobacco held tight between your cheek and gums. Now, speak out of the front of your mouth without dripping or making a mess on something. The sounds you make will be akin to the reserved, conservative sounds of Appalachian speech.

The letter *i* is almost always pronounced short and flat, as "ah." For example, instead of pronouncing each *i* in this statement as "eye," say "ah": "I like fried rice and ice cream. Mike, turn out the lights; good night."

Other vowels get special treatment. The short *e* adjacent to *r* makes thresh into "thrash"; wrestle, "wrastle"; service, "sarvice"; learn, "larn." The short *e* becomes short *i*, as in "git" for get; "pin" (pen), "min" (men). The short *o* sometimes becomes short *e* as in "fer" (for). The diphthong *oi* becomes *i* as in "bile" (boil), "jine" (join), "pint" (point), "hist" (hoist), and "rile" (royal), just as it did in the eighteenth century.[8]

The letter *y* gets added when the word ends in *a*, such as "opry" for opera, "sody" for soda, "Sary" for Sara or Sarah. When I first moved to Knoxville, Tennessee, in 1967, a filling station attendant referring to high-octane gasoline floored me with, "Fill 'er up with extry?"

Glossary of Appalachian Terms Collected 1910–79

Abide. Tolerate, agree. "I cain't abide with his cussin'."
Acrost. Across. "Let's go acrost the river."
Afeared (Old English). Afraid. "I'm afeared that Jack might be lost."
Afore. Before. "I want to git home afore dark."
Aigs. Eggs. "Sary Jane, git the aigs outter the hen house."
Aim. Plan. "I aim to settle down and git a job one of these days."
Airy. Any. "I ain't seen ary single blessed hawk this year."
Allers. Always. "My sister's allers late."
Aplenty. Enough, sufficient. "He's got rocks aplenty out thar in the old field."
Arn. Iron. "Momma's got to arn Sary's dress."
Arter. After. "He went arter the pig what got away."
Ast. Asked. "I ast you to do as I say."
Aw (outcry, exclamation). "Aw shucks, I bunged up my thumb with a hammer."

Backer. Tobacco. "Backer prices this year are goin' to be low."
Backerds. Backward. "Jim put the mule in the barn backerds. He shouldn't done that!"
Bait. Large portion of food. "Grandpa ate a bait of fried taters last night."

Barn. Born. "Sister was barn down in Monroe County."
Beholden. Obliged. "I am beholden to you fer that hound you gave me."
Biggety. Snobbish. "She got biggety after winnin' the Kentucky lottery."
Bile. Boil. "Bile that cabbage down, boys, bile that cabbage down."
Blowed. Blew. "She blowed out the candles on her birthday cake."
Bodacious. Much, big, a lot. "I'm bodaciously tuckered out."
Briarpatch child. A child born out of wedlock. "Bresh colt" means the same thing in the Ozarks.
Brickle, brickly. Brittle or fragile. Originated in Scotland, where "bruckle" is still used.
Britches. Pants. "Scratch Britches shore is a good name for that thicket. I tore my britches there."

Chanct. Chance. "I'll holp the first chanct I git."
Cheer. Chair. "Bring up a cheer and sit a spell."
Chunk. Throw. "We used to chunk rocks in the river."
Clobber. Hit someone or something. "When Joe lost the mule, Pa was so mad he was ready to clobber him."
Clum or clomb. Climbed. "That coon clum that old tree faster than Doc ever seed."
Cracklins. Fried pork rind. "Cracklins an' cornbread an' buttermilk, now that's a right good supper."
Crick. Creek. "Cratis caught a big trout in a deep pool in the crick."
Cur. Mongrel dog. "A cur cain't foller a cold track. He jus' runs by sight."
Cussedness. Meanness. "Hits jus' pure cussedness that makes Jack kick the dawg."

Dast. Dare. "She don't dast leave in the dark of night."
Dinner. The midday meal, often the largest meal of the day. "Lunch" is not used, and the evening meal is called "supper."
Drap. Drop. "He drank ever blesset drap of that moonshine."
Drug. Dragged. "Look what the cat drug in."

Endurabler. Longer lasting. "Poplar logs in houses are a heap endurabler."
Ever. Every, as in directions. "I looked for Nell ever whar in ever direction."

Far. Fire. "Dan put out the far in the woods."
Fer, Fur. Far. "They live a fer piece up the holler."
Fer. For. "This is fer the baby."
Fitten. Fit. "Cold grits ain't fitten to eat."
Furriner. Foreigner or stranger. Used to designate someone from outside the local community. "He's a furriner from South Carolina."

Gant lot. A pen used for to reduce the grass diet of wild cattle that have been foraging in the forest before market. The animals slimmed down and traveled better after some time in the gant lot. The cattle are "penned up to git gant and nimble."[9]
Git. Get. "Go git me my rifle."
Git shet uv. To get rid of; to get shed (or shut) of. "She cain't git shet uv that no account husband uv hern."

Haint. Ghost. "Watch out for haints on Halloween night."
Hern. Hers. "That plate is hern."
Hisn. His. "But the cup is hisn."
Hidy. Hello. A contraction of "hi" and "howdy."
Hit (Old English). It. "Hit don't matter no more. I lost hit and that's that."
Holler. Hollow or valley. "They live a fer piece up the holler."
Holp. Helped. "Mamaw holp with the dishes."
Hundert. Hundred. "That log cabin is a hundert years old."

Idee. Idea. "I ain't go no idee whar he went to."
If'n. If. "If'n you'ns go to town, git Sary some heavy black thread."

Jine. Join. "Will jus' wanted to jine the gall-danged air force."
Jus'. Just. "Jus' wait here."

Keepin' company. Dating, courting. "Your brother Jess is keepin' company with Annie Mae."
Kiver. Cover for a bed. "Pull up the kiver, hits goin' to git cold tonight."
Kilt. Killed. "Davy Crockett kilt a bar."

Larn (learn). To teach. "We are goin' to have to larn him a lesson."
Leave be. Let alone. "Don't touch that rattler; leave it be."
Lible. Likely, apt to. "That mad dawg's lible to bite yew."

Mater. Tomato. "Grainger County farmers grow good maters."
Mess. Enough for a meal. "She cooked a mess of greens for supper."
Middlin'. So-so; about average or midway. "We asked the old man how he was feeling. 'Fair to middlin', I reckon,' he replied."
Moonshine, moonshiner. Untaxed, illegal corn whiskey; the maker of such liquor. Also called blockade likker, blockader, white lightning, mountain dew, popskull.

Nanners. Bananas. "Granny allers took nanners with her on long bus trips."
Nary. Not any. "Those folks up the holler got nary a bite to eat."
Nekkid. Naked. "Skinny dippin' means you swim nekkid."
Newfangled. New, modern technology. "Hobart came in with a newfangled 'lectric saw t'other day."
Nope. No. "Nope, we cain't go to church cawz hits a-rainin'."
Nuff sed. Enough said. "Nuff sed, hits a gullywarsher."

Onc't. Once, at one time. "Old man Carver was struck by lightnin' onc't, and that wuz all hit took."
Ort. Ought. "He ort not go in the river; hits a-floodin'."
Overhauls. Overalls. "My favorite overhauls are the Pointer Brand."

Passle. A large amount. "She gathered a passle of black walnuts this afternoon."
Peaked (two syllables). Weak and sickly. "The preacher shore looked peaked after his operation."
Peart. Feeling well. "Aunt May was right peart at church today."
Piddle. To play with, fool with. "My cousin likes to piddle with antique tractors."
Pizen. Poison.
Plumb. Completely. "She was plumb wore out."
Poke. Bag or sack. Scottish origin, still used in the border country between Scotland and England.[10]
Popskull. Bad moonshine, poisoned liquor.[11]

Quair. Queer. "They sung a purty good song, but the words wuz quair."
Quar. Choir. "He hopes the quar don't sing hit agin."

Rat cheer. Right here. "Come on in and sit a spell, rat cheer."
Reckon. Believe. "I reckon I might go to church Sunday."

Retard. Retired. "Her husband Calvin retard about a month ago."
Retch. Reach. "She couldn't retch the lightbulb to change it, so she stayed in the dark."
Rosnears. Roasting ears of corn.
Roughness. Any kind of rough fodder such as cornstalks.[12]

Sallet. Salad.
Scoot. To move or slide something. Scottish origin.[13]
Shootin' arn. Shooting iron, gun. "Horace went barr huntin' with a new shootin' arn."
Smack dab. Right there, located exactly. "Nellie fell smack dab in the creek."
Smidgen. Little tiny bit. "Use jus' a smidgen of black powder or you'ns could git blowed up."
Soon start. An early start. "We got to git a soon start if you want barr brains an' liver for supper."[14]
Suggin. Pouch or small carryall bag.

Tain't. It is not. "Tain't as cold this winter as last year."
Taters. Potatoes. "He likes mashed taters and gravy."
Tetched. Touched, crazy. "That feller walks like he's tetched in the head."
They's. There is. "They's a cold wind a-blowin'."
Tow sack. Cloth bag or sack. "They allers carry a tow sack when they dig taters."

Urine. Yours. "That frog ain't mine. He must be urine."

Varmint. Animal. "The skunk is one bad varmint."
Vittles. Food. "Won't you'ns come in and have some vittles with us fer supper?"

Was a year ago. Last year. "Mable came by, was a year ago last May."
Wasted. Used up, spent. "Whar's that pan of biscuits that was here? We'uns wasted em."
Whar. Where. "Whar are the corn growin' at?"
Whup. Whip. "Jack is gonna whup up on you behind the barn."
Wommpyjawed. Slanted or crooked. "That roof of yern, hits plumb wommpyjawed." "Anniegogglin" means the same thing in the Ozarks.

Wrench. Rinse. "She wrenched the warsh today." The use of "wrench" or "rench" for rinse is widespread throughout Northern Ireland in all counties that contributed to the Scotch-Irish migrations in the eighteenth century.[15]

Wropt. Wrapped. "She wropt the cut with wet leaves."

Y'all. You all, all of you. "Y'all come back now, you hear."

Ye. You. "If'n ye hurry, ye might not get so wet in the rain."

You'ns. You ones. "How's you'ns doin' tonight?"

You'nses. Yours. "Which mailbox is you'nses?"

Young'uns. Young ones, children. "Jim's sister has a whole houseful of young'uns."[16]

Language in any form is at the heart and soul of a culture. The French seem to love their language as much as their food. I am not sure that Appalachian folk are so much in love with theirs. They learn it, they use it, but they do not seem to have much consciousness about what they are saying. They simply say things in the traditional ways they were taught. Tradition lays the blueprint for speech patterns, and isolation preserves them. As television, radio, and educational influences permeate the region and modify folk speech, "Elizabethan English" as Appalachian speech might vanish from the Southern Highlands. I remember a newspaper cartoon I saw more than thirty years ago. A graduate student with a tape recorder is interviewing an Appalachian farmer out in a cornfield. The student says, "Now, repeat after me: 'Whar are the corn growin' at?' Otherwise you're going to ruin my entire thesis."

Epilogue

In our Appalachian journey, we have witnessed traditional folk traits in the architecture of the region, in ways of making a living, in foodways, in folk remedies and beliefs, and in folk music and crafts, among many forms of Appalachian culture. Tradition continues to be the common denominator that makes things folk. Despite good intentions to bring about change in Appalachia, isolation has been a major force in preserving Appalachia's folk culture, and it still can be internalized as personal cultural isolation. I know people who remain culturally isolated by choice. They have lived along major highways and watched the world pass by. They have electricity and access to modern society but have remained steadfast in their comfortable, familiar folk culture. Physical isolation was overcome through outside forces in much of the region, yet cultural isolation prevailed. That was then, but now the changes arrive much too fast for most folks to get out of the way.

There was a time when I believed that rural Appalachia was a well-preserved living museum and that isolation and tradition surely would protect it. I was wrong. The vanishing of the landscape heritage discussed in chapter 4 is upon us. The region needs an active means of preserving traditional folk culture and folk traits. The preservation must be done carefully so as not to contaminate the setting and its artifacts. There are not enough protected outdoor museums that represent Appalachia. John Rice Irwin's Museum of Appalachia near Norris, Tennessee, at exit 122 off I-75, and the Museum of American Frontier Culture near Staunton, Virginia, at the junction of I-81 and I-64, both represent the levels of preservation that are required to maintain even the smallest glimpse of the folk past. These wonderful private efforts are on a par with many of the outstanding outdoor folk museums in Europe. In the Great Smoky Mountains National Park and along the Blue Ridge Parkway, the National Park Service has several fine outdoor exhibits and pioneer farmstead collections. But it could do much more. It *needs* to do much

Passing down traditions from one generation to another is the folk way of life in Appalachia. Bluegrass legend Ralph Stanley signs an autograph for a little girl who is holding an autographed picture signed just up the line by a bluegrass musician of another generation, Stanley's son, Ralph Stanley Jr. (Photograph by J. Rehder, 2002)

more. On the North Carolina Piedmont there is Old Salem, a well-developed living museum of eighteenth- and nineteenth-century German Moravian culture that models itself after the preservation ideal of colonial Williamsburg, Virginia. I spent my high school years in the late 1950s in Winston-Salem, when little or nothing of Old Salem had yet been "discovered" as worth preserving. Whew! I am so glad that someone had the foresight to safeguard Old Salem, restore its architecture, and demonstrate its ancient ways of life. Would that more could be accomplished before more of our North American folk landscapes vanish.

In the core chapters of this book, we observed many forms of Appalachian folk culture. I wish I had time and space to examine other pieces of the puzzle. Still, I hope that our exploration of Appalachian folkways and the region's landscape heritage has lent a decidedly soul-like grace to the subject for you, as it has for me.

Notes

Chapter 1. The Real Appalachia

1. Evans [1942] 1963, 5.
2. Whisnant 1981 [1994], 266–67; also in Higgs, Manning, and Miller 1995, 192–93.
3. Cabeza de Vaca [1542] 1994, 4–8.
4. Raitz, Ulack, and Leinbach 1984, 9–35.
5. Guyot 1861.
6. Powell 1895.
7. Thornbury 1965, 221.
8. Mood 1951, 74.
9. U.S. Department of Agriculture 1935.
10. Mangus 1940, 3–5.
11. Bradshaw 1992, 1–52.
12. Campbell [1921] 1969, 19–21.
13. Campbell [1921] 1969, 10–11.
14. Kniffen 1965.
15. Glassie 1968, 39; Jordan-Byschkov 2002.
16. Jordan-Bychkov and Domosh 2003, 9; Jordan-Bychkov 2003, 7.
17. Newton 1974.
18. Terry G. Jordan-Bychkov, letter to John B. Rehder, December 4, 2002.
19. Raitz, Ulack, and Leinbach 1984.
20. Salstrom 1995, 76–79.
21. Miller 1968; Shoemaker 1955; Clevinger 1938, 1942.
22. Gritzner 2002.
23. Shifflett 1991, 72; Eller 1982, 174.
24. Shapiro 1978, 256–57.
25. Dorson 1983, xi–xii, xvi, 115, 323–40.
26. Sharp [1932] 1966, xx.
27. Semple 1911; Huntington 1907.
28. Semple [1901] 1995, 174.
29. Eller 1982, xiii. 6–7, 8, 11, 16, 120; Shifflett 1991; Whisnant [1981] 1994; Shackelford and Weinberg 1988.
30. Mary Franks, personal communication, Knoxville, Tennessee, July 2002.
31. Rehder, Morgan, and Medford 1979; Wilhelm 1977.
32. Brown 1999, 92; Fetterman 1967; Halperin 1990; Hicks 1976.
33. Shapiro 1978, 18, 20–30; Batteau 1990; Batteau 1983.
34. Frost [1899] 1995.
35. Frost [1899] 1995, 92–93.

36. Frost [1899] 1995; Weller 1965.
37. Fox [1901] 1995, 123.
38. Fox [1901] 1995, 122–23.
39. Harney [1873] 1995, 48.
40. Semple [1901] 1995, 151–52.
41. Kephart [1913] 1984, 286–88.
42. Ford 1967; Ford [1967] 1991; Foster 1988.
43. Ergood 1991; Dickey 1970.
44. Jones 1994.
45. Shapiro 1978, ix.
46. Shapiro 1978, 263–64.

Chapter 2. The Shape of Appalachia

1. Raitz, Ulack, and Leinbach 1984, 39–84; Fenneman 1938.
2. Appalachian Trail Conference 1967, 2–33.
3. Murfree [1885] 1970, 1.
4. Moore 1994, 150–59; Mayfield 1984.
5. Carolyn White, personal communication, Knoxville, Tennessee, 2000.
6. Thornbury 1965, 106–7; Dykeman 1955; Lineback 1999, no. 214.
7. Sutton and Sutton 1985, 81–85.
8. Rehder 1996.
9. *Knoxville News Sentinel,* June 5, 2002, A1, A8.
10. Stupka 1964, 22–29.
11. Stupka 1964, 44–46.
12. Allen [1886b] 1995, 62–63.
13. Allen [1886b] 1995, 65.
14. Allen [1886b] 1995, 72.
15. Wise 1996, 45, 49–50.
16. Observations made at noon on Sunday, May 19, 2002.

Chapter 3. Ethnicity and Settlement

1. Rehder 1992, 99–100; Jordan-Bychkov 2003, 3–22.
2. Moody, Martin, and Byrne 1982, 222–65; Dunaway 1944, 4–15, 28; McWhiney 1988, xxi–xiii; Rehder 1992, 95; Rehder 1997a; Blethen and Wood 2001.
3. Cunningham 1987, xv–xvii; Berthoff 1986; Evans 1965, 1966, 1969.
4. Dickson 1966, 23–64.
5. Graham 1956, 185–88.
6. Hanna [1902] 1968, 1:83–84.
7. McDonald and McDonald 1980; Purvis 1984, 98.
8. Akenson 1984.
9. Purvis 1984, 98.
10. Rehder 1997b.
11. McWhiney 1988, xxi; Rehder 1992, 96–99; Caudill 1969; Berthoff 1986.

12. Purvis 1984, 98.
13. Jordan and Kaups 1989, 19–37, 54; Purvis 1984, 98.
14. Kilar 1997.
15. Jordan 1985, 119–24.
16. Jordan-Bychkov and Domosh 1999, 349.
17. McDonald and McDonald 1980, 198; Purvis 1984, 98.
18. Kilar 1997.
19. Fries 1922–69.
20. Prévos 1997; Golden 1988.
21. Kennedy 1997, xviii.
22. Kennedy 1997; Gallegos 1997; Price 1951.
23. Kennedy 1997, 1–3.
24. Kennedy 1997, 137.
25. Kennedy 1997, 114–19; Gallegos 1997, 27–68.
26. Kennedy 1997, 120–24; Quinn 1952, 251–55.
27. Gallegos 1997, 20–26.
28. Kennedy 1997, 99–102, 164; Price 1951.
29. Pollitzer and Brown 1969; Gallegos 1997, 75; Guthrie 1990, 14.
30. Pollitzer 1972.
31. Guthrie 1990.
32. Kennedy 1997, 139.
33. Price 1971, 10–12, 15–17.
34. Gallegos 1997, 61.
35. Campbell [1921] 1969, 28, 67; Davidson [1946] 1978, 135.
36. Hudson 1976.
37. Finger 1984, 3–4.
38. Finger 1984, 16.
39. Finger 1984; Neely 1991.
40. Rehder 1999, 54–55.
41. Montell 1993, 15–19, 21–22; Montell 1970; Turner and Cabbell 1985.
42. Eller 1982, 169, 168–72.
43. Campbell [1921] 1969, 23–24; Jordan and Kaups 1989, 20–22, 233–46; Rehder 1992, 97–99.
44. Jordan and Kaups 1989, 53–59.
45. Rouse 1995, 25, 26, 135.
46. Dunaway 1944, 50–71; Leyburn 1962, 184–236; Campbell [1921] 1969, 22–49; Hanna [1902] 1968, 2: 60–93; Kennedy 1997; Otto and Anderson 1982.
47. Rouse 1995, 93; Brissot and Mitchell 1993.
48. Weld [1796] 1998, 127, 129; Lemon 1966, [1972] 2002.
49. Rouse 1995, 135–36.
50. Works Projects Administration [1939] 1986, 45, 91.
51. Hanna [1902] 1968, 2:25–30.
52. Davidson [1946] 1978, 167–81.
53. Davidson [1946] 1978, 149–66.

54. Rehder 1992, 99; Finger 1984.

55. Mitchell 1972, 470–72; Jordan and Kaups 1989, 7–13.

56. Raitz 1996; National Geographic Society 1988, 192–93; Brown 1948, 184–86.

Chapter 4. Folk Architecture and the Cultural Landscape

1. Wilhelm 1978.

2. Kniffen 1965, 551; McIntire 1958; Zelinsky 1992, 13–14; Rehder 1999, 29–30, 61.

3. Meinig 1986, 240–42; Johnson 1976, 25–26; Thrower 1966, 25–29, 118, 127; Hart 1998, 145–48.

4. Glassie 1972, 418–20; Evans 1965, 1966, 1969 1974.

5. Rehder 1992, 102.

6. Allen [1886b] 1995, 69–70.

7. Semple [1901] 1995, 153.

8. Frost [1899] 1995, 93, 95.

9. Kephart [1913] 1984, 30–31.

10. Ashworth [1913] 1995, 196.

11. McCleary 2000, 92.

12. Shurtleff 1939.

13. Glassie 1963.

14. Kniffen 1965.

15. Kniffen and Glassie 1966, 40–56.

16. O'Malley and Rehder 1978.

17. Jordan 1985.

18. Morgan 1990.

19. Jordan and Kaups 1989; Jordan-Bychkov 2003.

20. Kniffen 1969; Kniffen and Glassie 1966.

21. Evans 1974; Brunskill 1978, 201.

22. Hanna [1902] 1968, 1:519–31; Naismith 1985, 20, 24.

23. Hanna [1902] 1968, 2:25–27.

24. Forman [1948] 1967, 9–13, 15.

25. Glassie 1975b, 124–25; Morrison 1952, 135–39.

26. Jordan and Kaups 1989, 135–78, 247–53; Jordan 1985, 3–14; Wright 1958; Mercer 1926, 583; Weslager 1969, 148–202.

27. Kniffen and Glassie 1966, 59; Glassie 1978, 401–2; Bucher 1962, 14; Wertenbaker 1938, 298–303.

28. Jordan and Kaups 1989, 135–78; Jordan 1985, 41–48.

29. Kniffen and Glassie 1966, 58–59; Glassie 1978, 401–2; Wertenbaker 1938, 298–303.

30. Jordan and Kaups 1989, 36–37, 53–63, 79–82.

31. Wigginton 1972, 53–114.

32. Del Scruggs, personal communication, Morgan County, Tennessee, 2001.

33. Jordan and Kaups 1989, 162–65; Jordan 1978, 158.
34. Morgan 1990, 27.
35. Wiggington 1972, 108–14.
36. Kniffen 1969; Kniffen and Glassie 1966; Jordan and Kaups 1989, 135–62.
37. Kniffen 1969; Kniffen and Glassie 1966; Jordan and Kaups 1989, 135–62; Jordan 1985, 55, 57, 91–93; Jordan-Bychkov 2003, 24–30.
38. Kniffen and Glassie 1966.
39. Reding 2002, 41–51.
40. Mann 2002, iv, 88–92.
41. Kniffen 1979.
42. Wilson 1970, 21–28, 26–31.
43. Wilson 1975, 77.
44. Morgan and Medford 1980, esp. 152–56.
45. Ambrosia 1979, 10, 14.
46. Morgan 1990, 28–29, 34–36.
47. Smith 1979, 89–165; Bishir 1990.
48. Addy 1898, 17, 66–69, 195–215.
49. Forman [1948] 1967, 9–13, 15.
50. Glassie 1978, 404–6; Glassie 1963, 9; Wilson 1971, 9–10; Wilson 1975, 54–55.
51. Glassie 1978, 404–13.
52. Wilson 1975, 27–28.
53. Morgan and Medford 1980, 152–56; Reding 2002, 41–51.
54. Ambrosia 1979, 10.
55. Morgan and Medford 1980, 152–56.
56. Dykes 1968; John C. Kemmer III, interview, Grassy Cove, Tennessee, June 8, 2002.
57. Riedl, Ball, and Cavender 1976; Jordan and Kaups 1989, 209–10; Jordan-Bychkov 2003, 36–40.
58. Wilson 1975, 78.
59. Morgan and Medford 1980.
60. Smith 1979, 89–165; Bishir 1990.
61. Rehder 1992, 107–8.
62. Kniffen 1965, 561; Carlisle 1982, 46–76.
63. Wilson 1975, 79.
64. Ambrosia 1979, 14.
65. Smith 1979, 89–165.
66. Carlisle 1982, 46–76.
67. Morgan 1986.
68. Montell and Morse 1976, 21.
69. Morgan 1990, 31.
70. Scofield 1936; Wright 1958; Jordan 1985, 146, 149; Jordan and Kaups 1989, 179–96, 249.
71. Kniffen 1965, 561–63.

72. Jordan and Kaups 1989, 179–96.
73. Wilson 1975, 32–43.
74. Hulan 1975, 42; Ferris 1980, 1986; Wilson and Ferris 1989.
75. Glassie 1968, 89–99.
76. Wright 1958; Jordan and Kaups 1989, 179–96, 249; Jordan-Bychkov 2003, 36–40.
77. Kniffen 1965, 553–57; Jordan 1985, 30–31; Glassie 1968, 64–69; Francaviglia 1979.
78. Kniffen 1936; Kniffen 1965, 553.
79. Kniffen 1965, 555.
80. O'Malley 1972, chap. 3; O'Malley 1977.
81. Rehder 1989.
82. Ambrosia 1979, 14.
83. O'Malley and Rehder 1978.
84. Smith 1979, 89–165.
85. John Kivett, interview. New Tazwell, Tennessee, February 24, 2001.
86. Kniffen 1965, 555; Jordan 1985, 199; Pillsbury and Kardos 1970; Pillsbury 1976, 1977; letter from Terry G. Jordan-Bychkov, December 4, 2002.
87. O'Malley and Rehder 1978; Newton and Pulliam-DiNapoli 1977.
88. Bucher 1962; Lay 1982; www.oldsalem.org.
89. Morgan 1990, 98–107.
90. Morgan 1990, 98–107; Williams 1995, 70–74.
91. Noble 1984, vol. 2; Schimmer and Noble 1984.
92. Semple [1901] 1995, 153.
93. Jordan 1985, 146, 149.
94. Glassie 1968, 89; Jordan-Bychkov 2003, 43–60.
95. Glassie 1968, 89, 228; Jordan 1985, 113, 149; Jordan-Bychkov 1998.
96. Ensminger 1992; Long 1972; Noble and Cleek 1995.
97. Ensminger 1992, 1.
98. Jordan 1985, 35–37, 98–101, 149; Ensminger 1992, 1–50.
99. Moffett and Wodehouse 1993, 3–4, 8–9.
100. Moffett and Wodehouse 1993, 9.
101. Moffett and Wodehouse 1993, 9.
102. Moffett and Wodehouse 1993, 4.
103. Moffett and Wodehouse 1993, xv, 86–114.
104. Kniffen 1965.
105. Morgan 1997; Ensminger 1992.
106. Moffett and Wodehouse 1993, 16.
107. Glassie 1964, 23.
108. Rehder, Morgan, and Medford 1979, 77.
109. Rehder, Morgan, and Medford 1979, 79.
110. Reding 2002, 41–51.
111. Jordan and Kaups 1989,105; Martin [1887] 1992, 7.
112. Venable 2000, 117–18.

113. U.S. Department of Agriculture 1872, 498–507; Jordan and Kaups 1989, 108–9; Cox 1978.

114. Glassie 1968, 227.

115. Brunhes 1952, plate 25, figs. 71 and 72.

116. Murray-Wooley and Raitz 1992; Allen [1886a], 375; Hart 1998, 184–87.

117. Jordan and Kaups 1989, 101; Campbell [1921] 1969, 251–52; Kephart 1984, 36; Semple [1901] 1995.

118. Hill 1978; Hart 1977, 148–66.

119. McCauley and Porter 2003.

120. Jeane 1969, 1978; Fielder 1982; Cobb 1978b; Jordan-Bychkov 2003, 71–81.

121. Montell 1993, 117–20.

122. Moore 1975.

123. Reding 2002, 54.

124. Weals 1991, 17; Madden and Jones 1977.

Chapter 5. Ways of Making a Living

1. Koons 2000, 3–20; Mitchell 1972, 1977.
2. Hilliard [1969] 1998, 82.
3. Black 1972.
4. Wiggington 1972, 212–19; Newton 1971.
5. Jordan and Kaups 1989, 211–19.
6. Kephart [1913] 1984, 76.
7. Kephart [1913] 1984, 99–101.
8. Hoyt Vanosdale, personal communication, Knoxville, Tennessee, 2001.
9. Works Projects Administration [1939] 1986, 439.
10. Richard Strange, personal communication, Knoxville, Tennessee, 2002.
11. Cobb 1978a.
12. Cobb 1978a, 45; Rostlund 1952, 101–4; 170; Shields 1967.
13. Landry 1998.
14. Wilhelm 1974, 243.
15. Price 1960.
16. Allen [1886b] 1995, 67.
17. Wilhelm 1974.
18. Montell 1993, 102; Lopes, Moser, and Perkinson 1996; Rafferty 1973.
19. Warner 2002; www.1stchineseherbs.com.
20. Tennessee Department of Environment and Conservation 2001.
21. Tennessee Department of Environment and Conservation 2001.
22. Price 1960.
23. Price 1960.
24. Price 1960; Allen [1886b] 1995, 67.
25. Stan Guffey, personal communication, Knoxville, Tennessee, 2002.
26. Boyer 2001.
27. Tennessee Department of Environment and Conservation 2001.

28. Nickens 2001; Brown 2002b.

29. Price 1960.

30. Lisa Roberts, unpublished field report and personal communications, 1984.

31. Grieve [1931] 2002b; www.botanical.com; Reeleder 2001.

32. Www.wildgrown.com.

33. Allen [1886b] 1995, 67; Price 1960.

34. Price 1960; www.wildgrown.com.

35. Price 1960; www.botanical.com.

36. Wilhelm 1974; www.botanical.com.

37. Eller 1982, 86; Moore 1990.

38. Eller 1982, 86, 89; Horn [1943] 1951, 108.

39. Eller 1982, 92.

40. Weals 1991, 4, 15.

41. Weals 1991, 4; Wilhelm 1974, 237.

42. Weals 1991, 4.

43. Great Smoky Mountains Natural History Association n.d.

44. Weals 1991, 85.

45. Weals 1991, 58–59, 90, 74–75.

46. Weals 1991, 77.

47. Stupka 1964, 16–18.

48. Eller 1982, 107; Weals 1991, 115; Great Smoky Mountains Natural History Association n.d.

49. Eller 1982, 96, 104–5.

50. Moore 1995, 140.

51. Shifflett 1991, 88; Caudill 1963.

52. Shifflett 1991, 85–87, 104.

53. Shifflett 1991, 89.

54. Maertens 1990, 19; Lambie 1954, 29–30; U.S. Department of Energy 2002.

55. Shifflett 1991, 85.

56. Brown 2002a.

57. Eller 1982, 165–74.

58. Shifflett 1991, 48.

59. Gillenwater 1972, 143.

60. U.S. Congress, Senate 1925, 1470; Eller 1982, 183–84.

61. Gillenwater 1972.

62. Gillenwater 1972, 70–80.

63. Gillenwater 1972, 68–92; Magnusson 1918, 186–95.

64. Kniffen 1965; Rehder 1999; Vlach 1986, 1993.

65. Gillenwater 1972, 81–83.

66. Gillenwater 1972, 83–84.

67. Gillenwater 1972, 83–86.

68. Gillenwater 1972, 86–90.

69. Rehder 1999, 100–116.
70. Gillenwater 1972, 132–36.
71. Gillenwater 1972, 135.
72. Tams 1963, 68; Gillenwater 1972, 135.
73. Gillenwater 1972, 133.
74. Gillenwater 1972, 136.
75. Big South Fork Scenic Railway 1994; National Park Service, Blue Heron interpretive exhibit, Blue Heron, Kentucky, 1989.
76. National Park Service, Blue Heron interpretive exhibit, Blue Heron, Kentucky, 1989.
77. Eller 1982, 184.
78. National Park Service, Blue Heron interpretive exhibit, Blue Heron, Kentucky, 1989.
79. Dabney 1974, 33–41.
80. Dabney 1974, xiv; Durand [1956] 1998, 121.
81. Durand [1956] 1998, 121–25.
82. Dabney 1974, 3.
83. Durand [1956] 1998, 120.
84. Dabney 1974, xiii.
85. Raine [1924] 1997, 131–32.
86. Nichols 1973.
87. Nichols 1973, 114, 44–82, 97–98.
88. Gabbard 2000.
89. Www.tbi.state.tn.us.
90. Www.whitehousedrugpolicy.gov; www.globalhemp.com; www.prevnet.org;www.tbi.state.tn.us.

Chapter 6. Foodways

1. Rehder 1999, 88–89.
2. Faye Goodman, personal communication, Wilmington, North Carolina, 1953; Dabney 1998, 115.
3. Faye Goodman, personal communication, Wilmington, North Carolina, 1953.
4. Egerton 1987, 191.
5. Dabney 1998, 115.
6. Egerton 1987, 11.
7. Dabney 1998, 99.
8. Dabney 1998, 112.
9. Dabney 1998, 100–101; Fowler 1995.
10. Dabney 1998, 112.
11. Todd Fox, unpublished field report and personal communication, Knoxville, Tennessee, 2001.
12. Dabney 1998, 188; Rehder, Morgan, and Medford 1979.
13. Bartram [1791] 1955, 283–84, 288–89.

14. Ben Mainor, personal communication. Knoxville, Tennessee, April 10, 2002.
15. Rehder 1999.
16. Dabney 1998, 431, 438–39.
17. Hilliard 1972.
18. Manthey 2002.
19. Farr 1983, 7.

Chapter 7. Folk Remedies and Belief Systems

Epigraph: Quoted in Price 1960, 3.
1. Price 1960, 1–2.
2. Wiggington 1972, 231–34.
3. Wiggington 1972, 230–48.
4. Montell 1993, 102; Lopes, Moser, and Perkinson 1996; Logsdon 1984, 1986.
5. Grieve [1931] 2002a.
6. Grieve [1931] 2002b; Reeleder 2001.
7. Lisa Roberts, unpublished field report and personal communication, Knoxville, Tennessee, 1984.
8. Wiggington 1972, 230–48; Montell 1993, 101–9.
9. Wiggington 1972, 230–44.
10. Leonard 1982, 474; McCauley 1995.
11. Leonard 1982, 20.
12. Jones 1999, 3–9.
13. Leonard 1982, 22.
14. Leonard 1982, 15, 26.
15. Brown and McDonald 2000.
16. Burton 1993, 24–25, 10, 204.
17. Burton 1993, 102, 161, 163; Brown and McDonald 2001; 101.
18. Brown and McDonald 2000, 231, 250; Burton 1993, 158.
19. Olson 1998b, 25.
20. Burton 1993, 32–40, 81; Leonard 1982, 482; Melton 1996, 636.
21. Burton 1993, 7.
22. Gillespie 1982, 205.
23. Jones, 1999, 150–51; Dorgan 1987, 114–15; Gillespie 1982, 351–69.
24. Dorgan 1987, 168–69; Jones 1999, 147–50, photographs between 166 and 167.
25. Williams 1995, 147, 143–50.
26. Crissman 1994, 151–55.
27. Montell 1993, 64.
28. Montell 1993, 94–95; Leonard 1982, 265–66.
29. Whilden and Cox 1973, 304–23; Crissman 1994.
30. Jeane 1969.
31. Montell 1993, 117–19.

Chapter 8. Folk Music, Folk Art, and Folk Festivals

1. Carney 1998, 129.
2. Olson 1998a, 83; Shapiro 1978, 252–58; Scarborough 1966.
3. Sharp [1932] 1966, xxi.
4. Sharp [1932] 1966, v–xi; Olson 1998a, 86; Wolfe 1977, 3–6, 11.
5. Sharp [1932] 1966, xv.
6. Sharp [1932] 1966, 385–417; Wolfe 1977, 3–6; Olson 1998a, 81; Bascom 1909.
7. Sharp [1932] 1966; Smith 1998, 137.
8. Sharp [1932] 1966, 21.
9. Sharp [1932] 1966, xxvii; Wolfe 1977, 10.
10. Miles [1905] 1975, 147; Wolfe 1977.
11. Jim Russell died on March 30, 2001; Irwin 1983.
12. Malone 1993, 36.
13. Conway 1995; Linn 1991.
14. Conway 1995, 120–21.
15. Williams 1995, 48.
16. Seeger 1958.
17. Carolyn White, personal communication, Kingsport, Tennessee, 1999.
18. Carolyn White, personal communications, Kingsport and Knoxville, Tennessee, 1999–2002.
19. Wolfe 1977, 6.
20. Sharp [1932] 1966, xx.
21. Jackson [1933] 1965; Jackson 1943.
22. Jackson [1933] 1965, 15–19.
23. Jackson 1943, 42, ix–xi.
24. Jackson [1933] 1965, 11–25; Lomax 1998.
25. White and King [1859] 1968; Swan [1867] 1978.
26. Jackson [1933] 1965, 25; Wolfe 1977, 11–12.
27. Forbis 1991, 18, 68, 143, 604.
28. Brose and John C. Campbell Folk School 1995; Lomax 1998; Jones 1984.
29. Brown 1999b, E1–E8.
30. Rosenberg 1985, 231–33.
31. Wolfe 1993.
32. Wolfe 1993.
33. Green 1976, 203, 211; Carney 1979.
34. Wolfe 1977, 54–56.
35. Wolfe 1977, 59, 67–70.
36. Green 1976, 203–7; Wolfe 1977, 76–78.
37. Wolfe 1977, 99–100.
38. Green 1976, 43–44, 167.
39. Parton 1999; Parton 1994; Sharp [1932] 1966, ii, 129–36, 229–30.
40. McDill 1994.
41. Cordle 1999.

42. Carney 1998, 139.

43. Carney 1998, 141–43.

44. Rosenberg 1985, 10–14; Wolfe 1977, 82–85; Carney 1974; Carney 1996; Carney 1998.

45. Smith 1964.

46. Www.gibson.com/products/Dobro(R), 1998.

47. Rosenberg 1985, 6–9.

48. Rosenberg 1985, 68–90, 95–165.

49. Rosenberg 1985, 406–20.

50. Www. country.com/sitesearch/ralph stanley; Gillam 1998.

51. Rosenberg 1985, 112.

52. Rooney 1971, 77; Ralph Rinzler quoted in Rosenberg 1985, 109.

53. Rosenberg 1985, 184–85.

54. Rinzler 1994.

55. Seeger 1990; Rosenberg 1985, 147–48.

56. Wolfe 1977, 61.

57. Ritchie 1955.

58. Brown 1999a; Adams 1995; Smith 1995, ix, xi.

59. Eaton [1937] 1973, 7–9.

60. Eaton [1937] 1973, 167.

61. Eaton [1937] 1973, 169; Law and Taylor 1991; Bullard 1976.

62. Irwin 1982, 43–44.

63. Wilson 2001, x–xi, xii.

64. Hall 1935; Clarke 1976; Eanes 1998; Ramsey and Waldvogel 1986; Waldvogel 1990.

65. Irwin 1984, 31.

66. Rinzler [1937] 1973, vii–lx; Becker 1998; Barker 1991.

67. Eaton [1937] 1973, 69–91.

68. John C. Campbell Folk School n.d.; www. folkschool.com.

69. John C. Campbell Folk School n.d.

70. John C. Campbell Folk School n.d.; www.folkschool.com.

71. Hindman Settlement School 1994, 5.

72. Hindman Settlement School 1994; Stoddart 2002.

73. Carolyn White, personal communication, Knoxville, Tennessee, 2002.

74. Hindman Settlement School 1994.

75. McCormick 1991; Cosby Ruritan Club 1999.

76. Www.fiddlersgrove.com.

77. George Brosi, interview, Norris, Tennessee, October 15, 2000.

Chapter 9. Folk Speech: Terms and Sayings

1. Dial 1969; Dial 1970; Fink 1974.

2. Kurath 1949; Kurath 1972.

3. McCrum, Cran, and McNeil, 1986; Sharp 1932; Orton 1974.

4. Campbell [1921] 1969.

5. Williams 1961; Williams 1962; Williams 1992.
6. Montgomery 1991, 177–91; Hall 1942; Hall 1972.
7. Hall 1972, 59.
8. Williams 1992, 13–16.
9. Kephart [1913] 1984, 93.
10. Glauser 1974, 95.
11. Kephart [1913] 1984, 122.
12. Kephart [1913] 1984, 112.
13. Carver 1987, 162.
14. Kephart [1913] 1984, 86.
15. Mather and Speitel 1977, 46.
16. Sources for this glossary are Kephart [1913] 1984; Williams 1991; Fink 1974; Hall 1972; and personal fieldwork since 1967.

Glossary

Barn Typology

single crib. A single unit barn, usually sixteen feet on a side

single crib with gear shed. A single unit barn with a cantilevered open shed on the side. The crib is usually a corncrib; the gear shed shelters farm equipment.

double crib. A barn with two cribs and a common roof over both. Each crib is about sixteen feet on a side. The passage between the cribs is a drive-through and a shelter for farm equipment.

four crib. A barn with four cribs (one on each corner) separated by two open passages that cross in the middle

transverse crib. A barn with two rows of multicrib units. Each side has three or more cribs. An open passageway goes between the two rows of cribs.

forebay barn. A large two-level barn with a distinctive overhang in the loft called a forebay that makes it easier to feed livestock in the feedlot below. The barn may also have an earth ramp in back to allow farm wagons to enter the loft on the second floor.

cantilever barn. Supported by two cribs as a foundation, the cantilever barn has massive, long wooden beams that support a large framed hayloft above.

House Typology

single pen. A small one-room cabin that is usually about sixteen feet square. It has a single front door and one outside gable-end chimney.

Cumberland (double pen). A two-room house with chimneys at the gable ends and two front doors. The house is found on the Cumberland Plateau of Tennessee and points westward toward the Nashville Basin, western Kentucky, and beyond.

saddlebag (double pen). A two pen house with a common central chimney that serves both pens. The house has a saddle roof and two front doors. The saddlebag follows much the same distribution as the Cumberland house.

dogtrot (double pen). A small two-room southern folk house that has an open passage (dogtrot) between the two pens. Door openings are usually in the dogtrot passageway. A common roof covers both pens and the passage. It is widespread from Tennessee to Missouri, Arkansas, Alabama, Mississippi, northern Louisiana, and Texas.

I-house. A two-story house that is one room deep and two rooms wide. The I-house has English origins and is a common rural house type throughout much of the United States.

four pen. A large two-story house with four rooms on each floor. It is the ultimate rural folk house in size and perceived economic importance, especially in Appalachia.

German three room. An eighteenth-century house with three rooms and a central chimney. Usually built of logs or stone, this now rare house of German origin was once more prevalent in Pennsylvania and the North Carolina Piedmont.

bungalow. A rectangular twentieth-century house with a front-facing gable, meaning that the front door is on the gable end. Similar in size to a double-wide trailer, the bungalow was the popular replacement for nineteenth-century log houses because of its balloon framing with low-cost sawmill lumber.

box house. A one-story, square, twentieth-century house with four rooms. The box house was the cheapest popular replacement for nineteenth-century log houses because of its usually rough-cut board-and-batten construction.

Notch Typology

saddle notch. A rounded cut made on the bottom side of a round log so that adjacent logs will overlap. Unlike most other notches, the saddle notch does not lock.

V notch. Viewed from the butt end, the V notch has an inverted V at the top of a hewn log. The log's sides and bottom are straight. The V notch looks like the gable end of a house.

saddle V. Viewed from the butt end, a saddle notch is pear shaped with a V shape on top and a rounded bottom. It is always used on round logs.

half dovetail. Viewed from the butt end, the half dovetail notch has a thirty- to forty-five-degree slope on the top of a hewn log. The sides and bottom are straight.

full dovetail. Viewed from the butt end, the full dovetail has forty-five-degree slopes cut on the top and bottom of a hewn log. The sides are vertical.

diamond. Viewed from the butt end, the diamond notch has slopes on all four sides at forty-five-degree angles to the squared hewn log. The diamond looks like an ace of diamonds in a card deck.

square. A square notch has four small right-angle corners that parallel the alignment of the hewn log. Viewed from the butt end, it looks like a small square on the end of a larger one. This notch does not lock.

semilunate crown on half logs. The semilunate crown, named for its half moon shape, is a rounded notch on split round logs (logs that have been split in half down the middle).

double notch. The double notch on hewn logs has squared notches cut on the top and bottom of the log, several inches in from the butt end. They are in the shape of the familiar toy Lincoln logs, but the double notch is rare in Appalachia.

Other Terminology

anticline. A convex geologic fold with an upward arch; an anticline is the opposite of a syncline, which is U shaped.

board-and-batten. A low-cost construction method for walls that uses wide vertical boards with two-inch wooden strips called battens nailed over the interstices between them. Between the 1880s and the 1940s, board-and-batten construction was popular in logging and coal camps and on farms that needed buildings constructed with cheap sawmill lumber.

cadastral pattern. The pattern of landownership. Cadastral surveys result in plat maps that show landownership for property tax registry.

cantilever. A construction technique that uses long horizontal beams to support a superstructure overhead. Cantilevers, thirty to forty feet long, are used in some barns to support a large hayloft.

cove. A small valley that extends into mountains or a fairly level limestone-floored valley surrounded by mountains. Cades Cove and Grassy Cove, Tennessee, and Burkes Garden, Virginia, are good examples.

cruck. Pairs of large curved timbers that form an A-shaped pattern for walls and roof in the earliest English houses in the Virginia colony.

culture hearth. A culture hearth is the center of a culture region where people share common culture traits. It is a focused place, usually with a deep history and a concentration of well-defined cultural characteristics. Examples are southeastern Pennsylvania and the North Carolina Piedmont for eighteenth-century Germans and Upper East Tennessee for a tertiary Scotch-Irish hearth along the Watauga River.

culture region. A spatial unit or area occupied by a group or groups of people who share a common culture.

dendrochronology. The science of analyzing growth rings from trees to determine past climatic conditions of wet and dry years based on ring width. Wider rings indicate wet years; narrow rings show drought periods. Dendrochronology is used to calculate the age of a tree when it was cut and thus to estimate a date for log construction.

fall line. The geologic boundary between the Atlantic Coastal Plain and the Piedmont Province. Rivers flowing across the fall line go over waterfalls and rapids because elevations abruptly change at the contact point between geologic units. Early settlements formed at heads of navigation and became the fall line cities of Richmond, Virginia; Raleigh and Fayetteville, North Carolina; Columbia, South Carolina; and Augusta and Macon, Georgia, among others.

folk architecture. Dwellings, outbuildings, and other structures built by a folk culture using traditional methods and styles and local materials. Folk architecture lacks professional architects and blueprints.

folk culture. A traditional, usually rural way of life among a relatively small, cohesive group of people who share common culture traits. The material objects in a folk culture are usually handmade.

folklife or folkways. All aspects of a folk culture involving material and nonmaterial culture traits.

folklore. Nonmaterial traits in a folk culture, usually expressed in oral traditions in tales and sayings

gable. The short side of a rectangular building; the area between wall and roof on a saddle- or gable-roofed building.

gable roof. See saddle roof

geomorphic provinces. Landform regions based on common geologic history, similar topography, and other general physical criteria

geomorphology. The science and study of landforms that focuses on the surface expression of topography and terrain such as mountains, plains, plateaus, deltas, dunes, mesas, and buttes, among many others

hearth. In an architectural sense, the stone or brick floor of a fireplace, where the warmth of a fire and coals radiates to warm people and to cook and heat food. A home's hearth, especially in winter, was a gathering point for a family. *See also* culture hearth.

karst topography. Landforms and landscapes that form in limestone geology where dissolved rocks create openings and depressions such as sinkholes, caves, and underground drainage. Named for the Karst district of limestone landscapes near the eastern Adriatic Sea.

material culture. All of the tangible, visible objects that a culture group makes or possesses. These may be buildings, tools, clothing, art objects, or thousands more material traits in the culture.

microclimate. The unique concentrated climate for a small area or ecological niche on the earth's surface. Mountains and valleys have different microclimates, with mountains having cooler temperatures and usually more rainfall than valley sites.

nonmaterial culture. All of the oral traditions of a culture in songs, stories, lore, beliefs, and ideas. Nonmaterial culture traits are intangible and invisible but are clearly present in the spoken word.

orogeny. The process of forming mountains. *Oro* refers to mountains, and *geny* to origin.

orographic rainfall. Rainfall that occurs on the windward sides of mountains. As moisture-laden winds are forced aloft on windward slopes, the air cools, condenses, forms clouds, and often produces rain. Such windward slopes always have more rainfall than leeward slopes.

saddle roof. Also called gable roof or pitched roof, the saddle roof is the most common roof type, with gables at each end of the building and equal slopes in front and in back that meet at an apex or roof ridge.

transhumance. The seasonal migration of livestock to fresh pastures. In the spring in mountain areas, transhumance migrations go from valley farms to mountain pastures. In autumn, herders and livestock return to the valleys.

vernacular. In an anthropological context *vernacular* refers to a native language in a particular culture group and place. Since the 1970s, *vernacular* has been used interchangeably with *folk,* especially in folk architecture.

Watauga culture hearth. An eighteenth-century concentration of Scotch-Irish settlers along the Watauga River in northeastern Tennessee. The area became a tertiary culture hearth from which Scotch-Irish settlers expanded to the south and west. Some of the Watauga settlements there became the present cities of Elizabethton, Johnson City, Bristol, and Kingsport.

wattle-and-daub. An old European construction technique for walls that uses interwoven saplings, vines, and twigs to form the wattle, which is then daubed with a mixture of clay and straw to fill the interstices

weatherboard. Also called clapboard or siding, weatherboarding consists of overlapping horizontal sawn planks fastened to the exterior walls of a building to protect it from the weather.

References

Adams, Sheila K. 1995. *Come Go Home with Me.* Chapel Hill: University of North Carolina Press.

Addy, Sidney O. 1898. *The Evolution of the English House.* London: Swan Sonnenschein.

Akenson, Donald H. 1984. "Why the Accepted Estimates of the Ethnicity of the American People, 1790, Are Unacceptable." *William and Mary Quarterly,* 3rd ser., 41 (1): 102–19.

Allen, James Lane. 1886a. "The Blue-Grass Region of Kentucky." *Harper's New Monthly Magazine* 72 (February): 365–82.

———. 1886b [1995]. "Through Cumberland Gap on Horseback." *Harper's Magazine* 73 (June): 50–66. Reprinted in *Appalachian Images in Folk and Popular Culture,* ed. W. K. McNeil, 2nd ed., 59–73. Knoxville: University of Tennessee Press.

Ambrosia, Vincent G. 1979. "Log Architecture in Union County, Tennessee: Dwellings, Outbuildings, and Notching Types." Manuscript. Knoxville: Department of Geography, University of Tennessee.

Appalachian Trail Conference. 1967. *Guide to the Appalachian Trail in Tennessee and North Carolina, Cherokee, Pisgah, and Great Smokies.* 2nd ed. Washington, DC: Appalachian Trail Conference.

Ashworth, John H. 1913 [1995]. "The Virginia Mountaineers." *South Atlantic Quarterly* 12:193–211. Reprinted in *Appalachian Images in Folk and Popular Culture,,* ed. W. K. McNeil, 2nd ed., 187–203. Knoxville: University of Tennessee Press, 1995.

Barker, Garry G. 1991. *The Handcraft Revival in Southern Appalachia, 1930–1990.* Knoxville; University of Tennessee Press.

Bartram, William. [1791] 1955. *Travels of William Bartram.* New York: Dover.

Bascom, Louise Rand. 1909. "Ballads and Songs of Western North Carolina." *Journal of American Folk-Lore* 22:238–50.

Batteau, Allen. 1990. *The Invention of Appalachia.* Tucson: University of Arizona Press.

———, ed. 1983. *Appalachia and America: Autonomy and Regional Dependence.* Lexington: University of Kentucky Press.

Becker, Jane S. 1998. *Selling Tradition: Appalachia and the Selling of an American Folk: 1930–1940.* Chapel Hill: University of North Carolina Press.

Berthoff, Rowland. 1986. "Celtic Mist over the South." *Journal of Southern History* 52:523–50.

Big South Fork Scenic Railway. 1994. *Souvenir Book: The Story of a Railway*

Excursion into the Blue Heron Mining Camp of the Big South Fork. Stearns, KY: Big South Scenic Railway.

Bishir, Catherine W. 1990. *North Carolina Architecture.* Chapel Hill: University of North Carolina Press.

Black, T. E. 1972. *T. E. Black's Lifetime Planting, Business and Fishing Guide.* Andalusia, AL: C. J. Black.

Blethen, H. Tyler, and Curtis W. Wood Jr., eds. 2001. *Ulster and North America: Transatlantic Perspectives on the Scotch-Irish.* Tuscaloosa: University of Alabama Press.

Boyer, Marj. 2001. *North Carolina Ginseng Dealers, 2001–2001.* Raleigh: North Carolina Department of Agriculture and Consumer Services.

Bradshaw, Michael. 1992. *The Appalachian Regional Commission: Twenty-five Years of Government Policy.* Lexington: University of Kentucky Press.

Brissot, Jacques Pierre, and Robert D. Mitchell. 1993. "Town and Country in Backcountry Virginia: Winchester and the Shenandoah Valley, 1730–1800," *Journal of Southern History* 59:691.

Brose, David, and John C. Campbell Folk School. 1995. *Meeting in the Air: Sacred Music of the Southern Appalachians.* Brasstown, NC: John C. Campbell Folk School.

Brown, Corie. 1999. "Deep in the Heart of Appalachia." *Newsweek,* November 29, 92.

Brown, Fred. 1999a. "Echoes from the Past: Sheila Adams Gives Voice to Her Mountain Ancestors." *Knoxville News-Sentinel,* January 24, 1999, E1, E5.

———. 1999b. "Sounds from the Soul: Old Harp Singing Is Harmony of History." *Knoxville News-Sentinel,* May 23, 1999, E1–E8.

———. 2002a. "Remembering Fraterville's Dead." *Knoxville News Sentinel,* May 20, 2002, A1, A7.

———. 2002b. "Dye Goes to Root of Ginseng Problem." *Knoxville News Sentinel,* November 24, 2002, B1, B6.

Brown, Fred, and Jeanne McDonald. 2000. *The Serpent Handlers: Three Families and Their Faith.* Winston-Salem, NC: John F. Blair.

Brown, Ralph H. 1948. *Historical Geography of the United States.* New York: Harcourt Brace and World.

Brunhes, Jean. 1952. *Human Geography.* Abr. ed. New York: Rand McNally.

Brunskill, R. W. 1978. *Illustrated Handbook of Vernacular Architecture.* London: Faber and Faber.

Bucher, Robert C. 1962. "The Continental Log House." *Pennsylvania Folklife* 12 (4): 14–19.

Bullard, Helen. 1976. *Crafts and Craftsmen of the Tennessee Mountains.* Falls Church, VA: Summit Press.

Burton, Thomas. 1993. *Serpent-Handling Believers.* Knoxville: University of Tennessee Press.

Cabeza de Vaca, Alvar Núñez. [1542] 1907. "The Narrative of Cabeza de Vaca." In *Spanish Explorers in the Southern United States, 1528–1543,* ed. Frederick

W. Hodge. New York: Charles Scribner's Sons. Reprinted in *Voices of the Old South,* ed. Alan Gallay, 4–8. Athens: University of Georgia Press, 1994.

Campbell, John C. [1921] 1969. *The Southern Highlander and His Homeland.* Lexington: University of Kentucky Press.

Cantrell, Brent. 1981. "Traditional Grave Structures on the Eastern Highland Rim." *Tennessee Folklore Society Bulletin* 67 (3): 93–103.

Carlisle, Ronald C. 1982. *An Architectural Study of Some Folk Structures in the Area of the Paintsville Lake Dam, Johnson and Morgan Counties, Kentucky.* Report to the U.S. Army Corps of Engineers, Huntington District, Huntington, WV.

Carney, George O. 1974."Bluegrass Grows All Around: The Spatial Dimensions of a Country Music Style." *Journal of Geography* 73:34–55.

———. 1979. "T for Texas, T for Tennessee: The Origins of American Country Music Notables." *Journal of Geography* 78:218–25.

———. 1996. "Western North Carolina: Culture Hearth of Bluegrass Music." *Journal of Cultural Geography* 16 (1): 65–87.

———, ed. 1998. *Baseball, Barns, and Bluegrass: A Geography of American Folklife.* Lanham, MD: Rowman and Littlefield.

Carver, Craig M. 1987. *American Regional Dialects: A Word Geography.* Ann Arbor: University of Michigan Press.

Caudill, Harry M. 1963. *Night Comes to the Cumberlands: A Biography of a Depressed Area.* Boston: Little, Brown.

———. 1969. "Anglo-Saxon vs. Scotch Irish: Round 2." *Mountain Life and Work* 45:18–19.

Clarke, Mary Washington. 1976. *Kentucky Quilts and Their Makers.* Lexington: University of Kentucky Press.

Clevinger, Woodrow R. 1938. "The Appalachian Mountaineers in the Upper Cowlitz Basin." *Pacific Northwest Quarterly* 29:115–34.

———. 1942. "Southern Appalachia Highlanders in Western Washington." *Pacific Northwest Quarterly* 33:3–25.

Cobb, James E. 1978a. "Historic Fish Traps on the Lower Holston River." *Tennessee Anthropologist* 3 (1): 33–58.

———. 1978b. "Supplementary Information on Gravehouses in Tennessee." *Tennessee Anthropological Association Newsletter* 3 (6): 4–7.

Conway, Cecelia. 1995. *African Banjo Echoes in Appalachia.* Knoxville: University of Tennessee Press.

Cordle, Larry. 1999. "Murder on Music Row." Nashville, TN: Shell Point Records.

Cosby Ruritan Club. 1999. *46th Annual Ramp Festival Program.* Cosby, TN: Cosby Ruritan Club.

Cox, William E. 1978. *Hensley Settlement: A Mountain Community.* N.p.: Eastern National Park and Monument Association.

Crissman, James K. 1994. *Death and Dying in Central Appalachia: Changing Attitudes and Practices.* Urbana: University of Illinois Press.

Cunningham, Rodger. 1987. *Apples on the Flood: Minority Discourse and Appalachia.* Knoxville: University of Tennessee Press.

Dabney, Joseph Earl. 1974. *Mountain Spirits: A Chronicle of Corn Whiskey.* Asheville, NC: Bright Mountain Books.

———. 1998. *Smokehouse Ham, Spoon Bread, and Scuppernong Wine: The Folklore and Art of Southern Appalachian Cooking.* Nashville, TN: Cumberland House.

Davidson, Donald. [1946] 1978. *The Tennessee: The Old River.* Knoxville: University of Tennessee Press.

Dial, Wylene P. 1969. "The Dialect of the Appalachian People." *West Virginia Quarterly* 30:463–71.

———. 1970. "Folk Speech Is English, Too." *Mountain Life and Work* 46 (March): 15–17.

Dickey, James. 1970. *Deliverance.* Boston: Houghton Mifflin.

Dickson, R. J. 1966. *Ulster Emigration to Colonial America, 1718–1775.* London: Routledge and Kegan Paul.

Dorgan, Howard. 1987. *Giving Glory to God in Appalachia: Worship Practices of Six Baptist Subdenominations.* Knoxville: University of Tennessee Press.

Dorson, Richard M. 1983. *Handbook of American Folklore.* Bloomington: University of Indiana Press.

Dunaway, Wayland P. 1944. *The Scotch-Irish of Colonial Pennsylvania.* Chapel Hill: University of North Carolina Press.

Durand, Loyal, Jr. 1956 [1998]. "Mountain Moonshining in East Tennessee." *Geographical Review* 46:168–81. Reprinted in *Baseball, Barns and Bluegrass,* ed. George O. Carney, 120–28. Lanham, MD: Rowman and Littlefield.

Dykeman, Wilma. 1955. *The French Broad.* Knoxville: University of Tennessee Press.

Dykes, Jim. 1968. "Cumberland Plateau's Grassy Cove Is World of Its Own." *Knoxville News-Sentinel,* November 3, D8.

Eanes, Ellen Fickling. 1988. *North Carolina Quilts.* Chapel Hill: University of North Carolina Press.

Eaton, Allen H. [1937] 1973. *Handicrafts of the Southern Highlands.* New York: Dover.

Egerton, John. 1987. *Southern Food: At Home, on the Road, in History.* New York: Alfred A. Knopf.

Eller, Ronald D. 1982. *Miners, Millhands, and Mountaineers: Industrialization of the Appalachian South, 1880–1930.* Knoxville: University of Tennessee Press.

Ensminger, Robert F. 1992. *The Pennsylvania Barn: Its Origin, Evolution, and Distribution in North America.* Baltimore: Johns Hopkins University Press.

Ergood, Bruce. 1991. "Toward a Definition of Appalachia." In *Appalachia: Social Context, Past and Present,* ed. Bruce Ergood and Bruce E. Kuhre, 3rd ed. 39–48. Dubuque, IA: Kendall Hunt.

Ergood, Bruce, and Bruce Kuhre. 1991. *Appalachia: Social Context, Past and Present.* 3rd ed. Dubuque, IA: Kendall Hunt.

Evans, E. Estyn. [1942] 1963. *Irish Heritage: The Landscape, the People, and Their Work.* Dundalk, Ireland: Dundalgan Press.

———. 1965. "Cultural Relics of the Ulster-Scots in the Old West of North America." *Ulster Folklife* 11:33–38.

———. 1966. "Culture and Land Use in the Old West of North America." *Heidelberger Studien zur Kulturgeographie* 15:72–80.

———. 1969. "The Scotch-Irish: Their Cultural Adaptation and the Heritage in the American Old West." In *Essays in Scotch-Irish History,* ed. E. R. R. Green, 69–86. London: Routledge and Kegan Paul.

———. 1974. "Folk Housing in the British Isles in Materials Other Than Timber." In *Man and Cultural Heritage: Papers in Honor of Fred B. Kniffen,* ed. H. J. Walker and W. G. Haag, 53–64. Baton Rouge: Louisiana State University School of Geoscience.

Farr, Sidney Saylor. 1983. *More Than Moonshine: Appalachian Recipes and Recollections.* Pittsburgh: University of Pittsburgh Press.

Fenneman, Nevin M. 1938. *Physiography of the Eastern United States.* New York: McGraw-Hill.

Ferris, William R., Jr. 1980. "The Dog Trot: A Regional Home and Its Builder." *Perspecta: The Yale Architectural Journal* 17:68–73.

———. 1986. "The Dogtrot: A Mythic Image in Southern Culture." *Southern Quarterly* 25:72–85.

Fetterman, John. 1967. *Stinking Creek.* New York: Dutton.

Fielder, George F. (Nick), Jr. 1982. "Gravehouses: Mortuary Folk Architecture." *Courier* (Tennessee Historical Commission) 21 (1): 4–5.

Finger, John R. 1984. *The Eastern Band of Cherokees, 1819–1900.* Knoxville: University of Tennessee Press.

Fink, Paul. 1974. *Bits of Mountain Speech.* Jonesborough, TN: Paul Fink.

Forbis, Wesley, ed. 1991. *Baptist Hymnal.* Nashville, TN: Convention Press.

Ford, Thomas R. [1967] 1991. "The Passing of Provincialism." Reprinted in *Appalachia: Social Context Past and Present,* ed. Bruce Ergood and Bruce E. Kuhre, 3rd ed., 80–103. Dubuque, IA: Kendall Hunt.

———, ed. 1967. *The Southern Appalachian Region.* Lexington: University of Kentucky Press.

Forman, Henry Chandlee. [1948] 1967. *The Architecture of the Old South: The Medieval Style, 1585–1850.* New York: Russell and Russell.

Foster, Stephen William. 1988. *The Past Is Another Country: Representation, Historical Consciousness, and Resistance in the Blue Ridge.* Berkeley: University of California Press.

Fowler, Damon Lee. 1995. *Classical Southern Cooking.* New York: Crown.

Fox, John. 1901 [1995]. "The Southern Mountaineer." *Scribner's* 29:387–99, 556–70. Reprinted in *Appalachian Images in Folk and Popular Culture,* ed. W. K. McNeil, 2nd ed., 121–44. Knoxville: University of Tennessee Press, 1995.

Francaviglia, Richard V. 1979. *The Mormon Landscape: Creation and Perception of a Unique Image in the American West.* New York: AMS Press.

Fries, Adelaide L. 1922–69. *Records of the Moravians in North Carolina.* Raleigh: Edwards and Broughton.

Frost, William Goodell. 1899 [1995]. "Our Contemporary Ancestors in the Southern Mountains." *Atlantic Monthly* 83:311–19. Reprinted in *Appalachian Images in Folk and Popular Culture,* ed. W. K. McNeil, 2nd ed., 91–106. Knoxville: University of Tennessee Press, 1995.

Gabbard, Alex. 2000. *Return to Thunder Road: The Story Behind the Legend.* Lenoir City, TN: Gabbard Publications.

Gallegos, Eloy J. 1997. *The Melungeons: The Pioneers of the Interior of the Southeastern United States, 1526–1997.* Knoxville, TN: Villagra Press.

Gillam, Martin. 1998. "The Life and Times of Ralph Stanley." Nashville, TN: Network Enterprises. Videotape.

Gillenwater, Mack H. 1972. "Cultural and Historical Geography of Mining Settlements in the Pocohontas Coal Field of Southern West Virginia, 1880 to 1930." Ph.D. diss., Department of Geography, University of Tennessee.

Gillespie, Paul F., ed. 1982. *The Foxfire Book,* vol. 7. Garden City, NY: Anchor Books.

Glassie, Henry. 1963. "The Appalachian Log Cabin." *Mountain Life and Work* 39 (4): 5–14.

———. 1964. "The Smaller Outbuildings of the Southern Mountains." *Mountain Life and Work* 40 (1): 21–25.

———. 1968. *Pattern in the Material Folk Culture of the Eastern United States.* Philadelphia: University of Pennsylvania Press.

———. 1972. "Eighteenth-Century Cultural Process in Delaware Valley Folk Building." *Winterthur Portfolio.* Reprinted in *Common Places: Readings in Vernacular Architecture,* ed. Dell Upton and John Michael Vlach, 394–425. Athens: University of Georgia Press, 1986.

———. 1975, *Folk Housing in Middle Virginia.* Knoxville: University of Tennessee Press.

———. 1978. "The Types of the Southern Mountain Cabin. In *The Study of American Folklore: An Introduction,* ed. Jan Brunvand, 2nd ed., 390–420. New York: Norton.

Glauser, Beat. 1974. *The Scottish-English Linguistic Border: Lexical Aspects.* Bern: Francke.

Golden, Richard M. 1988. *The Hueguenot Connection.* Dordrecht, The Netherlands: Kluwer Academic.

Graham, Ian Charles C. 1956. *Colonists from Scotland: Emigration to North America, 1707–1783.* Ithaca, NY: Cornell University Press for American Historical Association.

Great Smoky Mountains Natural History Association. n.d. *Tremont Logging History.* Gatlinburg, TN: Great Smoky Mountains Natural History Association.

Green, Douglas B. 1976. *Country Roots: The Origins of Country Music.* New York: Hawthorn Books.
Grieve, M. [1931] 2002a. "Boneset." In *A Modern Herbal.* www.botanical.com.
———. [1931]. 2002b. "Goldenseal." In *A Modern Herbal.* www.botanical.com.
Gritzner, Charles F. 2002. "Restoring 'Culture' and Related Terms to Their Rightful Geographic Meaning." *Journal of Geography* 101:222–23.
Guthrie, James L. 1990. "Melungeons: Comparisons of Gene Distributions of Those of Worldwide Populations." *Tennessee Anthropologist* 5 (1): 13–22.
Guyot, Arnold. 1861. "On the Appalachian Mountain System." *American Journal of Science and Arts* 31:157–87.
Hall, Carrie A. 1935. *The Romance of the Patchwork Quilt in America.* N.p.: Claxton.
Hall, Joseph S. 1942. *The Phonetics of Smoky Mountain Speech.* American Speech Reprints and Monographs no. 4. New York: King's Crown Press.
———. 1972. *Sayings from Old Smoky.* Asheville, NC: Cataloochee Press.
Halperin, Rhoda. 1990. *The Livelihood of Kin: Making Ends Meet "The Kentucky Way."* Austin: University of Texas Press.
Hanna, Charles A. [1902] 1968. *The Scotch Irish.* Vols. 1 and 2. Baltimore: Genealogical Publishing Company.
Harney, Will Wallace. 1873 [1995]. "A Strange Land and Peculiar People. *Lippincott's* 12:429–38. Reprinted in *Appalachian Images in Folk and Popular Culture,* ed. W. K. McNeil, 2nd ed., 45–58. Knoxville: University of Tennessee Press, 1995.
Hart, John F. 1977. "Land Rotation in Appalachia." *Geographical Review* 67:148–66.
———. 1998. *The Rural Landscape.* Baltimore: Johns Hopkins University Press.
Hicks, George L. 1976. *Appalachian Valley.* Prospect Heights, IL: Waveland Press.
Hill, Stephan L. 1978. "Land Rotation in Washington County, Tennessee." Master's thesis, Department of Geography, University of Tennessee.
Hilliard, Sam B. 1969 [1998]. "Hog Meat and Cornpone: Food Habits in the Ante-bellum South." *Proceedings of the American Philosophical Society* 113 (1): 1–13. Reprinted in *Baseball, Barns, and Bluegrass,* ed. George O. Carney, 79–101. Lanham, MD: Rowman and Littlefield.
———. 1972. *Hog Meat and Hoecake: Food Supply in the Old South, 1840–1860.* Carbondale: Southern Illinois University Press.
Hindman Settlement School. 1994. *The Mountain Echo* 13 (1): 1–8.
Horn, Stanley F. [1943] 1951. *This Fascinating Lumber Business.* Indianapolis: Bobbs-Merrill.
Hudson, Charles. 1976. *The Southeastern Indians.* Knoxville: University of Tennessee Press.

Hulan, Richard. 1975. "Middle Tennessee and the Dogtrot House." *Pioneer America* 7 (2): 37–47.

Huntington, Ellsworth. 1907. *The Pulse of Asia.* Boston: Houghton Mifflin.

Irwin, John Rice. 1982. *Baskets and Basket Makers in Southern Appalachia.* Atglen, PA: Schiffer.

———. 1983. *Musical Instruments of the Southern Appalachian Mountains.* Atglen, PA: Schiffer.

———. 1984. *A People and Their Quilts.* Atglen, PA: Schiffer.

Jackson, George Pullen. [1933] 1965. *White Spirituals in the Southern Uplands: The Story of the Fasola Folk, Their Songs, Singing and "Buckwheat Notes."* New York: Dover.

———. 1943. *White and Negro Spirituals: Their Life Span and Kinship.* Locust Valley, NY. J. J. Augustin.

Jeane, D. Gregory. 1969. "The Traditional Upland South Cemetery." *Landscape* 18 (2): 39–41.

———. 1978. "The Upland South Cemetery." *Journal of Popular Culture* 11:895–903.

John C. Campbell Folk School. n.d. Brochure. Brasstown, NC: John C. Campbell Folk School.

Johnson, Hildegard Binder. 1976. *Order upon the Land.* New York: Oxford University Press.

Jones, Loyal. 1984. *Minstrel of the Appalachians: The Story of Bascom Lamar Lunsford.* Boone, NC: Appalachian Consortium Press.

———. 1994. *Appalachian Values.* Ashland, KY: Jesse Stuart Foundation.

———. 1999. *Faith and Meaning in the Southern Uplands.* Urbana: University of Illinois Press.

Jordan, Terry G. 1978. *Texas Log Buildings: A Folk Architecture.* Austin: University of Texas.

———. 1985. *American Log Buildings: An Old World Heritage.* Chapel Hill: University of North Carolina Press.

Jordan, Terry G., and Matti Kaups. 1989. *The American Backwoods Frontier: An Ethnic and Ecological Interpretation.* Baltimore: Johns Hopkins University Press.

Jordan-Bychkov, Terry G. 1998. "Transverse Crib Barns, the Upland South, and Pennsylvania extended." *Material Culture* 30 (2): 1–31.

———. 2003. *The Upland South: The Making of an American Folk Region and Landscape.* Santa Fe, NM: Center for American Places.

Jordan-Bychkov, Terry G., and Mona Domosh. 2003. *The Human Mosaic: A Thematic Introduction to Cultural Geography.* 9th ed. New York: W. H. Freeman.

Kennedy, N. Brent. 1997. *The Melungeons: The Resurrection of a Proud People.* Macon, GA: Mercer University Press.

Kephart, Horace. [1913] 1984. *Our Southern Highlanders.* Knoxville: University of Tennessee Press.

Kilar, Jeremy. W. 1997. "Germans." In *American Immigrant Cultures: Builders of a Nation,* ed. David Levinson and Melvin Ember, 315–25. New York: Simon and Schuster Macmillan..

Kimball, Fiske. [1922] 1966. *Domestic Architecture of the American Colonies and of the Early Republic.* New York: Dover.

Kniffen, Fred B. 1936. "Louisiana House Types." *Annals of the Association of American Geographers* 26:179–93.

———. 1965. "Folk Housing: Key to Diffusion." *Annals of the Association of American Geographers* 55 (4): 549–77.

———. 1969. "On Corner-Timbering." *Pioneer America* 1 (1): 1–8.

———. 1979. "The Geographer's Craft: Why Folk Housing?" *Annals of the Association of American Geographers* 69 (1): 59–63.

Kniffen, Fred B., and Henry Glassie. 1966. "Building in Wood in the Eastern United States: A Time-Place Perspective." *Geographical Review* 56 (1): 40–66.

Koons, Kenneth. 2000. "The Staple of Our Country: Wheat in the Regional Farm Economy of the Nineteenth-Century Valley of Virginia." In *After the Back Country: Rural Life in the Great Valley of Virginia, 1800–1900,* ed. Kenneth E. Koons and Warren R. Hofstra, 3–20. Knoxville: University of Tennessee Press.

Kurath, Hans. 1949. *A Word Geography of the Eastern United States.* Ann Arbor: University of Michigan Press.

———. 1972. *Studies in Area Linguistics.* Bloomington: Indiana University Press.

Lambie, Joseph T. 1954. *From Mine to Market: The History of Coal Transportation on the Norfolk and Western Railway.* New York: New York University Press.

Landry, Bill. 1998. "Heartland Tonight." Knoxville, TN: WBIR Television. Videotape.

Law, Rachel Nash, and Cynthia W. Taylor. 1991. *Appalachian White Oak Basketmaking: Handing Down the Basket.* Knoxville: University of Tennessee Press.

Lay, K. Edward. 1982. "European Antecedents of Seventeenth and Eighteenth Century Germanic and Scotch-Irish Architecture in America." *Pennsylvania Folklife* 32 (1): 2–43.

Lemon, James T. 1966. "The Agricultural Practices of National Groups in Eighteenth-Century Southeastern Pennsylvania." *Geographical Review* 56:467–96.

———. [1972] 2002. *The Best Poor Man's Country: A Geographical Study of Early Southeastern Pennsylvania.* Baltimore: Johns Hopkins University Press.

Leonard, Bill J. 1982. "Historical Overview and Appendix." In *The Foxfire Book,* vol. 7, ed. Paul F. Gillespie. Garden City, NY: Anchor Books.

Leyburn, James G. 1962. *The Scotch Irish: A Social History.* Chapel Hill: University of North Carolina Press.

Lineback, Neal G. 1999. "How Old the New?" *Geography in the News,* no. 214. Southern Pines: NC: Karo Hollow Press.

Linn, Karen. 1991. *That Half Barbaric Twang: The Banjo in American Popular Culture.* Urbana: University of Illinois Press.

Logsdon, Charles W. 1984. "The Pharmacology of Folk Medicinal Use of Panax in the Southern Appalachians: An Overview with Suggested Applications for Further Study." *Tennessee Anthropologist* 9 (1): 66–79.

———. 1986. "Plant Medicines of the Southern Appalachians, I, Hematic Herbs ('The Tonics of Spring')." *Tennessee Anthropologist* 11 (1): 55–68.

Lomax, Alan. 1998. *Southern Journey, and Glory Shown Around: More All Day Singing from The Sacred Harp.* Vol. 10. Liner notes. Cambridge, MA: Rounder Records. Rounder CD 1710.

Long, Amos, Jr. 1972. *The Pennsylvania German Family Farm: A Regional Architecture and Folk Cultural Study of an American Agricultural Community.* Publications, vol. 6. Breinigsville, PA: Pennsylvania German Society.

Lopes, Danielle, Joan Moser, and Annie Louise Perkinson. 1996. *Appalachian Folk Medicine: Native Plants and Healing Traditions.* Asheville, NC: Warren Wilson College Press.

Madden, Robert R., and T. Russell Jones. 1977. *Mountain Home: The Walker Family Farmstead, Great Smoky Mountains National Park.* Washington, DC: U.S. Department of the Interior, National Park Service.

Maertens, Thomas Brock, Jr. 1990. "The Relationships of Climate and Terrain to Maintenance of Way on the Norfolk and Southern Railroad between Norfolk, Virginia, and Portsmouth, Ohio." Ph.D. diss., Department of Geography, University of Tennessee.

Magnusson, Leifur. 1918. "Company Housing in the Anthracite Region of Pennsylvania." *Monthly Labor Review* 10:186–95.

Malone, Bill C. 1993. *Singing Cowboys and Musical Mountaineers.* Athens: University of Georgia Press.

Mangus, Arthur R. 1940. *Rural Regions of the United States.* Washington, DC: Government Printing Office.

Mann, David F. 2002. "The Dendroarchaeology of the Swaggerty Blockhouse, Cocke County, Tennessee." Master's thesis. Department of Geography, University of Tennessee.

Manthey, Tanya. 2002. "Food as a Cultural Marker: A Sketch and the Development and Significance of Soul Food." Senior thesis, College Scholars Program, University of Tennessee.

Martin, George A. [1887] 1992. *Fences, Gates, and Bridges: A Practical Manual.* Chambersburg, PA: Alan C. Hood.

Mather, J. Y., and H. H. Speitel, eds. 1977. *The Linguistic Atlas of Scotland.* Vol. 2. Hamden, CT: Archon Books.

Mayfield, Michael Wells. 1984. "Variations in Streamflow among Watersheds of the Cumberland Plateau, Tennessee." Ph.D. diss., Department of Geography, University of Tennessee.

McCauley, Deborah Vansau. 1995. *Appalachian Mountain Religion: A History.* Urbana: University of Illinois Press.

McCauley, Deborah Vansau, and Laura E. Porter. 2003. *Mountain Holiness: A Photographic Narrative.* Knoxville: University of Tennessee Press.

McCleary, Ann E. 2000. "Forging a Regional Identity: Development of Rural Vernacular Architecture in the Central Shenandoah Valley, 1790–1850." In *After the Back Country: Rural Life in the Great Valley of Virginia, 1800–1900,* ed. Kenneth E. Koons and Warren R. Hofstra, 92–110. Knoxville: University of Tennessee Press.

McCormick, M. J. 1991. "In Search of the Wild Ramp." *Herb Companion.* www.foodfinder.com/ramp.

McCrum, Robert, William Cran, and Robert McNeil. 1986. *The Story of English.* New York: Viking Press.

McDill, Bob. 1994. *Gone Country.* New York: Polygram International Publishing, Ranger Bob Music (ASCAP).

McDonald, Forrest, and Ellen Shapiro McDonald. 1980. "The Ethnic Origins of the American People, 1790." *William and Mary Quarterly,* 3rd ser., 37:179–99.

McIntire, William G. 1958. *Prehistoric Indian Settlements of the Changing Mississippi Delta.* Baton Rouge: Louisiana State University Press.

McNeil, W. K., ed. 1995. *Appalachian Images in Folk and Popular Culture.* 2nd ed. Knoxville: University of Tennessee Press.

McWhiney, Grady. 1988. *Cracker Culture: Celtic Ways in the Old South.* Tuscaloosa: University of Alabama Press.

Meinig, D. W. 1986. *The Shaping of America: A Geographical Perspective on 500 Years of History.* New Haven: Yale University Press.

Melton, L. Gordon. 1996. "Church of God with Signs Following." *Encyclopedia of American Religions.* 5th ed. Washington, DC: Gale Research.

Mercer, Henry C. 1926. "The Origin of Log Houses in the United States." *Papers, Bucks County Historical Society* 5:568–83.

Miles, Emma Bell. [1905] 1975. *The Spirit of the Mountains.* Knoxville: University of Tennessee Press.

Miller, E. Joan Wilson. 1968. "The Ozark Culture Region as Revealed by Traditional Materials." *Annals of the Association of American Geographers* 58 (1): 51–77.

Mitchell, Robert D. 1972. "The Shenandoah Valley Frontier." *Annals of the Association of American Geographers* 62 (3): 461–86.

———, 1977. *Commercialism and Frontier: Perspectives on the Early Shenandoah Valley.* Charlottesville: University Press of Virginia.

Moffett, Marian, and Lawrence Wodehouse. 1993. *East Tennessee Cantilever Barns.* Knoxville: University of Tennessee Press.

Montell, William Lynwood. 1970. *The Saga of Coe Ridge: A Study in Oral History.* Knoxville: University of Tennessee Press.

———. 1993. *Upper Cumberland Country.* Jackson: University Press of Mississippi.

Montell, William Lynwood, and Michael Lynn Morse. 1976. *Kentucky Folk Architecture.* Lexington: University of Kentucky Press.

Montgomery, Michael. 1991. "The Roots of Appalachian English: Scotch-Irish or British Southern?" *Journal of the Appalachian Studies Association* 3:177–91.

Mood, Fulmer. 1951. "The Origin, Evolution, and Application of the Sectional Concept, 1750–1900." In *Regionalism in America,* ed. Merrill Jensen, 5–98. Madison: University of Wisconsin Press.

Moody, T. W., F. X. Martin, and F. J. Byrne. 1982. *A New History of Ireland: A Chronology of Irish History to 1976.* Vol. 8. Oxford: Clarendon Press.

Moore, Harry L. 1994. *A Geologic Trip across Tennessee by Interstate 40.* Knoxville: University of Tennessee Press.

Moore, Marat. 1995. "Because the Earth Is Dark and Deep." In *Appalachia Inside Out,* vol.1, ed. Robert J. Higgs, Ambrose N. Manning, and Jim Wayne Miller, 135–45. Knoxville: University of Tennessee Press.

Moore, Tyrel G., Jr. 1975. "The Role of Ferry Crossings in the Development of the Transportation Network in East Tennessee, 1790–1974." Master's thesis, Department of Geography, University of Tennessee.

———. 1990. "Development and Change in Appalachian Kentucky's Economy: 1870–1890." *Southeastern Geographer* 30:121–39.

Morgan, John T. 1986. "The Decline of Log House Construction in Blount County, Tennessee." Ph.D. diss., Department of Geography, University of Tennessee.

———. 1990. *The Log House in East Tennessee.* Knoxville: University of Tennessee Press.

———. 1997. "The Cantilever Barn in Southwest Virginia." In *Diversity and Accommodation: Essays on the Cultural Composition of the Virginia Frontier,* ed. Michael J. Puglisi, 275–94. Knoxville: University of Tennessee Press.

Morgan, John [T.], and Joy Medford. 1980. "Log Houses in Grainger County, Tennessee." *Tennessee Anthropologist* 5 (2): 137–58.

Morrison, Hugh. 1952. *Early American Architecture.* New York: Dover.

Murfree, Mary Noailles. [1885] 1970. *The Prophet of the Great Smoky Mountains.* New York: AMS Press.

Murray-Wooley, Carolyn, and Karl Raitz. 1992. *Rock Fences of the Bluegrass.* Lexington: University Press of Kentucky.

Naismith, Robert J. 1985. *Buildings of the Scottish Countryside.* London: Victor Gollancz.

National Geographic Society. 1988. *Historical Atlas of the United States.* Washington, DC: National Geographic Society.

Neely, Sharlotte. 1991. *Snowbird Cherokees: People of Persistence.* Athens: University of Georgia Press.

Newton, Milton B. 1971. "The Annual Round in the Upland South." *Pioneer America* 3 (2): 63–73.

———. 1974. "Cultural Preadaptation and the Upland South." In *Man and Cultural Heritage: Papers in Honor of Fred B. Kniffen,* Geoscience and Man, vol. 5, ed. H. J. Walker and W. G. Haag, 143–54. Baton Rouge: School of Geoscience, Louisiana State University.

Newton, Milton B., Jr., and Linda Pulliam-DiNapoli. 1977. "Log Houses as Public Occasions: A Historical Theory." *Annals of the Association of American Geographers* 67:360–83.

Nichols, Edward Ray. 1973. "The Geography of Moonshining in Tennessee." Master's thesis, Department of Geography, University of Tennessee.

Nickens, T. Edward. 2001. "Catching Bandits in the Smokies." *National Wildlife* 29, no. 2 (February–March): 34–39.

Noble, Allen G. 1984. *Wood, Brick, and Stone: The North American Settlement Landscape.* 2 vols. Amherst: University of Massachusetts Press.

———, ed. 1992. *To Build in a New Land: Ethnic Landscapes in North America.* Baltimore: Johns Hopkins University Press.

Noble, Allen G., and Richard K. Cleek. 1995. *The Old Barn Book: A Field Guide to North American Barns and Other Farm Structures.* New Brunswick, NJ: Rutgers University Press.

Olson, Ted. 1998a. *Blue Ridge Folklore.* Jackson: University Press of Mississippi.

———. 1998b. "They Shall Take Up Serpents." *Christian History* 17 (2): 25.

O'Malley, James Ross. 1972. "The 'I' House: An Indicator of Agricultural Opulence in Upper East Tennessee." Master's thesis, Department of Geography, University of Tennessee.

———. 1977. "The 'I' House: An Indicator of Agricultural Attainment in the Southern Appalachian Valley." In *West Virginia and Appalachia: Selected Readings,* ed. Howard G. Adkins, Steve Ewing, and Chester E. Zimolzak, 105–13. Dubuque, IA: Kendall-Hunt.

O'Malley, James R., and John B. Rehder. 1978. "The Two-Story Log House in the Upland South." *Journal of Popular Culture* 11:904–15.

Orton, Harold. 1974. *A Word Geography of England.* New York: Seminar Press.

Otto, J. S., and N. E. Anderson. 1982. "The Diffusion of Upland South Folk Culture: 1790–1840." *Southeastern Geographer* 22:89–98.

Parton, Dolly. 1994. *Dolly: My Life and Other Unfinished Business.* New York: HarperCollins.

———. 1999. *The Grass Is Blue.* Liner notes. Durham, NC: Sugar Hill Records. SUG-CD-3900.

Pillsbury, Richard. 1976. "The Construction Materials of the Rural Folk Housing of the Pennsylvania Culture Region." *Pioneer America* 8 (2): 98–106.

———. 1977. "Pattern in the Folk and Vernacular House Forms of the Pennsylvania Culture Region." *Pioneer America* 9 (2): 12–31.

Pillsbury, Richard, and Andrew Kardos. 1970. *A Field Guide to the Folk Architecture in the Northeast.* Hanover, NH: Dartmouth College.

Pollitzer, William S. 1972. "The Physical Anthropology and Genetics of Marginal People in the Southeastern United States." *American Anthropologist* 74 (3): 719–34.

Pollitzer, William S., and William H. Brown. 1969. "Survey of Demography, Anthropology, and Genetics of the Melungeons of Tennessee: An Isolate Hybrid Origin in Process of Dissolution." *Human Biology* 41:388–99.

Powell, John Wesley. 1895. *Physiographic Regions of the United States.* New York: American Book Company.

Prévos, André J. M. 1997. "French." In *American Immigrant Cultures: Builders of a Nation,* ed. David Levinson and Melvin Ember, 286–91. New York: Simon and Schuster Macmillan.

Price, Edward T. 1951. "The Melungeons: A Mixed Blood Strain of the Southern Appalachians." *Geographical Review* 41:256–71.

———. 1960. "Root Digging in the Appalachians: The Geography of Botanical Drugs." *Geographical Review* 50:1–20.

Price, Henry R. 1971. *Melungeons: The Vanishing Colony of Newman's Ridge.* Rogersville, TN: Hancock County Drama Association.

Purvis, Thomas L. 1984. "The European Ancestry of the United States Population, 1790." *William and Mary Quarterly,* 3rd ser., 41 (1): 85–101.

Quinn, David B., ed. 1952. *The Roanoke Voyages.* Vol. 1, ser. 204. London: Hakluyt Society.

Rafferty, Milton D. 1973. "The Black Walnut Industry: The Modernization of a Pioneer Custom." *Pioneer America* 5 (1): 23–32.

Raine, James Watt. [1924] 1997. *The Land of Saddle-Bags: A Study of Mountain People of Appalachia.* Lexington: University of Kentucky Press.

Raitz, Karl B., ed. 1996. *The National Road.* Baltimore: Johns Hopkins University Press.

Raitz, Karl B., and Richard Ulack, with Thomas R. Leinbach. 1984. *Appalachia: A Regional Geography.* Boulder, CO: Westview Press.

Ramsey, Bets, and Marikay Waldvogel. 1986. *The Quilts of Tennessee: Images of Domestic Life prior to 1930.* Nashville, TN: Rutledge Hill Press.

Reding, William M. 2002. "Assessment of Spatial and Temporal Patterns in Log Structures in East Tennessee." Master's thesis, Department of Geography, University of Tennessee.

Reeleder, R. D. 2001 "Goldenseal: *Hydrastis canadensis.*" Southern Crop Protection and Food Research Centre. Ottawa: Agriculture and Agri-Food Canada.

Rehder, John B. 1992. "The Scotch-Irish and English in Appalachia." In *To Build in a New Land: Ethnic Landscapes in North America,* ed. Allen G. Nobel, 95–118. Baltimore: Johns Hopkins University Press.

———. 1996."Mapping Needleleaf Forests in the Great Smoky Mountains National Park. In *Raster Imagery in Geographic Information Systems,* ed. Stan Morain and Shirley Lopez Baros, 431–36. Santa Fe: OnWord Press.

———. 1997a. "Scotch-Irish." In *American Immigrant Cultures,* ed. David Levinson and Melvin Ember, 767–73. New York: Simon and Schuster Macmillan.

———. 1997b. "Scots." In *American Immigrant Cultures,* ed. David Levinson and Melvin Ember, 773–79. New York: Simon and Schuster Macmillan.

———. 1999. *Delta Sugar: Louisiana's Vanishing Plantation Landscape.* Baltimore: Johns Hopkins University Press.

Rehder, John B., John Morgan, and Joy L. Medford. 1979. "The Decline of Smokehouses in Grainger County, Tennessee." *West Georgia College Studies in the Social Sciences* 18:75–83.

Rehder, Karen. 1989. "Observations of Folk Houses as Found in the Eastern Tennessee Region." Unpublished data in Special Collections, John C. Hodges Library, University of Tennessee.

Riedl, Norbert F., Donald B. Ball, and Anthony P. Cavender. 1976. *A Survey of Traditional Architecture and Related Material Folk Culture Patterns in the Normandy Reservoir, Coffee County, Tennessee.* Department of Anthropology Report of Investigations, no. 17. Knoxville: University of Tennessee and Tennessee Valley Authority.

Rinzler, Ralph. [1937] 1973. "Preface." In *Handicrafts of the Southern Highlands,* by Allen H. Eaton. New York: Dover.

———. 1994. *Doc Watson and Clarence Ashley: The Original Folkways Recordings: 1960–1962.* Liner notes. Washington, DC: Smithsonian/Folkways Recordings. CD SF 40029/30.

Ritchie, Jean. 1955. *Singing Family of the Cumberlands.* New York: Oxford University Press.

———. 1997. *Folk Songs of the Southern Appalachians.* 2nd ed. Lexington: University of Kentucky Press.

Rooney, James. 1971. *Bossmen: Bill Monroe and Muddy Waters.* New York: Dial Press.

Rosenberg, Neil V. 1985. *Bluegrass: A History.* Urbana: University of Illinois Press.

Rostlund, Erhard. 1952. *Freshwater Fish and Fishing in Native North America.* University of California Publications in Geography, vol. 9. Berkeley: University of California Press.

Rouse, Parke, Jr. 1995. *The Great Wagon Road.* Richmond, VA: Dietz Press.

Salstrom, Paul. 1995. "Newer Appalachia as One of America's Last Frontiers." In *Appalachia in the Making: The Mountain South in the Nineteenth Century,* ed. Mary Beth Pudup, Dwight B. Billings, and Altina L. Waller, 76–102. Chapel Hill: University of North Carolina Press.

Scarborough, Dorothy. 1966. *A Song Catcher in the Southern Mountains: American Folk Songs of British Ancestry.* New York: AMS Press.

Schimmer, James R., and Allen G. Noble. 1984. "The Evolution of the Corn Crib." *Pioneer America Society Transactions* 7: 21–33.
Scofield, Edna. 1936. "The Evolution and Development of Tennessee Houses." *Journal of the Tennessee Academy of Science* 11 (4): 229–40.
Seeger, Charles. 1958. "The Appalachian Dulcimer." *Journal of American Folklore* 71:40–51.
Seeger, Mike. 1990. *American Banjo: Three Finger and Scruggs Style.* Liner notes. Washington, DC: Smithsonian /Folkways Records. CD SF 40037.
Semple, Ellen Churchill. [1901] 1995. "The Anglo Saxons of the Kentucky Mountains: A Study in Anthropogeography." *Geographical Journal* 17:588–623. Also in *Bulletin of the American Geographical Society* 42 (1910): 561–94. Reprinted in *Appalachian Images in Folk and Popular Culture,* ed. W. K. McNeil, 2nd ed. 145–74. Knoxville: University of Tennessee Press, 1995.
———. 1911. *Influences of Geographic Environment.* New York: Henry Holt.
Shackelford, Laurel, and Bill Weinberg. 1988. *Our Appalachia.* Lexington: University of Kentucky Press.
Shapiro, Henry D. 1978. *Appalachia on Our Mind: The Southern Mountains and Mountaineers in the American Consciousness, 1875–1920.* Chapel Hill: University of North Carolina Press.
Sharp, Cecil J. [1932] 1966. *English Folksongs from the Southern Appalachians.* 2nd ed. London: Oxford University Press.
Shields, Wayne F. 1967. "The Chariton River Fish Trap." *Missouri Historical Review* 61 (4): 489–96.
Shifflett, Crandall A. 1991. *Coal Towns: Life, Work, and Culture in Company Towns of Southern Appalachia, 1880–1960.* Knoxville: University of Tennessee Press.
Shoemaker, Floyd C. 1955. "Missouri's Tennessee Heritage." *Missouri Historical Review* 49:127–42.
Shurtleff, Harold R. 1939. *The Log Cabin Myth: A Study of the Dwellings of the English Colonists in North America.* Cambridge: Harvard University Press.
Simmons, Morgan. 2003. "Task Force Racing Clock in Battle vs. Hemlock Foe." *Knoxville News Sentinel,* August 18, B1, B6.
Smith, Betty N. 1998. *Jane Hicks Gentry: A Singer among Singers.* Lexington: University of Kentucky Press.
Smith, H. McKelden. 1979. *Architectural Resources: An Inventory of Historic Architecture.* Raleigh: North Carolina Department of Cultural Resources, Division of Archives and History.
Smith, L. Mayne. 1964. "Bluegrass Music and Musicians." Master's thesis, Indiana University.
Smith, Lee. 1995. "Introduction." In *Come Go Home with Me,* by Sheila Adams. Chapel Hill: University of North Carolina Press.
Stoddart, Jess, 2002. *Challenge and Change in Appalachia: The Story of the Hindman Settlement School.* Lexington: University Press of Kentucky.

Stupka, Arthur. 1964. *Trees, Shrubs, and Woody Vines of the Great Smoky Mountains National Park.* Knoxville: University of Tennessee Press.

Sutton, Ann, and Myron Sutton. 1985. *Eastern Forests.* Audubon Nature Guides. New York: Alfred A. Knopf.

Swaim, Doug, ed. 1978. *Carolina Dwelling: Towards Preservation of Place.* Raleigh: North Carolina State University.

Swan, M. L. [1867] 1978. *The New Harp of Columbia.* Knoxville: University of Tennessee Press.

Tams, W. P., Jr. 1963. *The Smokeless Coal Fields of West Virginia.* Morgantown, WV: West Virginia University Foundation.

Tennessee Department of Environment and Conservation. 2001. *Tennessee Ginseng Program: American Ginseng in Tennessee.* Nashville: Tennessee Department of Environment and Conservation.

Thornbury, William D. 1965. *Regional Geomorphology of the United States.* New York: John Wiley.

Thrower, Norman J. G. 1966. *Original Survey and Land Subdivision.* Chicago: Rand McNally.

"Tiny Beetles Put in Great Smokies to Save Hemlocks." 2002. *Knoxville News-Sentinel,* June 5, A1, A8.

Turner, William H., and Edward J. Cabbell, eds. 1985. *Blacks in Appalachia.* Lexington: University of Kentucky Press.

U.S. Congress, Senate. 1925. *Report of the United States Coal Commission.* Sen. Doc. 195, 68th Cong., 2nd sess. Washington, DC: Government Printing Office.

U.S. Department of Agriculture. 1872. "Statistics of Fences in the United States." *Report of the Commissioner of Agriculture for the Year 1871.* Washington, DC: Government Printing Office.

———. 1935. *Economic and Social Problems and Conditions of the Southern Appalachians.* Misc. Pub. 205. Washington: U.S. Department of Agriculture.

U.S. Department of Energy, Energy Information Administration. 2002. *U.S. Coal Supply and Demand. 2002 Review.*

Upton, Dell, ed. 1986. *America's Architectural Roots: Ethnic Groups That Built America.* Washington, DC: Preservation Press.

Upton, Dell, and John Michael Vlach, eds. 1986. *Common Places: Readings in American Vernacular Architecture.* Athens: University of Georgia Press.

Venable, Sam. 2000. *Mountain Hands: A Portrait of Southern Appalachia.* Knoxville: University of Tennessee Press.

Vlach, John Michael. 1986. "The Shotgun House: An African Architectural Legacy." In *Common Places: Readings in American Vernacular Architecture,* ed. Dell Upton and John Michael Vlach, 58–78. Athens: University of Georgia Press.

———. 1993. *Back of the Big House: The Architecture of Plantation Slavery.* Chapel Hill: University of North Carolina Press.

Waldvogel, Marikay. 1990. *Soft Covers for Hard Times: Quiltmaking and the Great Depression.* Nashville, TN: Rutledge Hill Press.

Warner, John. 2002. *John E. Warner's Medical Catalog.* Crossville, TN: Warner Herb Company.

Weals, Vic. 1991. *Last Train to Elkmont: A Look Back at Life on Little River in the Great Smoky Mountains.* Knoxville, TN: Olden Press.

Weld, Isaac, Jr. [1796] 1998. "East and West of the Blue Ridge." Reprinted in *The Height of Our Mountains,* ed. Michael P. Branch and Daniel J. Philippon, 126–31. Baltimore: Johns Hopkins University Press.

Weller, Jack. E. 1965. *Yesterday's People: Life in Contemporary Appalachia.* Lexington: University of Kentucky Press.

Wertenbaker, Thomas J. 1938. *The Founding of American Civilization: The Middle Colonies.* New York: Charles Scribner's Sons.

Weslager, C. A. 1969. *The Log Cabin in America: From Pioneer Days to the Present.* New Brunswick, NJ: Rutgers University Press.

Whilden, Betse, and Karen Cox. 1973. "Old Time Burials." In *The Foxfire Book,* vol. 2, ed. Eliot Wiggington, 304–23. Garden City, NY: Anchor Books.

Whisnant, David. [1981] 1994. *Modernizing the Mountaineer: People, Power, and Planning in Appalachia.* Rev. ed. Knoxville: University of Tennessee Press.

White, B. F., and E. J. King. [1859] 1968. *The Sacred Harp.* Nashville, TN: Broadman Press.

Wigginton, Eliot, ed. 1972. *The Foxfire Book.* Vol. 1 Garden City, NY: Anchor Books.

———. 1973. *The Foxfire Book.* Vol. 2. Garden City, NY: Anchor Books.

Wilhelm, Gene, Jr. 1974. "The Mullein: Plant Piscicide of the Mountain Folk Culture." *Geographical Review* 64 (2): 235–52.

———. 1977. "Appalachian Isolation: Fact or Fiction." In *An Appalachian Symposium,* ed. Jerry Williamson, 77–91. Boone, NC: Appalachian State University Press.

———. 1978. "Folk Settlements in the Blue Ridge Mountains." *Appalachian Journal* 5 (2): 204–45.

Williams, Cratis D. 1961. "The Content of Mountain Speech." Part 4 of "Mountain Speech." *Mountain Life and Work* 37 (4): 13–17.

———. 1962. "Mountaineers Mind Their Manners." Part 5 of "Mountain Speech." *Mountain Life and Work* 38 (2): 19–25.

———. 1992. *Southern Mountain Speech.* Berea, KY: Berea College Press.

Williams, Michael Ann. 1995. *Great Smoky Mountains Folklife.* Jackson: University Press of Mississippi.

Wilson, Charles Reagan, and William R. Ferris Jr., eds. 1989. *Encyclopedia of Southern Culture.* Chapel Hill: University of North Carolina Press.

Wilson, Eugene M. 1970. "The Single Pen House in the South." *Pioneer America* 2 (1): 21–28.

———. 1971. "Some Similarities between American and European Folk Houses." *Pioneer America* 3 (2): 8–14.

———. 1975. *Alabama Folk Houses.* Montgomery: Alabama Historical Commission.

Wilson, Kathleen Curtis. 2001. *Textile Art from Southern Appalachia: The Quiet Work of Women.* Johnson City, TN: Overmountain Press.

Wise, Kenneth. 1996. *Hiking Trails of the Great Smoky Mountains: A Comprehensive Guide.* Knoxville: University of Tennessee Press.

Wolfe, Charles K. 1977. *Tennessee Strings: The Story of Country Music in Tennessee.* Knoxville: University of Tennessee Press.

———. 1993. *The Carter Family: Anchored in Love. Their Complete Victor Recordings, 1927–1928.* Liner notes. Cambridge, MA: Rounder Records. Rounder CD 1064.

Works Projects Administration. [1939] 1986. *The WPA Guide to Tennessee.* Reissued Knoxville: University of Tennessee Press.

Wright, Martin. 1958. "The Antecedents of the Double-Pen House Type." *Annals of the Association of American Geographers* 48:109–17.

Zelinsky, Wilbur. 1992. *The Cultural Geography of the United States: A Revised Edition.* Englewood Cliffs, NJ: Prentice Hall.

Web Sites

www.1stchineseherbs.com
www.botanical.com
www.ralphstanley.net
www.fiddlersgrove.com
www.folkschool.com
www.gibson.com/products/oai/dobro/story.html
www.globalhemp.com
www.oldsalem.org
www.prevnet.org
www.tbi.state.tn.us
www.whitehousedrugpolicy.gov
www.wildgrown.com

Index

John Burkhardt Rehder is professor of geography at the University of Tennessee in Knoxville, where he has been a scholar for more than thirty-six years. He was born in Wilmington, North Carolina, and grew up in the "projects" on flat land with swamps and coastal marshes. He has a B.A. in geography from East Carolina College and an M.A. and Ph.D. in geography from Louisiana State University. In 1967 he came to the mountains to live and work in southern Appalachia. Only seven years out of high school, Rehder became an assistant professor and the first cultural geographer at the University of Tennessee. One among his more than sixty publications, *Delta Sugar: Louisiana's Vanishing Plantation Landscape,* published in 1999 by the Johns Hopkins University Press, won the prestigious Abbott Lowell Cummings Award in 2000 for the most outstanding book on vernacular architecture in North America.